AF251646

SECOND HARMONIC AND SUM-FREQUENCY SPECTROSCOPY

BASICS AND APPLICATIONS

SECOND HARMONIC AND SUM-FREQUENCY SPECTROSCOPY
BASICS AND APPLICATIONS

Yuen-Ron Shen

University of California, Berkeley, USA

World Scientific

NEW JERSEY · LONDON · SINGAPORE · BEIJING · SHANGHAI · HONG KONG · TAIPEI · CHENNAI · TOKYO

Published by

World Scientific Publishing Co. Pte. Ltd.

5 Toh Tuck Link, Singapore 596224

USA office: 27 Warren Street, Suite 401-402, Hackensack, NJ 07601

UK office: 57 Shelton Street, Covent Garden, London WC2H 9HE

Library of Congress Control Number: 2022942257

British Library Cataloguing-in-Publication Data
A catalogue record for this book is available from the British Library.

SECOND HARMONIC AND SUM-FREQUENCY SPECTROSCOPY
Basics and Applications

ISBN 978-981-126-227-2 (hardcover)
ISBN 978-981-126-228-9 (ebook for institutions)
ISBN 978-981-126-229-6 (ebook for individuals)

For any available supplementary material, please visit
https://www.worldscientific.com/worldscibooks/10.1142/13022#t=suppl

Desk Editor: Joseph Sebastian Ang

Typeset by Stallion Press
Email: enquiries@stallionpress.com

To Hsiaolin T. Shen and my family.

Preface

This book is an updated version of an earlier one titled "Fundamentals of Sum Frequency Spectroscopy" (FSFS) published by Cambridge University Press in 2016. Over the past 6 years, the field of second-harmonic/sum-frequency generation (SHG/SFG) spectroscopy has made strides in a number of areas, both technically and scientifically. It also clearly shows that further development of the technique is warranted, and the scope of applications can be broadened. In preparation of this new book, attempt was made to include more important recent advances in the field. I must however apologize in advance if some noteworthy works appear to have been neglected.

This book puts more emphasis on physical concepts and description. Complex mathematical derivations are purposely avoided. Readers interested in more rigorous theories are referred to the literature. There are however places where equations are helpful for illustration. They are marked by shading and can be skipped by readers not interested in theory. Because of the book page limit, discussion on each topic is necessarily brief and focuses on ideas and concepts. For details, the readers are again referred to the relevant literature. The reference list of each chapter of the book is also intentionally kept short; the older references on a topic can be found in the earlier FSFS book. Selected recent review articles on various topics are quoted at the end of the reference list. The readers are reminded that descriptions in references and review articles are not necessarily all correct. This is particularly true of issues that have not yet been resolved. In most chapters, a section of summary and prospect is included at the end. I should note that there were many mistakes and printing errors in the FSFS book; hopefully in this new book they have all been removed and few new ones would appear.

Some parts of this new book are significantly different from the old one. A more careful discussion on experimental setups and precautions is provided (Chapter 3) and comparison with other techniques is made if possible. A perspective on bulk characterization by SHG/SFG spectroscopy is given (Chapter 4). Recent extension of SF vibrational spectroscopy studies of catalysis and surface reactions on nanoparticles is delineated (Chapter 6). Importance of probing surface structures of both media at an interface for interfacial studies is emphasized (Chapters 7–9). New schemes of SF vibrational spectroscopy (SFVS) to properly interrogate buried water interfaces, including electrochemical interfaces, are introduced (Chapter 8). Capability of chiral SF spectroscopy to probe achiral structure and chiral molecular arrangement at interfaces of chiral materials is described (Chapter 5). Use of SFVS to study industrial polymer problems is discussed (Chapter 9). Applications of SHG/SFG and SH/SF bio-imaging to real bio-systems are delineated (Chapters 10 and 11). A detailed general comment on the approach to analyze ultrafast surface dynamics is presented (Chapter 11). The book ends with an overview on the prospect of the field (Chapter 12).

I owe my sincere thanks to many of my former group members and collaborators I have had the fortune to work with at Berkeley. They made seminal contributions to the initiation of SHG/SFG spectroscopy as a viable surface analytical tool and its subsequent development at all stages. The name list can be found in the preface of the FSFS book, and will not be reproduced here. In recent years, I have benefited from discussions and collaborations with Wei-Tao Liu, Chuan-Shan Tian, and Yu-Chieh Wen. I would also like to thank Zhan Chen for his careful reading and expert comments on Chapters 9 and 10 of the book. Finally, I am indebted to my wife, Hsiaolin Shen, for her overall general assistance.

Contents

Chapter 1

Introduction

Sum-frequency generation (SFG) is a wave mixing process in which two input waves at angular frequencies ω_1 and ω_2 interact in a medium and generate an output at $\omega_3 = \omega_1 + \omega_2$. (We label ω as frequency throughout the book.) Second harmonic generation (SHG) and difference frequency generation (DFG) are special cases of SFG with $\omega_1 = \omega_2$ and $\omega_1 = -\omega_2$, respectively. In optics, SFG is one of the simplest and earliest discovered nonlinear optical effects, but it is also among the most useful ones. The most important application of SFG (together with SHG and DFG) is in frequency conversion of light. It provides an effective means to extend the spectral range of coherent light sources from terahertz to vacuum ultraviolet and has been instrumental in the spectacular advances of laser-related science and technology in the past half century.

Understanding how optical processes originate from material responses to light allows in turn optical characterization of materials through measurement of material responses. Knowing material responses over a wide range of optical frequencies provides vast amount of information about structural and functional properties of materials. Soon after the renaissance period, during which linear optical phenomena were comprehended, linear optical spectroscopy was developed to study materials. Together with advance of quantum mechanics, it readily became an effective tool to probe quantum states and optical transitions of matter and to learn about quantum nature of materials. However, known linear optical processes were limited, and so was their capability to characterize materials.

The situation was radically changed with the advent of lasers. Being coherent and directional with possible high spectral purity, strong peak intensity, and ultrashort temporal pulse width, lasers have revolutionized the optical spectroscopy field. They have not only greatly improved the capacity and sensitivity of known optical spectroscopic techniques of the pre-laser era, but also created a plethora of new techniques based on newly discovered nonlinear optical processes. Such techniques opened the door to studies in many hitherto unexplored areas of material science. SHG/SFG, in particular, has been developed into a versatile spectroscopic tool for material studies that is capable of yielding information not attainable by linear optical spectroscopy. This book is designed to provide an updated comprehensive introduction to SF spectroscopy and applications.[1]

1.1. Linear and Nonlinear Optical Responses from Materials

We give in this section a brief description of the principles of SHG/SFG; more details of the basic theory of SHG/SFG will be addressed in Chapter 2. Optical effects resulting from light interaction with matter in a medium are characterized by the intrinsic response coefficients of the medium. In the case of linear optics, the incoming field at frequency ω impinging on a medium induces a polarization (or oscillating dipoles per unit volume) at ω linearly proportional to the field in the medium. The proportional constant is the linear response coefficient known as the linear susceptibility $\chi^{(1)} = (\varepsilon - 1)/4\pi$ in cgs units, where ε is the optical dielectric constant and $n = \sqrt{\varepsilon}$ is the refractive index. [In SI units, $\chi^{(1)} = (\varepsilon/\varepsilon_0 - 1)$ and $n = \sqrt{\varepsilon/\varepsilon_0}$ with $\varepsilon_0 = 8.85 \times 10^{-12}$ F/m.] If the input fields are sufficiently strong, the nonlinear material response to the fields becomes important. SFG is a second-order nonlinear optical process originating from nonlinearly induced polarization proportional to the product of two input fields at frequencies ω_1 and ω_2; the response coefficient is the second-order nonlinear susceptibility $\chi^{(2)}$. Optical characterization of materials is through measurement of ε and $\chi^{(2)}$, as they not only represent macroscopic optical properties of a material but also carry microscopic structural information about the material.

Generally speaking, symmetry of response coefficients of a material reflects the structural symmetry of the material and resonant features of response coefficients reveal its microscopic structural properties. For elucidation, we first consider the case of $\chi^{(1)}$ or ε. The optical dielectric constant $\overset{\leftrightarrow}{\varepsilon}$ is a rank-2 tensor that has 9 tensor elements in the Cartesian coordinates. However, with proper choice of the coordinates, depending on the structural symmetry of a material, and under the so-called electric-dipole approximation, some of the tensor elements are zero and others related to one another, resulting in only a few nonvanishing, independent elements. For example, one can generally choose the coordinates to diagonalize $\overset{\leftrightarrow}{\varepsilon}$ with only three nonvanishing diagonal elements. For materials with isotropic or cubic symmetry, the three elements are equal; for materials with uniaxial and biaxial symmetries, two of the three and none of the three are equal, respectively. Each tensor element is a complex quantity $(\overset{\leftrightarrow}{\varepsilon} = \overset{\leftrightarrow}{\varepsilon}{}' + i\overset{\leftrightarrow}{\varepsilon}{}'')$ depending on the input light frequency; the real part governs refraction and the imaginary part describes absorption or gain (positive for absorption and negative for gain) of the optical field in a material. It is through the frequency dependence that ε bears the microscopic structural information about a material. The frequency dependence comes from optical transitions in the material. Optical absorption/gain or $\overset{\leftrightarrow}{\varepsilon}{}''(\omega)$ is resonantly enhanced when frequency ω approaches an optical transition; $\overset{\leftrightarrow}{\varepsilon}{}'(\omega)$ is related to $\overset{\leftrightarrow}{\varepsilon}{}''(\omega)$ by the famous Kramers–Kronig relation. Conversely, if $\overset{\leftrightarrow}{\varepsilon}{}'(\omega)$ and $\overset{\leftrightarrow}{\varepsilon}{}''(\omega)$ are known, we can extract from their frequency dependence characteristics of the optical transitions, and hence the molecular-level structural information about the material. This is usually achieved by comparing measured $\overset{\leftrightarrow}{\varepsilon}(\omega)$ with theoretically calculated $\overset{\leftrightarrow}{\varepsilon}(\omega)$. For example, band structures of semiconductors were determined this way in early days.

A similar description applies to $\overset{\leftrightarrow}{\chi}{}^{(2)}$ although it is more complex.[2] As a second-order response coefficient, $\overset{\leftrightarrow}{\chi}{}^{(2)}$ is a rank-3 tensor with 27 tensor elements in the Cartesian coordinates. Again, as a result of structural symmetry of a material, the number of independent nonvaishing elements of $\overset{\leftrightarrow}{\chi}^{(2)}$ in an appropriately chosen coordinate system can be quite limited. Materials of different symmetry classes have different sets of nonvanishing elements of $\overset{\leftrightarrow}{\chi}^{(2)}$, which can be used to identify the possible symmetry class a material belongs to. Same as $\overset{\leftrightarrow}{\varepsilon}(\omega)$, the tensor elements of $\overset{\leftrightarrow}{\chi}^{(2)}(\omega_3 = \omega_1 + \omega_2)$ of a material depend on input frequencies and are resonantly enhanced when ω_i approaches optical transitions of the material; if known, $\overset{\leftrightarrow}{\chi}^{(2)}(\omega_3)$ can be used to extract information about optical transitions and microscopic structure of the material. They are also complex quantities, although unlike $\overset{\leftrightarrow}{\varepsilon}(\omega)$, the imaginary part of $\overset{\leftrightarrow}{\chi}^{(2)}$ is not necessarily connected to absorption of the fields involved, but is an indication of how the specific optical transitions affect the second-order response through resonances.

1.2. Material Characterization by Sum Frequency Generation

Comparing $\overset{\leftrightarrow}{\varepsilon}$ and $\overset{\leftrightarrow}{\chi}^{(2)}$, one naturally expects the more complex $\overset{\leftrightarrow}{\chi}^{(2)}(\omega = \omega_1 + \omega_2)$ to carry more information about a material: two input beams with tunable frequencies, beam polarizations, and beam directions clearly provide much more possibility to probe a material than a single input beam with one tunable frequency, beam polarization, and beam direction. Measurement of $\overset{\leftrightarrow}{\varepsilon}(\omega)$ can only interrogate resonance with one optical transition at a time, but measurement of $\overset{\leftrightarrow}{\chi}^{(2)}(\omega = \omega_1 + \omega_2)$ allows not only interrogation of single resonances, but also double resonances with two coupled optical transitions. Unfortunately, theoretical calculation of $\overset{\leftrightarrow}{\chi}^{(2)}$ $(\omega = \omega_1 + \omega_2)$ is much more complex and difficult.

From symmetry consideration, $\overset{\leftrightarrow}{\chi}^{(2)}$ often has more nonvanishing tensor elements than $\overset{\leftrightarrow}{\varepsilon}$, permitting higher-level determination of structural symmetry of materials. We mention here a few examples: (More details will be given in Chapter 2.) $\overset{\leftrightarrow}{\chi}^{(2)}$ can identify two-fold and three-fold rotational symmetries, but $\overset{\leftrightarrow}{\varepsilon}$ can only identify two-fold rotational symmetry; $\overset{\leftrightarrow}{\chi}^{(2)}$ can distinguish trigonal symmetry from hexagonal symmetry, cubic structure from zincblende structure, etc., but ε cannot; $\overset{\leftrightarrow}{\chi}{}^{(2)}$ can detect polar orientation of a material, but $\overset{\leftrightarrow}{\varepsilon}$ cannot; $\overset{\leftrightarrow}{\chi}{}^{(2)}$ can probe anti-ferromagnetization, but ε cannot. There exists however a serious limitation on $\overset{\leftrightarrow}{\chi}{}^{(2)}$; it vanishes under electric-dipole approximation for materials with inversion symmetry that constitute a large class of materials including liquids, polymers, amorphous solids, and others. This is seen from a simple symmetry argument. As a rank-3 tensor, $\chi^{(2)}$ must have all its elements change sign when the coordinates are inverted, but for a material with inversion symmetry, all elements of $\overset{\leftrightarrow}{\chi}{}^{(2)}$ should remain unchanged upon inversion of coordinates. We then have $\overset{\leftrightarrow}{\chi}{}^{(2)} = -\overset{\leftrightarrow}{\chi}{}^{(2)} = 0$. This limitation on $\overset{\leftrightarrow}{\chi}{}^{(2)}$ seems disappointing, but it actually facilitates a very useful application of SFG, i.e., SF spectroscopy for surface studies. Inversion symmetry is necessarily broken at a surface or interface. Since SFG from bulk is suppressed, SFG from surface or interface stand out, allowing it to become an effective spectroscopic tool for surface characterization. Surface SF spectroscopy will be the main topic of the book.

Surfaces and bulk are intertwined. While objects generally appear in bulk form, surfaces support the bulk form and protect the bulk. Surfaces and bulk of the same material often have very different properties. Physical properties of a material are largely controlled by the bulk, but chemical properties (reactivity and functionality) are usually dominated by surfaces. A bulk can have its structure and properties changed by temperature, pressure, doping, etc., but surfaces may have their structure and properties also varied through interaction with the environment.

The basic idea of SFG (or second-order nonlinear optical processes in general) for surface studies is that $\overset{\leftrightarrow}{\chi}{}^{(2)}$ vanishes in media with inversion symmetry, but is electric-dipole allowed at surfaces or interfaces. As in the bulk case, $\overset{\leftrightarrow}{\chi}{}^{(2)}$ from a surface or interface is expected to carry structural information about the surface or interface. [We note in passing that SFG as a surface probe may be extended to crystals without inversion symmetry. As mentioned earlier, there exist vanishing elements in the $\overleftrightarrow{\chi}^{(2)}$ tensor for the bulk of a crystal of a certain symmetry class, but because of the structural difference of bulk and surface, the same elements of the surface $\overleftrightarrow{\chi}^{(2)}$ may not vanish. Such surface $\overleftrightarrow{\chi}^{(2)}$ elements, if accessible by SFG with selective beam geometry and polarization with respect to the crystal orientation, can be used to characterize the surface.] Over the past decades, surface SFG has been developed successfully into a most viable surface analytical tool and has impacted almost all areas of surface science, as we shall see in later chapters of the book.

1.3. Modern Surface Science and Surface Probes

Surface science is one of the most important branches of physical science as it plays an essential role in all aspects of our modern life. In geoscience, important aspects of oceanography, plate tectonics, glacier sliding, soil formation, mineral dissolution, nutrient transportation in water, and so on all involve surfaces. In environmental and atmospheric sciences, a large part of studies deal with water interfaces. In life science, biological processes usually start at biointerfaces. In our daily life, surface problems like cleaning, adhesion, friction, lubrication, material corrosion, and others are what we frequently encounter. In modern industry, development relies heavily on knowledge of surface science. It is through microscopic understanding of surfaces and interfaces of metals, semiconductors, and insulators that electronic device technology has been able to rapidly advance in the past decades to the current status. The same can be said about the chemical industry, which has benefited from knowledge of surface catalysis and electrochemistry.

Surface science in early years focused more on macroscopic thermodynamic properties of surfaces. Modern surface science beginning in 1960s has striven to probe and understand surface structures and properties at the molecular level.[3] The dream is to understand surface structures and properties of materials in given environment and be able to control them. However, even after more than half a century, the field is still far from being mature because surface and interface are much more difficult to study than bulk. Wofgang Pauli once

said: "God made the bulk; the surface was invented by the devil."
A surface generally composed of only a few atomic or molecular
layers covering a bulk has its structure depending not only on the
bulk, but also on the environment it is exposed to. Many possible
variations of structure and properties of a surface make the surface
study difficult but interesting. Moreover, surface probing requires
techniques that have monolayer sensitivity and ability to distinguish
against signals from bulk. Surface preparation is also a problem
as surfaces can be easily contaminated resulting in their structure
and property changes. Theoretical studies of surfaces are not easy
either as they have to take into account the effect of environment
the surfaces are in.

It was realized very early that to obtain reliable molecular-level
information about a surface, we must first be able to prepare the
surface reproducibly. To avoid surface contamination and influence
of environment on the surface, a sample needs to be situated
in ultrahigh vacuum, which was developed in the 1960s together
with techniques such as electron diffraction and X-ray spectroscopy
for surface characterization. This led to the earlier boom of the
semiconductor electronic industry.[3] Yet surfaces of the same material
in vacuum and in real environment can be very different. Many real,
relevant surface problems, for example, surface catalysis, cannot be
addressed with samples in vacuum. Surfaces of volatile liquid cannot
even be situated in vacuum. We need techniques that can probe
surface microscopic structure *in situ*, but this has not generally been
possible. To study a surface in real environment, we often have to be
satisfied with results of measurement that are reproducible, and try
to understand and learn about the surface from the results.

Reference [4] provides an exhaustive list of techniques that
have been used to study surfaces and interfaces. They can be
roughly divided into four categories. (1) Particle-based techniques:
Most common ones are thermal desorption mass spectroscopy, low
energy electron diffraction, photoemission, Auger electron spectroscopy, electron energy loss spectroscopy, surface scanning electron
microscopy, Rutherford backscattering, secondary ion spectroscopy,
and He-scattering. If incoming particles are of sufficiently low energy,

they can only interact with the surface layer of a material, and are therefore highly surface-specific, but they cannot be used to probe buried interfaces. Generally, the particle-based techniques require samples in high vacuum in order to avoid particles scattered by ambient gas molecules. Many of them are not able to identify surface species and surface composition. (2) X-ray-based techniques: Usual ones are surface X-ray diffraction, X-ray photoemission spectroscopy, and surface X-ray absorption spectroscopy. With total reflection geometry, the X-ray penetration depth into a material can be limited to a few nm, and hence X-ray probes can be considered quasi-surface-specific. To distinguish surface and bulk contributions, data analysis usually relies on models. X-rays can pass through thin ambient gas, and micron-thick liquid or solid. Accordingly, with proper arrangement, X-ray techniques can be employed to study gas/solid interfaces and buried interfaces. A major advantage of X-ray spectroscopy is that it is capable of identifying surface species and their local environment, as well as surface composition. (3) Close-contact scanning microscopic techniques: Development of scanning tunneling microscopy (STM), atomic force microscopy (AFM), and their auxiliaries was the most important advance of surface science technology in the past half century. They are highly surface-specific and have capability of mapping a surface with atomic resolution; STM in particular can also identify local surface species and composition. They are best fit with sample surfaces in ultrahigh vacuum although AFM with limited spatial resolution is often used to image topography of surface under ambient atmosphere. There is obvious difficulty to apply STM and AFM to buried interfaces. (4) Optical techniques: Well-known optical techniques that have been employed to probe surfaces are reflection-absorption spectroscopy, Raman spectroscopy, and ellipsometry. They have submonolayer sensitivity. Compared with techniques in the other three categories, they are simpler and easier to use. They are however not surface-specific and have to rely on spectral difference to distinguish surface from bulk and learn about surface structure, thus limiting their applications. SFG spectroscopy is a relatively new addition to the arsenal. Being surface-specific, it greatly expands the scope of other optical techniques and opens up many unique opportunities for novel surface studies. It should be remarked that there is no single technique in surface science that can provide all the needed information about a surface. One always hopes to be able to apply as many techniques as possible on the same surface and obtain complementary information about the surface from them.

1.4. Sum Frequency Spectroscopy for Surfaces and Interfaces

As a coherent laser spectroscopy technique, SFG has a number of advantages over other techniques for surface studies. In addition to surface specificity, it has high spectral and spatial resolution, as well as submonolayer sensitivity. The output signal is highly directional, allowing effective discrimination against background radiation by spatial filtering, and hence the use of the technique for *in situ*, non-destructive, remote sensing of sample surfaces in hostile environment. With ultrashort input light pulses, it can also monitor surface dynamics with fs time resolution. Most importantly, the technique is applicable to all interfaces accessible by light, making it most versatile among all existing surface analytical tools.

We describe here an early experiment on fused silica/liquid interfaces that was designed to demonstrate surface specificity and monolayer sensitivity of SFG and its ability to probe buried interfaces.[5] The experimental arrangement is sketched in the inset of Fig. 1.1. Two input beams were incident from the silica side, one fixed at a visible frequency and the other tunable in the infrared for vibrational excitation. The SF output in the visible was detected in the reflected direction. Three interfaces were probed: silica/hexadecane[HD, $H_3C(CH_2)_{14}CH_3$], octadecyltrichlorosilane (OTS, $CH_3(CH_2)_{17}SiCl_3$)-covered silica/HD, and OTS-covered silica/CCl_4. The molecular structures of HD and OTS are given in the inset of Fig. 1.1. Both HD and OTS have long alkyl (CH_2) chains. The CH stretch spectra of the three interfaces taken by surface SF vibrational spectroscopy (SFVS) are presented in Fig. 1.1. It is seen that despite the presence of HD liquid, the spectrum of silica/HD is very weak and has hardly any discernible features,

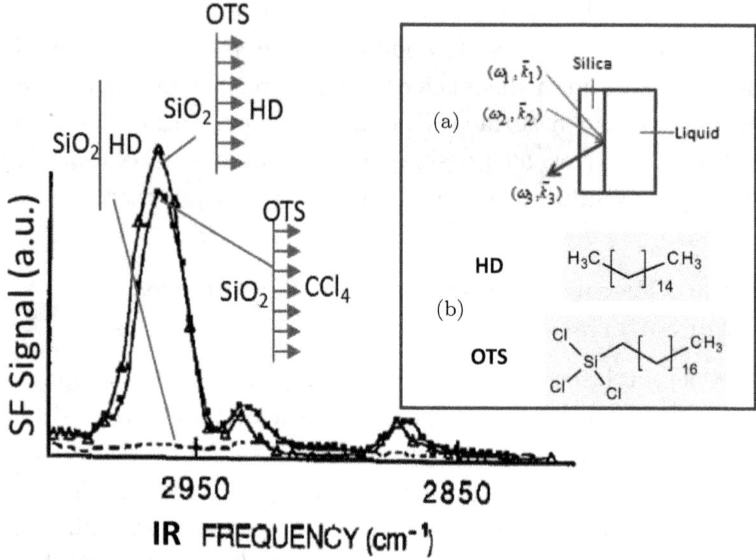

Fig. 1.1. SF vibrational spectra in the CH stretching region for three inter-facial systems: SiO$_2$/Hexadecane(HD), SiO$_2$/Octadecyltrichlorosilane(OTS)-monolayer/HD, and SiO$_2$/OTS-monolayer/CCl$_4$. Insets describe (a) the input/output beam geometry and (b) molecular structures of HD and OTS. (after Ref. [5])

although infrared and Raman measurements of the same interfacial system would produce spectra overwhelmed by CH$_2$ stretch modes. This is because the orientation distribution of HD molecules is random in bulk liquid with inversion symmetry and also nearly so at the interface. The spectrum of OTS-covered silica/HD shows three prominent CH$_3$ modes that can all be assigned to the OTS monolayer on silica. Again, despite its abundant presence, the HD liquid does not contribute to the observed CH stretch spectrum. The CH$_2$ modes are missing because the OTS chains are in all-trans conformation resulting in a CH$_2$ distribution around the chain that has near inversion symmetry. That the spectrum comes from the OTS monolayer is further confirmed by seeing the same spectrum from OTS-covered silica/CCl$_4$. This experiment clearly demonstrates that SFVS is capable of probing buried interfaces with high surface specificity and monolayer sensitivity.

1.5. Early Development of Second Harmonic and Sum Frequency Generation as Viable Surface Analytical Tools

Establishing SHG/SFG as viable surface probes took some time and effort. We give a brief description here on the history of the development. A more detailed account can be found in Ref. [1] and references therein. SHG/SFG were discovered right after the ruby laser was invented in 1960. Theoretical understanding of the process soon led to the realization that SHG can be used to probe structural symmetry of materials. SHG by reflection from a boundary surface of a medium was worked out theoretically and experimentally, and the existence of a surface layer with nonlinearities different from the bulk was recognized. Research on SHG around that time, however, was mainly focused on understanding the SHG process in different media instead of employing SHG for material characterization. Although submonolayer sensitivity of SHG was observed in experiments on atomic and molecular adsorption on metals, no in-depth study was pursued, and the works went by without being noticed.

Development of SHG as surface probe was triggered by the discovery of surface-enhanced Raman scattering (SERS) in 1974.[6] It was realized that like SERS, reflected SHG from a rough metal surface could enjoy a local-field enhancement of four orders of magnitude, and can be used to monitor adsorption and desorption of molecules at a roughened silver electrode during an electrolytic cycle in an electrochemical cell.[7] From the observed signal level, it was realized that even without local field enhancement, SHG could have enough sensitivity to detect a submonolayer of molecules on any substrate. This was soon demonstrated in experiments monitoring adsorption of molecules on various surfaces under ambient conditions including liquid interfaces. The theory of SHG (also applicable to SFG) was refined to better understand the role of a surface in the process. To assure that the technique was compatible with basic surface science studies at the time, SHG measurements of adsorption and desorption kinetics of atoms and molecules on well-defined metal and semiconductor surfaces in ultrahigh vacuum (UHV) were carried out. That SHG could monitor surface reconstruction of a crystal was also demonstrated. Applications of SHG were then extended to measurements of orientation of adsorbed molecules, orientation dynamics, molecular adsorption at liquid–solid interfaces, charged water interfaces, surfaces of colloidal particles, etc. SHG spectroscopy on adsorbed molecules was performed to show that it was possible to probe electronic transitions of adsorbed molecules, but it was realized that vibrational transitions would be more informative on surface structures. In principle, one could use tunable infrared input for SHG to probe vibrational spectra in the IR region. The problem is that tunable IR pulses are usually not intense enough and IR detectors not sensitive enough, making surface IR SHG spectroscopy not practical.

It is obvious that SFG with two independently adjustable input beams has more flexibility and viability than SHG. With one input beam tunable in the infrared and the other fixed in the visible, the SFG output in the visible ought to be able to generate vibrational spectra with submonolayer sensitivity. This was first demonstrated in 1987.[8] The technique relies on the availability of tunable coherent IR sources. Earlier works of surface SF vibrational spectroscopy (SFVS) were carried out with a ps pulsed Nd:YAG laser and an associated laser-pumped $LiNbO_3$ parametric amplifier that generated a tunable IR beam from ~ 2.65 μm (3780 cm^{-1}) to $\sim 4\mu$m (2500 cm^{-1}) covering the spectral range of CH, OH, and NH stretching vibrations. Soon after parametric amplifiers became commercially available, SFG spectroscopy for surface studies blossomed. A series of experiments was able to reveal the capacity, versatility, and unique-ness of SFG as a surface vibrational spectroscopy tool for different areas of surface science. It was shown that SFVS could be used to study adsorbed molecules at nearly all interfaces with focus on vibrational spectra of molecular subgroups. Exploration of interfaces of neat liquids and solids that had not been possible before became feasible with SFVS. Among them are the interfaces of two most important materials, water and polymers. The technique was also extended to biological molecules and membranes. In areas relevant to technology, SFVS also found many useful applications: monitoring electrochemical reactions *in situ*, investigating catalytic reactions in real atmosphere, tracking interfacial structures of junctions during operation, and others. Two significant technical advances of SFG spectroscopy following advancement of laser technology should be mentioned. One is the development of the so-called broadband scheme of SFVS that uses broadband fs IR pulses mixed with narrowband visible pulses to generate SF vibrational spectra. It significantly improves the signal-to-noise ratio of SFVS and facilitates ultrafast surface dynamics studies. The other is the development of phase-sensitive SF spectroscopy that allows measurement of both real and imaginary parts of surface nonlinear response coefficient, providing complete information on surface resonance characteristics. We should also mention that the continuous success of SFVS has relied very much on being able to correctly interpret observed spectra to glean information on surface structures. Fortunately, theoretical progress in this respect, particularly in terms of molecular dynamics (MD) simulations, has been truly remarkable, and has greatly helped promoting surface studies of SFVS. For theory on SFVS with MD simulations, we refer readers to the recent book by Morita.[9]

Over the past three decades, surface SF vibrational spectroscopy has been developed into a most viable surface analytical tool. Because of its ability to probe almost all types of interfaces, it has opened up unique research opportunities in many disciplines and in areas not easily accessed by other techniques. We should however note that SF spectroscopy is not limited to surfaces and interfaces. In the following chapters, we shall describe applications of SF spectroscopy to characterization of both surfaces and bulks, although more emphasis will be on surfaces to illustrate how SF spectroscopy has become a unique and viable surface probe.

References

1. This book is an updated, version of the existing book: Shen, Y. R.: *Fundamentals of Sum-Frequency Spectroscopy.* Cambridge University Press: Cambridge, 2016.
2. More details on nonlinear optical susceptibilities can be found in Shen, Y. R.: *The Principles of Nonlinear Optics.* J. Wiley: New York, 1984; Chapters 2.

3. Gatos, H. C.: Semiconductor Electronics and the Birth of the Modern Science of Surfaces. *Surf. Sci.* 1994, 299/300, 1–23.

4. Somorjai, G. A.; Li, Y.: *Introduction to Surface Chemistry and Catalysis.* 2nd ed. J. Wiley: New York, 2010.

5. Guyot-Sionnest, P.; Superfine, R.; Hunt, J. H.; Shen, Y. R.: Vibrational Spectroscopy of a Silane Monolayer at Air Solid and Liquid Solid Interfaces Using Sum-Frequency Generation. *Chem. Phys. Lett.* 1988, 144, 1–5.

6. Fleischmann, M.; Hendra, P. J.; Mcquilla, A. J.: Raman-Spectra of Pyridine Adsorbed at a Silver Electrode. *Chem. Phys. Lett.* 1974, 26, 163–166.

7. Chen, C. K.; Decastro, A. R. B.; Shen, Y. R.: Surface-Enchanced second Harmonic Generation. *Phys. Rev. Lett.* 1981, 46, 145–148.

8. Zhu, X. D.; Suhr, H.; Shen, Y. R.: Surface Vibrational Spectroscopy by Infrared-Visible Sum Frequency Generation. *Phys. Rev. B.* 1987, 35, 3047–3050.

9. Morita, A.: *Theory of Sum-Frequency Generation Spectroscopy.* Springer Nature Singapore, 2018.

Chapter 2

Basic Theory of Sum Frequency Generation

To see how sum frequency generation (SFG) arises from nonlinear interaction of light with matter and how it can be used to characterize materials, we need some basic understanding of the process. The basic theory of SFG was worked out in 1962 in the classical paper on nonlinear optics by Bloembergen and coworkers[1] (Bloembergen 1962).We give a sketch of the theory here with more focus on the role of surface/interface of media. Emphasis is on physical concept and understanding; only the most crucial equations will be presented. Details of the theory are somewhat complicated, but can be found elsewhere.[2–4]

2.1. Sum Frequency Generation from a Semi-infinite Medium

In elementary description of linear optics, reflection and transmission of a light beam at an interface are approximated by an infinite plane wave incident on a plane interface ($z = 0$) between two media. As depicted in Fig. 2.1(a), the two media, I and II, are both assumed to be semi-infinite with, respectively, optical dielectric constants $\overset{\leftrightarrow}{\varepsilon}_I$ and $\overset{\leftrightarrow}{\varepsilon}_{II}$. Description of SFG in Fig. 2.1(b) follows the same layout; for simplicity, it is usually assumed that medium I is linear and medium II nonlinear, characterized by a tensorial nonlinear susceptibility $\overset{\leftrightarrow}{\chi}^{(2)}$. In both Figs. 2.1(a) and 2.1(b), there should exist an interfacial layer of finite thickness ($z = 0^-$ to 0^+) with z-dependent $\overset{\leftrightarrow}{\varepsilon}$ and $\overset{\leftrightarrow}{\chi}^{(2)}$ due to structural transition from medium I to medium II. For the SFG case, we consider two input beams at

15

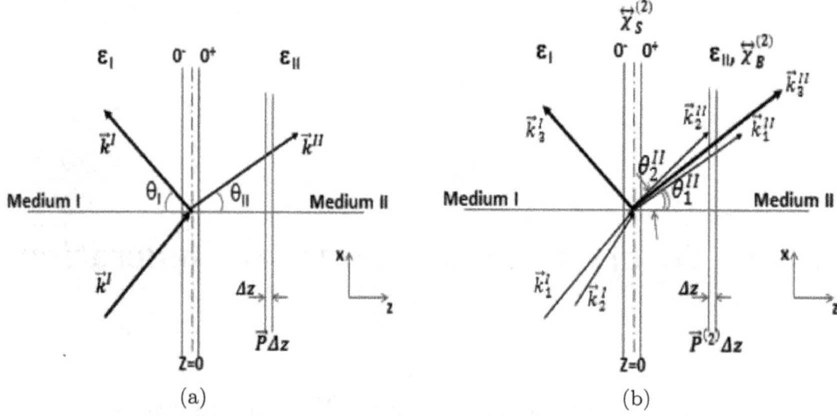

Fig. 2.1.   (a) Beam geometry for linear transmission and reflection at an interface $(z = 0^-$ to $0^+)$ between media I and II with optical dielectric constants, ε_I and ε_{II}, respectively. (b) Beam geometry for sum frequency generation at the interface described in (a), but medium II is nonlinearly active with nonlinear susceptibility $\overset{\leftrightarrow}{\chi}_B^{(2)}$. The input beams at ω_1 and ω_2 have wave vectors $\vec{k}_1^{II}$ and $\vec{k}_2^{II}$ in medium II, respectively, and the SF outputs in transmission and reflection have wave vectors $\vec{k}_3^{II}$ in medium II and $\vec{k}_3^{I}$ in medium I (after Ref. [4]).

frequencies ω_1 and ω_2 from the medium I side with wave vectors $\vec{k}_1^I$ and $\vec{k}_2^I$: They refract into medium II with wave vectors $\vec{k}_1^{II}$ and $\vec{k}_2^{II}$, and induce a nonlinear polarization at the sum frequency $\omega_3 = \omega_1 + \omega_2$ and wave vector $\vec{k}_1 + \vec{k}_2$ by wave mixing in medium II. The induced polarization radiates and generates SF output in both transmitted and reflected directions with corresponding wave vectors $\vec{k}_3$ governed by the boundary condition that the components of $\vec{k}_1 + \vec{k}_2$ and $\vec{k}_3$ along the interface must be matched, i.e., $\vec{k}_{3\parallel}^I = \vec{k}_{3\parallel}^{II} = (\vec{k}_1^I + \vec{k}_2^I)_\parallel = (\vec{k}_1^{II} + \vec{k}_2^{II})_\parallel$, and accordingly, $k_{3\perp}^I = \sqrt{(k_3^I)^2 - (k_{3\parallel}^I)^2}$ and $k_{3\perp}^{II} = \sqrt{(k_3^{II})^2 - (k_{3\parallel}^{II})^2}$ with $(k_3^I)^2 = (\omega_3^2/c^2)\varepsilon^I(\omega_3)$ and $(k_3^{II})^2 = (\omega_3^2/c^2)\varepsilon^{II}(\omega_3)$. [We insert here a note of caution. For absorbing media, $k_\parallel$ must be real, but $k_\perp = k'_\perp + i\beta$ is complex with $k'_\perp = (\omega/c)\varepsilon' \cos\theta$ and $\beta = \omega\varepsilon''/2c\sqrt{\varepsilon'} \cos\theta$, where $\varepsilon = \varepsilon' + i\varepsilon''$ and θ is the angle between $\vec{k}'$ and $\hat{z}$.]

Formally, the SF output field from the system of Fig. 2.1(b) can be found by solving the Maxwell equations with the induced nonlinear polarization in medium II as the source.[2,3] Here, we present a more physical approach.[4] We know all optical effects must originate from optical field-induced polarization in media. For example, linear optical transmission, refraction, and reflection at the interface of two media are the result of interference between the incoming wave and the radiated wave from induced polarization in the media. As described in Fig. 2.1(a), the incoming field $\vec{E}^{II}(\omega) = \vec{\mathcal{E}}^{II}e^{i\vec{k}^{II}\cdot\vec{r}-i\omega t}$ in medium II induces a stack of polarization sheets, $\vec{P}(\omega,\vec{k}_z^{II})dz = \overset{\leftrightarrow}{\chi}_{II}^{(1)} \cdot \vec{E}^{II}(\omega)dz = [(\overset{\leftrightarrow}{\varepsilon}_{II} - 1)/4\pi] \cdot \vec{E}^{II}(\omega)dz$, from $z = 0$ to ∞ in medium II, where $\overset{\leftrightarrow}{\chi}_{II}^{(1)} = [(\overset{\leftrightarrow}{\varepsilon}_{II} - 1)/4\pi]$ is the linear susceptibility in cgs units (but $\overset{\leftrightarrow}{\chi}_{II}^{(1)} = (\overset{\leftrightarrow}{\varepsilon}_{II} - 1)$ in SI units). The expression for electric-dipole (ED) radiation from a polarization sheet can be found from elementary electricity and magnetism textbooks:[5] The radiation field, $\vec{E}_R(\omega,\vec{k},\vec{r},z')$, with wave vector $\vec{k}$ and frequency ω, observed at position $\vec{r}$ from a polarization sheet of $\vec{P}\left(\omega,\vec{k}_s,z'\right)\Delta z' = |\vec{P}|\Delta z'e^{i\vec{k}_s\cdot\vec{r}'-i\omega t}$ at z' in a uniform medium $(z' > 0)$, is given by $\vec{E}_R(\omega,\vec{k},\vec{r},z') = \frac{i2\pi\omega^2}{c^2k_z}i\hat{k} \times [i\hat{k} \times \vec{P}(\omega,\vec{k}_s,z')]\Delta z'e^{i\vec{k}\cdot(\vec{r}-z'\hat{z})}$ in cgs units. It has been shown that the Fresnel coefficient for transmission from medium I to medium II can be deduced from interference between the incoming field and the radiation field from all polarization sheets in medium II, and the Fresnel coefficient for reflection solely from radiation of the polarization sheets.[4,6] The interface layer acts as a transition layer that allows the incoming beam with wavevector $\vec{k}^I$ in medium I to change over to the transmitted beam with wavevector $\vec{k}^{II}$ and vice versa.

The same approach can be used to find SFG in transmission and reflection with reference to Fig. 2.1(b).[2] It is now the stack of induced nonlinear polarization sheets, $\vec{P}^{(2)}\left(\omega_3,\vec{k}_s,z\right)dz$ with $\vec{P}^{(2)}\left(\omega_3,\vec{k}_s,z\right) = \overset{\leftrightarrow}{\chi}^{(2)}(\omega_3,\vec{k}_s,z) : \vec{E}_1^{II}(z,\omega_1)\vec{E}_2^{II}(z,\omega_2)$, responsible for SFG. We assume here electric dipole (ED) approximation for the nonlinear response, and the interfacial layer infinitesimally thin with a surface nonlinear susceptibility $\overset{\leftrightarrow}{\chi}_{SD}^{(2)}$ as the nonlinear response coefficient. (Possible complications arising beyond the ED approximation will be briefly commented upon later.) Intergration of radiation fields from all induced polarization sheets yields the SF output field. The result is given in the following. Detailed derivation will be omitted, but can be found in Ref. [4].

In the forward propagating direction, the SF output field generated by surface and bulk nonlinear polariztions in medium II is given by[4] (in cgs units)

$$\vec{E}_t^{II}(\omega_3, z) = \vec{E}_{St}^{II}(\omega_3, z) + \vec{E}_{Bt}^{II}(\omega_3, z)$$

$$= \frac{i2\pi\omega_3^2}{c^2 k_{3z}^{II}} (\overset{\leftrightarrow}{\chi}_{SD}^{(2)} - \frac{\overset{\leftrightarrow}{\chi}_{BD}^{(2)}}{i\Delta k_{t,z}^{II}}) : \vec{\mathcal{E}}_1^{II} \vec{\mathcal{E}}_2^{II} e^{i(\vec{k}_3^{II} \cdot \vec{r} - \omega_3 t)}, \tag{2.1}$$

if $|\Delta k_{t,z}^{II}| \neq 0$ and $z >> |1/\Delta k_{t,z}^{II}|$. (In SI units, the right-hand side should be divided by 4π.) Here, the subindex B refers to bulk, $\overset{\leftrightarrow}{\chi}_{SD}^{(2)}$ and $\overset{\leftrightarrow}{\chi}_{BD}^{(2)}$ denote the surface and bulk ED nonlinear susceptibilities, respectively, and $\Delta k_{t,z}^{II} = k_{1z}^{II} + k_{2z}^{II} - k_{3z}^{II}$ is the phase mismatch of SFG in the forward direction. In practice, if medium II is of finite length l (with $l >> |1/\Delta k_{t,z}^{II}|$) and embedded in medium I (e.g., air), and all fields are to be referenced to those in medium I, we need to use Fresnel transmission coefficients to connect fields in media I and II. The transmitted SF field at $z = l$ in medium I, following Eq. (2.1) and neglecting SFG from the output interface layer, appears as (in cgs units) (divided by 4π in SI units)

$$\vec{E}_t^I(\omega_3, z = l) = \vec{E}_{St}^I(\omega_3, z = l) + \vec{E}_{Bt}^I(\omega_3, z = l)$$

$$= \frac{i2\pi\omega_3^2}{c^2 k_{3z}^I} \overset{\leftrightarrow}{F}_t^{I-II}(\omega_3) : (\overset{\leftrightarrow}{\chi}_{SD}^{(2)} - \frac{\overset{\leftrightarrow}{\chi}_{BD}^{(2)}}{i\Delta k_{t,z}^{II}}) : \overset{\leftrightarrow}{F}_t^{I-II}(\omega_1)$$

$$\overset{\leftrightarrow}{F}_t^{I-II}(\omega_2) : \vec{\mathcal{E}}_1^I \vec{\mathcal{E}}_2^I e^{i(k_{3z}^{II} l + k_{3x}^{II} l - \omega_3 t)}, \tag{2.2}$$

where $\overset{\leftrightarrow}{F}_t^{I-II}(\omega_3)$ is the Fresenl transmission coefficient with components given by

$$F_{t,xx}^{I-II} = \frac{2\varepsilon_I k_z^{II}}{\varepsilon_{II} k_z^I + \varepsilon_I k_z^{II}}, \quad F_{t,yy}^{I-II} = \frac{2k_z^I}{k_z^I + k_z^{II}}, \quad F_{t,zz}^{I-II} = \frac{2\varepsilon_I k_z^I}{\varepsilon_{II} k_z^I + \varepsilon_I k_z^{II}} \tag{2.3}$$

The reflected SF output field in medium I has a similar expression given by (in cgs units)(divided by 4π in SI units)

$$\vec{E}_r^I(\omega_3, z) = \vec{E}_{Sr}^I(\omega_3, z) + \vec{E}_{Br}^I(\omega_3, z)$$

$$= \frac{i2\pi\omega_3^2}{c^2 k_{3z}^I} \overset{\leftrightarrow}{F}_t^{I-II}(\omega_3) : [\overset{\leftrightarrow}{\chi}_{SD}^{(2)} - \overset{\leftrightarrow}{\chi}_{BD}^{(2)}/i\Delta k_{r,z}^{II}] : \overset{\leftrightarrow}{F}_t^{I-II}(\omega_1)$$

$$\overset{\leftrightarrow}{F}_t^{I-II}(\omega_2) : \vec{\mathcal{E}}_1^I \vec{\mathcal{E}}_2^I e^{i(-k_{3z}^I z - \omega_3 t)} \tag{2.4}$$

The magnitude of the phase mismatch $\Delta k_{r,z}^{II} = k_{1z}^{II} + k_{2z}^{II} + k_{3z}^{II}$, for reflected SFG is much larger than $\Delta k_{t,z}^{II}$ for transmitted SFG. We note that Δk_z^{II} is complex if medium II is absorbing. In strongly absorbing media, $1/|\Delta k_z^{II}|$ could be limited by the absorption length. If medium II has no inversion symmetry, then $\overset{\leftrightarrow}{\chi}_{BD}^{(2)} \neq 0$, and $\overset{\leftrightarrow}{\chi}_{SD}^{(2)}$ should be negligible compared to $|\overset{\leftrightarrow}{\chi}_{BD}^{(2)}/\Delta k_z^{II}|$ in the above equations.

If medium II has inversion symmetry, then $\overset{\leftrightarrow}{\chi}_{BD}^{(2)} = 0$ and we need to go beyond ED approximation, replacing $\overset{\leftrightarrow}{\chi}_{BD}^{(2)}$ in Eqs. (2.1), (2.2), and (2.4) by $\overset{\leftrightarrow}{\chi}_{BQ}^{(2)}$, which denotes bulk nonlinear susceptibility from electric-quadrupole (EQ) (including magnetic-dipole) contribution (to be discussed in the next section), and the surface nonlinearity $\overset{\leftrightarrow}{\chi}_{SD}^{(2)}$ could be comparable to $\overset{\leftrightarrow}{\chi}_{BQ}^{(2)}/i\Delta k_z^{II}$. It turns out that for transmitted SFG, even if $\overset{\leftrightarrow}{\chi}_{BD}^{(2)} = 0$, surface contribution from $\overset{\leftrightarrow}{\chi}_{SD}^{(2)}$ is usually still negligible in comparison to $|\overset{\leftrightarrow}{\chi}_{BQ}^{(2)}/\Delta k_{tz}^{II}|$ because $\Delta k_{t,z}^{II}$ is small. For reflected SFG, $\Delta k_{r,z}^{II}$ is orders of magnitude larger than $\Delta k_{t,z}^{II}$ and we often have surface contribution dominate over bulk contribution. This is the reason why reflected SFG is effective as a surface-specific probe. There could also be EQ contribution to SFG from the interfacial layer. We shall briefly discuss the role of EQ nonlinearities in the next section, but later throughout the book, we shall always consider EQ contribution to SFG negligible unless specifically noted.

In practice, each input beam has its own finite beam diameter and intensity profile. To find SFG output, geometric optics approximation is used. The SF output power is given by integration of SF intensity over area,

$$\mathcal{P}_{t,r}^{I}(\omega_3) = \int \frac{cn_I}{2\pi} |E_{t,r}^{I}(\omega_3, x, y)|^2 dA \propto \int |E^I(\omega_1, x, y)$$
$$E^I(\omega_2, x, y)|^2 dA. \tag{2.5}$$

With pulse inputs, the SF output energy per pulse is given by

$$W_{t,r}^{I}(\omega_3) = \int \frac{cn_I}{2\pi} |E_{t,r}^{I}(\omega_3, x, y)|^2 dA dt \propto \int |E^I(\omega_1, x, y, t)$$
$$E^I(\omega_2, x, y, t)|^2 dA dt. \tag{2.6}$$

(In SI units, the right-hand side of Eqs. (2.5) and (2.6) should be multiplied by $4\pi\varepsilon_0$ with $\varepsilon_0 = 8.854 \times 10^{-12}\text{F/m}$.)

For material characterization using SFG spectroscopy, we are interested in deducing from measurement the nonlinear susceptibilities: $\overset{\leftrightarrow}{\chi}_{BD}^{(2)}$ or $\overset{\leftrightarrow}{\chi}_{BQ}^{(2)}$ for the bulk and $\overset{\leftrightarrow}{\chi}_{SD}^{(2)}$ for the interface. In Secs. 2.3–2.6, we shall discuss the general characteristics of $\overset{\leftrightarrow}{\chi}_{D}^{(2)}$, i.e., its symmetry, resonance behavior, relation to nonlinear polarizability of constituent molecules, structural and microscopic information of materials. We shall also describe how the tensor elements of $\overset{\leftrightarrow}{\chi}_{D}^{(2)}$ can be extracted from measurements. More emphasis will be on the subject related to surface SFG spectroscopy for surface/interface characterization. The following section can be skipped for readers not interested in complex mathematic description.

2.2. Electric-Quadrupole Nonlinear Response of a Medium[2–4]

If the EQ response is taken into account, the field-induced polarization in a medium should appear as $\vec{P} = \vec{P}_D - \nabla \cdot \overset{\leftrightarrow}{Q}$, where $\vec{P}_D$ is the ED polarization and $\overset{\leftrightarrow}{Q}$ is the EQ polarization.[4] In the linear response case, $\vec{P}_D^{(1)} = \overset{\leftrightarrow}{\chi}_D^{(1)} \cdot \vec{E} + \overset{\leftrightarrow}{\chi}_{Q1}^{(1)} \cdot \nabla\vec{E}$ and $\overset{\leftrightarrow}{Q}^{(1)} = \overset{\leftrightarrow}{\chi}_{Q2}^{(1)} \cdot \vec{E}$. For $\vec{E}(\omega) = \vec{\mathcal{E}}e^{i\vec{k}\cdot\vec{r}-i\omega t}$ in a bulk medium, we can combine the two and write

$$\vec{P}^{(1)}(\omega, \vec{k}) = \overset{\leftrightarrow}{\chi}_D^{(1)}(\omega) \cdot \vec{E} + \overset{\leftrightarrow}{\chi}_Q^{(1)}(\omega, \vec{k}) : i\vec{k}\vec{E}. \tag{2.7}$$

In the second-order nonlinear response case with two input fields $\vec{E}_1$ and $\vec{E}_2$, we have

$$\vec{P}_D^{(2)} = \overset{\leftrightarrow}{\chi}_D^{(2)} : \vec{E}_1\vec{E}_2 + \overset{\leftrightarrow}{\chi}_{q1}^{(2)} \cdot (\nabla\vec{E}_1)\vec{E}_2 + \overset{\leftrightarrow}{\chi}_{q2}^{(2)} \cdot \vec{E}_1(\nabla\vec{E}_2)$$
$$\overset{\leftrightarrow}{Q}^{(2)} = \overset{\leftrightarrow}{\chi}_{q3}^{(2)} \cdot \vec{E}_1\vec{E}_2 \tag{2.8}$$

where all $\overset{\leftrightarrow}{\chi}_{qi}^{(2)}$'s are fourth-rank tensors. For $\vec{E}_1(\omega) = \vec{\mathcal{E}}_1 e^{i\vec{k}_1\cdot\vec{r} - i\omega_1 t}$ and $\vec{E}_2(\omega) = \vec{\mathcal{E}}_2 e^{i\vec{k}_2\cdot\vec{r} - i\omega_2 t}$ in a bulk medium, we find

$$\vec{P}^{(2)}(\omega_3 = \omega_1 + \omega_2; \vec{k}_1, \vec{k}_2, \vec{k}_3) = \vec{P}_D^{(2)} - \nabla \cdot \overset{\leftrightarrow}{Q}^{(2)} =$$
$$(\overset{\leftrightarrow}{\chi}_{BD}^{(2)} + \overset{\leftrightarrow}{\chi}_{BQ}^{(2)}) : \vec{E}_1\vec{E}_2,$$
$$\overset{\leftrightarrow}{\chi}_{BQ}^{(2)}(\omega_3 = \omega_1 + \omega_2; \vec{k}_1, \vec{k}_2, \vec{k}_3) \equiv -i\vec{k}_3 \cdot \overset{\leftrightarrow}{\chi}_{q3}^{(2)} + i\vec{k}_1 \cdot \overset{\leftrightarrow}{\chi}_{q1}^{(2)}$$
$$+ i\vec{k}_2 \cdot \overset{\leftrightarrow}{\chi}_{q2}^{(2)} \tag{2.9}$$

The EQ response is however much weaker than the ED response if both are present; roughly speaking, $\overset{\leftrightarrow}{\chi}_{BQ}^{(2)}$ is kr times smaller than $\overset{\leftrightarrow}{\chi}_{BD}^{(2)}$ with r being the size of an electronic orbit in a molecule. In the visible range, $\overset{\leftrightarrow}{\chi}_{BD}^{(2)}$ is about $10^2 - 10^3$ times larger than $\overset{\leftrightarrow}{\chi}_{BQ}^{(2)}$ if $r < 1\text{nm}$ and therefore in media without inversion symmetry, the EQ contribution to SFG can generally be neglected. In media with inversion symmetry, $\overset{\leftrightarrow}{\chi}_{BD}^{(2)} = 0$ and $\overset{\leftrightarrow}{\chi}_{BQ}^{(2)}$ is solely responsible for SFG in bulk. In comparison with interface contribution from $\overset{\leftrightarrow}{\chi}_{SD}^{(2)}$ for transmitted SFG, we still have $|\overset{\leftrightarrow}{\chi}_{BQ}^{(2)}/i\Delta k_{tz}^{II}| >> |\overset{\leftrightarrow}{\chi}_{SD}^{(2)}|$ so that the interface contribution is negligible. For reflected SFG, however, because $|\Delta k_{rz}^{II}| \gg |\Delta k_{tz}^{II}|$, we may have $|\overset{\leftrightarrow}{\chi}_{BQ}^{(2)}/i\Delta k_{rz}^{II}|$ comparable to $|\overset{\leftrightarrow}{\chi}_{SD}^{(2)}|$, but in many cases discussed in the book, $|\overset{\leftrightarrow}{\chi}_{SD}^{(2)}|$ actually dominates.[4]

As discussed in Sec. 2.1, SFG from a medium with inversion symmetry measures $\overset{\leftrightarrow}{\chi}_{SD}^{(2)} - \frac{\overset{\leftrightarrow}{\chi}_{BQ}^{(2)}}{i\Delta k_z^{II}}$ with $\overset{\leftrightarrow}{\chi}_{BQ}^{(2)}$ given in Eq. (2.9). We can also write

$$\overset{\leftrightarrow}{\chi}_{SD}^{(2)} - \left[\overset{\leftrightarrow}{\chi}_{BQ}^{(2)}/i\Delta k_z^{II}\right] = \left(\overset{\leftrightarrow}{\chi}_{SD}^{(2)} - \hat{z}\cdot\overset{\leftrightarrow}{\chi}_{q3}^{(2)}\right) - \left[\overset{\leftrightarrow}{\chi}_{BBQ}^{(2)}/i\Delta k_z^{II}\right]$$

$$\text{with}\qquad \overset{\leftrightarrow}{\chi}_{BBQ}^{(2)} \equiv [\overset{\leftrightarrow}{\chi}_{q1}^{(2)}\cdot i\vec{k}_1^{II} + \overset{\leftrightarrow}{\chi}_{q2}^{(2)}\cdot i\vec{k}_2^{II} - i(\vec{k}_1^{II}+i\vec{k}_2^{II})\cdot\overset{\leftrightarrow}{\chi}_{q3}^{(2)}]$$

$$\tag{2.10}$$

where $\Delta k_z^{II} = \Delta k_{tz}^{II}$ for transmitted SFG and $\Delta k_z^{II} = \Delta k_{rz}^{II}$ for reflected SFG. In the above expression, $(\overset{\leftrightarrow}{\chi}_{SD}^{(2)} - \hat{z}\cdot\overset{\leftrightarrow}{\chi}_{q3}^{(2)})$ is independent of wave vectors, but describes a combination of two terms of ED and EQ origins. Actually, these two terms are intrinsically inseparable because of ambiguity in the common definition of ED versus EQ polarizations.[4] It is $\left[\overset{\leftrightarrow}{\chi}_{BBQ}^{(2)}/-i\Delta k_{tz}^{II}\right]$, not $\left[\overset{\leftrightarrow}{\chi}_{BQ}^{(2)}/-i\Delta k_{tz}^{II}\right]$, that usually dominates and is measured in transmitted SFG. The reflected SFG, on the other hand, measures $\overset{\leftrightarrow}{\chi}_{S}^{(2)} \equiv \overset{\leftrightarrow}{\chi}_{SD}^{(2)} + \left[(\overset{\leftrightarrow}{\chi}_{BBQ}^{(2)})_r/-i\Delta k_{rz}^{II}\right] - \hat{z}\cdot\overset{\leftrightarrow}{\chi}_{q3}^{(2)}$ and allows extraction of $(\overset{\leftrightarrow}{\chi}_{SD}^{(2)} - \hat{z}\cdot\overset{\leftrightarrow}{\chi}_{q3}^{(2)})$, because $(\overset{\leftrightarrow}{\chi}_{BBQ}^{(2)})_r$ can be deduced from $(\overset{\leftrightarrow}{\chi}_{BBQ}^{(2)})_t$ measured by transmitted SFG. It turns out that because $|\Delta k_{rz}^{II}|$ is large, the $\left[(\overset{\leftrightarrow}{\chi}_{BBQ}^{(2)})_r/-i\Delta k_{rz}^{II}\right]$ term is usually negligible in reflected SFG.

In explicit tensorial notation with the Cartesian coordinates, we have

$$(\chi_{BBQ}^{(2)})_{\alpha\beta\gamma} = \sum_{\kappa}\chi_{q1,\alpha(\bar{\kappa}\beta)\gamma}^{(2)}(ik_{1,\kappa}^{II}) + \chi_{q2,\alpha\beta(\bar{\kappa}\gamma)}^{(2)}(ik_{2,\kappa}^{II})$$
$$- i(k_{1,\kappa}^{II}+k_{2,\kappa}^{II})\chi_{q3,(\bar{\kappa}\alpha)\beta\gamma}^{(2)}] \tag{2.11}$$
$$(\hat{z}\cdot\overset{\leftrightarrow}{\chi}_{q3}^{(2)})_{\alpha\beta\gamma} = \chi_{q3,(\bar{z}\alpha)\beta\gamma}^{(2)}$$

where the bracketed sub-indices indicate the EQ field component with the one under a bar referring to the direction of the wave vector component; for example, $\chi_{q2,yy(\bar{z}z)}^{(2)}$ denotes the EQ nonlinear susceptibility associated with $E_{3y}(\omega_3)$ in response to $E_{1y}(\omega_1)[k_{2,z}^{II}E_{2z}(\omega_2)]$.

There is of course also an EQ contribution from an interface layer.[4] Because the optical fields must vary rapidly to change from $\vec{E}^I$ in medium I to $\vec{E}^{II}$ in medium II across an interface layer, the field gradient in the interface layer is dominated by field amplitude gradient in the surface normal direction $[\nabla\vec{E} = \nabla(\vec{\mathcal{E}}e^{i\vec{k}\cdot\vec{r}-i\omega t}) = (\nabla\vec{\mathcal{E}} + i\vec{k}\vec{\mathcal{E}})e^{i\vec{k}\cdot\vec{r}-i\omega t} \approx \nabla\vec{\mathcal{E}}e^{i\vec{k}\cdot\vec{r}-i\omega t}]$. Following Eqs. (2.8) and (2.9), we can express the induced second-order surface polarization of the interfacial layer as $\vec{P}_S^{(2)}(\omega_3) = (\overset{\leftrightarrow}{\chi}_{SD}^{(2)} + \overset{\leftrightarrow}{\chi}_{SQ}^{(2)}) : \vec{E}_1\vec{E}_2$, and we need to replace $\overset{\leftrightarrow}{\chi}_{SD}^{(2)}$ by $(\overset{\leftrightarrow}{\chi}_{SD}^{(2)} + \overset{\leftrightarrow}{\chi}_{SQ}^{(2)})$ in earlier expressions for SFG to take into account this interfacial EQ contribution. In later discussions on SFG for surface studies, we shall mainly focus on cases where $\overset{\leftrightarrow}{\chi}_{SD}^{(2)}$ dominates and comment on EQ contribution only when needed.

2.3. Symmetry and Characteristics of Surface Nonlinear Susceptibility

Symmetry of response coefficients of a medium generally reflects the structural symmetry of the medium. A response coefficient tensor can be greatly simplified from symmetry consideration. For a material belonging to a certain symmetry class, there is a corresponding set of symmetry operations such as inversion, reflection, rotation, etc., under which the material remains unchanged. With coordinates properly chosen to be along symmetry axes, each symmetry operation sets up a relation among the tensor elements of the response coefficient, thereby making some tensor elements vanish and others depend on one another. Materials with higher symmetry have less independent, non-vanishing elements.

Non-vanishing tensor elements of second-order ED nonlinear susceptibility $\overset{\leftrightarrow}{\chi}_{BD}^{(2)}$ for bulk materials of different symmetry classes can be found in books on nonlinear optics.[2] The form of surface nonlinear susceptibility $\overset{\leftrightarrow}{\chi}_{SD}^{(2)}$ is more complicated, depending on how the surface is oriented with respect to the bulk structure. If the surface normal is along a high symmetry axis of the bulk, then the form of $\overset{\leftrightarrow}{\chi}^{(2)}_{SD}$ can be obtained by amending $\overset{\leftrightarrow}{\chi}^{(2)}_{BD}$ with additional nonvanishing elements arising from the broken symmetry along the surface normal direction. Consider, for example, an isotropic material with inversion symmetry. The bulk $\overset{\leftrightarrow}{\chi}^{(2)}_{BD}$ vanishes, but $\overset{\leftrightarrow}{\chi}^{(2)}_{SD}$ has four independent, nonvanishing elements:

$$\chi^{(2)}_{SD,zzz}, \chi^{(2)}_{SD,zxx} = \chi^{(2)}_{SD,zyy}, \chi^{(2)}_{SD,xzx} = \chi^{(2)}_{SD,yzy},$$

$$\text{and } \chi^{(2)}_{SD,xxz} = \chi^{(2)}_{SD,yyz}, \tag{2.12}$$

with z along the surface normal. The same result applies to cubic and hexagonal crystals with inversion symmetry if z is along a symmetric axis of cubic crystals or along the uniaxis of hexagonal crystals. If the interfacial layer of a bulk material belonging to the above symmetric classes has additional broken mirror symmetry along x, the nonvanishing elements of $\overset{\leftrightarrow}{\chi}^{(2)}_{SD}$ become

$$\chi^{(2)}_{SD,zzz}, \chi^{(2)}_{SD,xxx}, \chi^{(2)}_{SD,xxz}, \chi^{(2)}_{SD,yyz}, \chi^{(2)}_{SD,zzx}, \chi^{(2)}_{SD,yyx}, \chi^{(2)}_{SD,zyy},$$

$$\chi^{(2)}_{SD,yzy}, \chi^{(2)}_{SD,zxx}, \chi^{(2)}_{SD,xzx}, \chi^{(2)}_{SD,xzz}, \chi^{(2)}_{SD,zxz}, \chi^{(2)}_{SD,xyy}$$

and $\chi^{(2)}_{SD,yxy}$.

On SFG experiment, it is important that we know how various tensor elements of $\overset{\leftrightarrow}{\chi}^{(2)}_{BD}$ or $\overset{\leftrightarrow}{\chi}^{(2)}_{SD}$ participate in the SFG process with chosen beam geometry and polarizations with respect to the material orientation. Knowing the tensorial form of $\overset{\leftrightarrow}{\chi}^{(2)}$ for a material allows us to design SFG experiment for desired output. Conversely, from the symmetry or nonvanishing elements of the rank-3 $\overset{\leftrightarrow}{\chi}^{(2)}$ tensor, two-fold and three-fold rotational symmetric axes, and more generally the symmetry class, of a material can be identified by SFG measurement. This allows SFG/SHG to be used to monitor structural phase transition of materials.

As discussed in Sec. 2.2, the SF output fields in transmission and reflection under ED approximation are given by $\vec{E}_{t,r}(\omega_3) \propto -\dfrac{\overset{\leftrightarrow}{\chi}^{(2)}_{BD}}{i(\Delta k_{t,r})_z} : \vec{E}(\omega_1)\vec{E}(\omega_2)$ in media without inversion symmetry and

$\vec{E}_r(\omega_3) \propto \overset{\leftrightarrow}{\chi}_{SD}^{(2)} : \vec{E}(\omega_1)\vec{E}(\omega_2)$ in media with inversion symmetry, respectively. In principle, n independent elements of $\overset{\leftrightarrow}{\chi}_{BD}^{(2)}$ or $\overset{\leftrightarrow}{\chi}_{SD}^{(2)}$ can be determined by a set of n measurements with different polarization combinations of the input/output fields. Each measurement may measure a linear combination of several tensor elements. For better characterization of materials, one would like to know as many independent elements of $\overset{\leftrightarrow}{\chi}_{BD}^{(2)}$ or $\overset{\leftrightarrow}{\chi}_{SD}^{(2)}$ as possible. In practice, however, measurement of a $\overset{\leftrightarrow}{\chi}^{(2)}$ spectrum is not very simple because $\overset{\leftrightarrow}{\chi}^{(2)}$ is complex and its determination requires information on the relative phase of $\vec{E}(\omega_3)$ and $\vec{E}(\omega_1)\vec{E}(\omega_2)$. We usually have to be satisfied with measurements of only a limited number of $\overset{\leftrightarrow}{\chi}^{(2)}$ elements. More details on extraction of complex $\overset{\leftrightarrow}{\chi}^{(2)}$ elements from measurements will be discussed in Chapter 3.

Similar to $\overset{\leftrightarrow}{\varepsilon}$ (or $\overset{\leftrightarrow}{\chi}^{(1)}$) in the linear case, $\overset{\leftrightarrow}{\chi}^{(2)}$ becomes complex as input or/and output frequencies approach resonances. In the single-resonant case, both $\overset{\leftrightarrow}{\varepsilon}$ and $\overset{\leftrightarrow}{\chi}^{(2)}$ (with ω_2 near resonances) have similar dispersive expressions showing resonant enhancement: We have, for discrete resonances,

$$
\begin{aligned}
\overset{\leftrightarrow}{\varepsilon}(\omega) &= \overset{\leftrightarrow}{\varepsilon}{}'(\omega) + i\overset{\leftrightarrow}{\varepsilon}{}''(\omega) \\
&= 1 + 4\pi\overset{\leftrightarrow}{\chi}^{(1)}(\omega) \\
&= \overset{\leftrightarrow}{\varepsilon}_{NR} - 4\pi\sum_q \frac{\overset{\leftrightarrow}{C}_q}{\omega - \omega_q + i\Gamma_q} \\
\overset{\leftrightarrow}{\varepsilon}{}'(\omega) &= \overset{\leftrightarrow}{\varepsilon}_{NR} - 4\pi\sum_q \frac{\overset{\leftrightarrow}{C}_q(\omega - \omega_q)}{(\omega - \omega_q)^2 + \Gamma_q^2} \\
\overset{\leftrightarrow}{\varepsilon}{}''(\omega) &= 4\pi\sum_q \frac{\overset{\leftrightarrow}{C}_q\Gamma_q}{(\omega - \omega_q)^2 + \Gamma_q^2}
\end{aligned}
$$

$$
\begin{aligned}
\overset{\leftrightarrow}{\chi}^{(2)}(\omega &= \omega_1 + \omega_2) \\
&= \mathrm{Re}\overset{\leftrightarrow}{\chi}^{(2)}(\omega) + i\mathrm{Im}\overset{\leftrightarrow}{\chi}^{(2)}(\omega) \\
&= \overset{\leftrightarrow}{\chi}_{NR}^{(2)} + \sum_q \frac{\overset{\leftrightarrow}{A}_q}{\omega_2 - \omega_q + i\Gamma_q} \\
\mathrm{Re}\overset{\leftrightarrow}{\chi}^{(2)} &= \overset{\leftrightarrow}{\chi}_{NR}^{(2)} + \sum_q \frac{\overset{\leftrightarrow}{A}_q(\omega_2 - \omega_q)}{(\omega_2 - \omega_q)^2 + \Gamma_q^2} \\
\mathrm{Im}\overset{\leftrightarrow}{\chi}^{(2)} &= -\sum_q \frac{\overset{\leftrightarrow}{A}_q\Gamma_q}{(\omega_2 - \omega_q)^2 + \Gamma_q^2}
\end{aligned}
$$

$$(2.13)$$

and for continuum resonances,

$$
\overset{\leftrightarrow}{\varepsilon}(\omega) = \overset{\leftrightarrow}{\varepsilon}_{NR} - 4\pi\int \frac{\overset{\leftrightarrow}{C}_q}{(\omega - \omega_q) + i\Gamma_q}\rho(\omega_q)d\omega_q
$$

$$
\overset{\leftrightarrow}{\varepsilon}{}''(\omega) = 4\pi\int \frac{\overset{\leftrightarrow}{C}_q\Gamma_q}{(\omega - \omega_q)^2 + \Gamma_q^2}\rho(\omega_q)d\omega_q \approx 4\overset{\leftrightarrow}{C}_q(\omega)\rho(\omega_q)
$$

$$\overset{\leftrightarrow}{\chi}^{(2)}(\omega = \omega_1 + \omega_2) = \overset{\leftrightarrow}{\chi}_{NR}^{(2)} + \int \frac{\overset{\leftrightarrow}{A}_q}{(\omega_2 - \omega_q) + i\Gamma_q}\rho(\omega_q)d\omega_q$$

$$\mathrm{Im}\,\overset{\leftrightarrow}{\chi}^{(2)} = -\int \frac{\overset{\leftrightarrow}{A}_q\Gamma_q}{(\omega_2 - \omega_q)^2 + \Gamma_q^2}\rho(\omega_q)d\omega_q$$

$$\approx -\overset{\leftrightarrow}{A}_q(\omega_2 = \omega_q)\rho(\omega_2 = \omega_q)/\pi \tag{2.14}$$

(In SI units, $\overset{\leftrightarrow}{\varepsilon}, \overset{\leftrightarrow}{\varepsilon}{}', \overset{\leftrightarrow}{\varepsilon}{}''$, and $\overset{\leftrightarrow}{\varepsilon}_{NR}$ in Eqs. (2.13) and (2.14) should be replaced by $\overset{\leftrightarrow}{\varepsilon}/\varepsilon_0, \overset{\leftrightarrow}{\varepsilon}{}'/\varepsilon_0, \overset{\leftrightarrow}{\varepsilon}{}''/\varepsilon_0$, and $\overset{\leftrightarrow}{\varepsilon}_{NR}/\varepsilon_0$ with $\varepsilon_0 = 8.854 \times 10^{-12}$F/m, and the final expression divided by 4π.). In Eqs. (2.13) and (2.14), $\overset{\leftrightarrow}{\varepsilon}_{NR}$ and $\overset{\leftrightarrow}{\chi}_{NR}^{(2)}$ are the nonresonant parts of $\overset{\leftrightarrow}{\varepsilon}$ and $\overset{\leftrightarrow}{\chi}^{(2)}$, ω_q and Γ_q are the frequency and damping constant of the qth resonant mode, respectively, and $\overset{\leftrightarrow}{C}_q$ and $\overset{\leftrightarrow}{A}_q$ are the corresponding resonant amplitude in $\overset{\leftrightarrow}{\varepsilon}$ and $\overset{\leftrightarrow}{\chi}^{(2)}$. For continuum resonance bands, $\rho(\omega_q)$ denotes the joint density of modes of the resonant transition at ω_q. It is seen from the above equations that both $\overset{\leftrightarrow}{\varepsilon}{}''$ and $\mathrm{Im}\,\overset{\leftrightarrow}{\chi}^{(2)}$ directly describe characteristics of resonances. The difference is that while the sign of $\overset{\leftrightarrow}{\varepsilon}{}''(\omega)$ denotes gain or loss of the field at ω (positive for gain and negative for loss), the sign of $\mathrm{Im}\,\overset{\leftrightarrow}{\chi}^{(2)}(\omega_2 = \omega_q)$ or $\overset{\leftrightarrow}{A}_q(\omega_q)$ denotes the polar orientation of the resonance at ω_q; since $\overset{\leftrightarrow}{\chi}^{(2)}$ is a response coefficient of a system with a net polar orientation, it changes sign when the polar orientation flips. This is a unique, powerful feature of SFG spectroscopy.

2.4. Surface Sum Frequency Spectroscopy

Optical spectroscopy is most effective for probing material properties and structures. Through measurement of resonant characteristics associated with optical transitions, microscopic structural information about materials can be gleaned. Having two independent input beams in measurement, SFG spectroscopy via $\overset{\leftrightarrow}{\chi}^{(2)}$ is more informative than other spectroscopy techniques. The discussion in this section is applicable to both bulk $\overset{\leftrightarrow}{\chi}^{(2)}_{BD}$ and surface $\overset{\leftrightarrow}{\chi}^{(2)}_{SD}$, but focus will be on $\overset{\leftrightarrow}{\chi}^{(2)}_{SD}$ because SFG spectroscopy has been more developed for surface characterization. To see how $\overset{\leftrightarrow}{\chi}^{(2)}_{D}$ carries information about microscopic structure of materials we need a microscopic description on $\overset{\leftrightarrow}{\chi}^{(2)}_{D}$.

On the molecular basis, a material is considered as an ensemble of molecules with a prescribed orientation distribution. (The microscopic theory of nonlinear optical responses of materials with crystalline band structure is much more complex, and has not yet been well developed for characterization of materials. However, even without detailed understanding, spectroscopic information provided by $\overset{\leftrightarrow}{\chi}^{(2)}_{SD}$ for such materials could still be useful.) The incoming fields induce on each molecule an electric dipole $\vec{p} = \vec{p}^{(1)} + \vec{p}^{(2)}$ with $\vec{p}^{(1)}(\omega) = \overset{\leftrightarrow}{\alpha}^{(1)}(\omega) : \vec{E}(\omega)$ for linear response and $\vec{p}^{(2)}(\omega_3) = \overset{\leftrightarrow}{\alpha}^{(2)}(\omega_3 = \omega_1 + \omega_2) : \vec{E}(\omega_1)\vec{E}(\omega_2)$ for SFG response, where $\overset{\leftrightarrow}{\alpha}^{(1)}$ and $\overset{\leftrightarrow}{\alpha}^{(2)}$ are the linear and second-order nonlinear polarizabilities, respectively. Here, we define $\overset{\leftrightarrow}{\alpha}^{(1)}$ and $\overset{\leftrightarrow}{\alpha}^{(2)}$ not as response coefficients of isolated molecules, but as response coefficients of molecules influenced by intermolecular interactions and local field effects from surrounding molecules. We neglect EQ contribution to $\vec{p}^{(1)}$ and $\vec{p}^{(2)}$. With N molecules per unit volume, the linear and nonlinear susceptibilities of the medium are given by $\overset{\leftrightarrow}{\chi}^{(1)} = N < \overset{\leftrightarrow}{\alpha}^{(1)} >$ and $\overset{\leftrightarrow}{\chi}^{(2)} = N < \overset{\leftrightarrow}{\alpha}^{(2)} >$, where the angular brackets denote an orientation average over molecules. To be more specific, we can express $\overset{\leftrightarrow}{\alpha}$ and $\overset{\leftrightarrow}{\chi}$ in explicit tensorial notation with molecular coordinates $(\hat{x}', \hat{y}', \hat{z}')$ for $\overset{\leftrightarrow}{\alpha}$ usually taken along molecular symmetry axes and lab coordinates $(\hat{x}, \hat{y}, \hat{z})$ for $\overset{\leftrightarrow}{\chi}$, along material symmetry axes. For $\overset{\leftrightarrow}{\chi}^{(2)}_{SD}$, we have explicitly

$$\chi^{(2)}_{SD,ijk} = N_S \sum_{\xi,\eta,\zeta} < (\hat{i} \cdot \hat{\xi})(\hat{j} \cdot \hat{\eta})(\hat{k} \cdot \hat{\zeta}) > \alpha^{(2)}_{\xi\eta\zeta}$$

$$= N_S \sum_{\xi,\eta,\zeta} \int (\hat{i} \cdot \hat{\xi})(\hat{j} \cdot \hat{\eta})(\hat{k} \cdot \hat{\zeta}) \alpha^{(2)}_{\xi\eta\zeta} f(\Omega) d\Omega \qquad (2.15)$$

where N_S is the surface density of molecules, ξ, η, and ζ refer to x', y', or z', i, j, and k to x, y, or z, and $f(\Omega)$ is the orientation distribution function with Ω denoting the solid angle. Measurement of M independent $\chi^{(2)}_{SD,ijk}$ elements from reflected SFG allows determination of m unknown $\alpha^{(2)}_{\xi\eta\zeta}$ elements and $M - m$ parameters that define $f(\Omega)$, assuming $M > m$. Information on molecular orientation at an interface is useful in many applications. More detailed discussion on how to find surface molecular orientation will be provided in Sec. 6.3.

As seen in Eq. (2.6) together with Eq. (2.4), if $\overleftrightarrow{\varepsilon}_I(\omega)$ and $\overleftrightarrow{\varepsilon}_{II}(\omega)$ are known and the Fresnel coefficients calculated, the reflected SF spectroscopic measurement with specific input/output polarizations on a medium with $\chi^{(2)}_{BD,ijk} = 0$ allows extraction of $|\chi^{(2)}_{SD,ijk}(\omega_3 = \omega_1 + \omega_2)|^2$, but not the complex $\chi^{(2)}_{SD,ijk}(\omega_3 = \omega_1 + \omega_2)$. Yet, like $\varepsilon''(\omega) = \mathrm{Im}\ \varepsilon(\omega)$ in linear spectroscopy, it is $\mathrm{Im}\ \chi^{(2)}_{SD}$ that carries direct information about surface resonances. We shall discuss in Sec. 3.4 how we can find $\mathrm{Im}\ \chi^{(2)}_{SD}$ by spectral fitting of $|\chi^{(2)}_{SD}(\omega_3 = \omega_1 + \omega_2)|^2$ with Eq. (2.13) in case of discrete resonances, or experimentally by phase-sensitive SF spectroscopy.

2.5. Information Content of Surface Nonlinear Susceptibility

Microscopic information carried by optical response coefficients comes from resonances associated with optical transitions as spectroscopy commonly does. Figures 2.2(a) and 2.2(b) describe possible resonance transitions of linear and SFG processes, respectively. We have $\overleftrightarrow{\chi}^{(1)} = N\langle\overleftrightarrow{\alpha}^{(1)}\rangle$ and $\overleftrightarrow{\chi}^{(2)} = N\langle\overleftrightarrow{\alpha}^{(2)}\rangle$, and the microscopic expression of $\overleftrightarrow{\alpha}^{(1)}$ and $\overleftrightarrow{\alpha}^{(2)}$ can be derived from the quantum mechanical perturbation theory:

$$\alpha^{(1)}_{\xi\eta}(\omega) = \sum_{g,n,n'} \left(-\frac{e^2}{\hbar}\right) \left[\frac{\langle g|r_\xi|n\rangle\langle n|r_\eta|g\rangle}{(\omega - \omega_{ng} + i\Gamma_{ng})} \right.$$

$$\left. - \frac{\langle g|r_\eta|n\rangle\langle n|r_\xi|g\rangle}{(\omega + \omega_{ng} + i\Gamma_{ng})} \right] \rho^0_{gg} \tag{2.16}$$

$$\alpha^{(2)}_{\xi\eta\zeta}(\omega_3 = \omega_1 + \omega_2) = \sum_{g,n,n'} \left(-\frac{e^3}{\hbar^2}\right)\left[\frac{\langle g|r_\xi|n\rangle\langle n|r_\eta|n'\rangle\langle n'|r_\zeta|g\rangle}{(\omega_3 - \omega_{ng} + i\Gamma_{ng})(\omega_2 - \omega_{n'g} + i\Gamma_{n'g})}\right.$$

$$\left. + \frac{\langle g|r_\xi|n\rangle\langle n|r_\zeta|n'\rangle\langle n'|r_\eta|g\rangle}{(\omega_3 - \omega_{ng} + i\Gamma_{ng})(\omega_1 - \omega_{n'g} + i\Gamma_{n'g})} + 6 \ \text{other terms}\right]\rho^0_{gg}$$

$$(2.17)$$

where ρ^0_{gg} is the population in the $<g|$ state, ω_{ng} and Γ_{ng} are the transition frequency from $<g|$ to $<n|$ and the associated damping constant, respectively. It is clear, from Eqs. (2.16) and (2.17), that $\chi^{(2)}_{\xi\eta\zeta}(\omega_3 = \omega_1+\omega_2)$ carries more spectral information than $\chi^{(1)}_{ij}(\omega)$.

For SFG, in the case of ω_2 approaching a resonance (sketched in Fig. 2.2(b.1)), Eq. (2.17) can be written as

$$\alpha^{(2)}_{\xi\eta\zeta}(\omega_3 = \omega_1 + \omega_2) = \alpha^{(2)}_{NR,\xi\eta\zeta} + \alpha^{(2)}_{R,\xi\eta\zeta}$$

$$\text{with} \ \ \alpha^{(2)}_{R,\xi\eta\zeta} = -\frac{(R_{\xi\eta})_{gn'}(\mu_\zeta)_{n'g}}{\hbar(\omega_2 - \omega_{n'g} + i\Gamma_{n'g})}\rho^0_{gg}$$

$$(\mu_\zeta)_{n'g} \equiv \langle n'|er_\zeta|g\rangle$$

$$(R_{\xi\eta})_{gn'} \equiv \sum_n \frac{\langle g|er_\xi|n\rangle\langle n|er_\eta|n'\rangle}{\hbar(\omega_3 - \omega_{ng} + i\Gamma_{ng})}$$

$$(2.18)$$

where $\alpha^{(2)}_{R,\xi\eta\zeta}$ and $\alpha^{(2)}_{NR,\xi\eta\zeta}$ denote the resonant and nonresonant terms of $\alpha^{(2)}_{\xi\eta\zeta}$, and $(\mu_\zeta)_{n'g}$ and $(R_{\xi\eta})_{gn'}$ are the one-photon and two-photon transition matrix elements between $\langle g|$ and $|n'\rangle$, respectively, for the transitions denoted in Fig. 2.2. If ω_{ng} is in the infrared range, one usually labels $(\mu_\zeta)_{n'g}$ and $(R_{\xi\eta})_{gn'}$ as IR and Raman transition elements. On a vibrational resonance at ω_{qg} , the following notations are often adopted:

$$\alpha^{(2)}_{R,\xi\eta\zeta} = -\left(\frac{1}{2\omega_{qg}}\right)\frac{(\partial\mu_\zeta/\partial Q)(\partial\alpha^{(1)}_{\xi\eta}/\partial Q)}{\hbar(\omega_2 - \omega_{qg} + i\Gamma_{qg})}\rho^0_{gg}$$

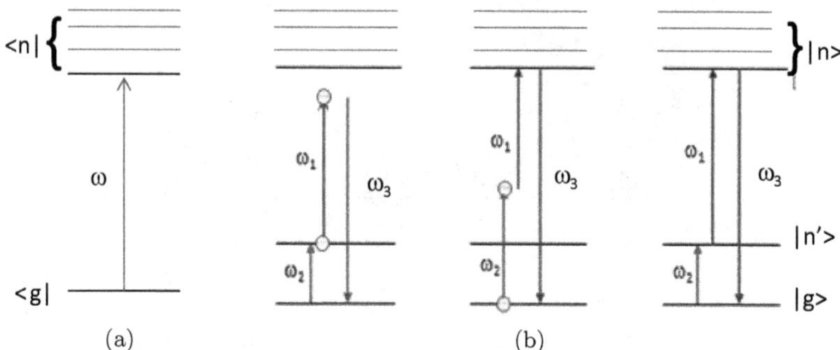

Fig. 2.2.   Description of resonant optical transitions in (a) linear optical process and (b) SFG processes in a medium with ground state |g>, and excited states |n> and |n'> . (b.1), (b.2), and (b.3) show ω_2 on resonance, $\omega_1 + \omega_2$ on resonance, and both ω_2 and $\omega_1 + \omega_2$ on resonances, respectively.

$$(\partial\mu_\zeta/\partial Q) \equiv \left(\frac{2\omega_{qg}}{\hbar}\right)^{1/2} \langle q|er_\zeta|g\rangle$$

$$(\partial\alpha_{\xi\eta}^{(1)}/\partial Q) \equiv \left(\frac{2\omega_{qg}}{\hbar}\right)^{1/2} \sum_n \frac{\langle g|er_\xi|n\rangle\langle n|er_\eta|q\rangle}{\hbar(\omega_3 - \omega_{ng} + i\Gamma_{ng})}$$

We note that unlike $\alpha^{(1)}$ for linear spectroscopy, $\alpha_R^{(2)}$ for SFG is nonvanishing only if the resonance transition is both one-photon and two-photon ED-active (or both IR and Raman active). Thus, compared to IR and Raman spectra of vibrations or phonons for a material, for example, the SFG spectrum often appears simpler.

Generally, a surface can be composed of multiple species, and accordingly the surface nonlinear susceptibility has the form $\overset{\leftrightarrow}{\chi}_{SD}^{(2)} = \sum_\sigma N_\sigma \langle \overset{\leftrightarrow}{\alpha}_\sigma^{(2)} \rangle$ with N_σ denoting the surface density of the σth surface species. Together with the microscopic expression of $\overset{\leftrightarrow}{\alpha}^{(2)}$ in Eq. (2.18), it is seen that spectroscopic measurement of $\chi_{SD}^{(2)}(\omega_3 = \omega_1 + \omega_2)$ should allow us to (1) identify surface species and their surface densities, (2) symmetry of surface molecular arrangement, (3) orientation of surface molecular species, (4) frequency, amplitude, and damping coefficient of resonance modes of each surface species, (5) population distribution in different states of each species, (6) spatial variation of surface composition on the ~10 μm scale, and finally, (7) temporal variation of the above quantities with ~ 10 fs time resolution. As shall be seen in later chapters, these specific features have promoted SFG spectroscopy to become a most viable surface analytical tool.

SFG spectroscopy with both input beams tunable in frequency can probe double resonances, as described in Fig. 2.2(b3) with $\omega_2 \sim \omega_{n'g}$ and $\omega_1 \sim \omega_{nn'}$. Similar to doubly-resonant Raman spectroscopy, doubly-resonant SFG spectroscopy yields additional information about coupling between two excited states $|n >$ and $|n' >$. The difference from Raman spectroscopy is that SFG is a coherent optical process that can provide information on coherent coupling among states. Doubly-resonant (or 2D) coherent spectroscopy has been actively developed into a technique for molecular spectroscopic studies in chemistry and bioscience in recent years.[7] Now, SFG facilitates such studies on surfaces and interfaces.

2.6. Signal Strength of Sum Frequency Generation from a Surface Layer

From Eq. (2.4) with $\overset{\leftrightarrow}{\chi}_{BD}^{(2)} = 0$, we can estimate the intensity of reflected SFG from an interface characterized by $\overset{\leftrightarrow}{\chi}_{SD}^{(2)}$. Let the input fields, $\vec{E}_1(\omega_1)$ and $\vec{E}_2(\omega_2)$, be polarized along $\hat{e}_1(\omega_1)$ and $\hat{e}_2(\omega_2)$ with incident angles θ_1 and θ_2 on a surface and their intensities be $I(\omega_i) = c\sqrt{\varepsilon(\omega_i)}|E(\omega_i)|^2/2\pi$. With $k_3^I/k_{3z}^I = \omega_3\sqrt{\varepsilon(\omega_3)}/ck_{3z}^I = \sec\theta_3$, the intensity of the reflected SF output exiting at angle θ_3 from the surface obtained from $\vec{E}_r^I(\omega_3)$ in Eq.(2.4) has the following expression:

$$I(\omega_3) = c\sqrt{\varepsilon(\omega_3)}|\hat{e}_3(\omega_3) \cdot \vec{E}_r^I(\omega_3)|^2/2\pi$$

$$= \frac{8\pi^3\omega_3^2\sec^2\theta_3}{c^3\sqrt{\varepsilon(\omega_3)\varepsilon(\omega_1)\varepsilon(\omega_2)}}|\hat{e}_3(\omega_3) \cdot \overset{\leftrightarrow}{\chi}_{S,eff}^{(2)} : \hat{e}_1(\omega_1)\hat{e}_2(\omega_2)|^2$$

$$\times I(\omega_1)I(\omega_2) \tag{2.19}$$

where $\overset{\leftrightarrow}{\chi}_{S,eff}^{(2)} \equiv \overset{\leftrightarrow}{F}^{I-II}(\omega_3) \cdot \overset{\leftrightarrow}{\chi}_{SD}^{(2)} : \overset{\leftrightarrow}{F}^{I-II}(\omega_1)\overset{\leftrightarrow}{F}^{I-II}(\omega_2)$. (In SI units, the final expression should be divided by $64\pi^3\varepsilon_0$.) Pulsed inputs are generally used for SF spectroscopy. For input pulses with overlapping area A and time duration T on a sample surface, the SF output energy per pulse is approximately given by

$$
\mathrm{W}(\omega_3) \approx \frac{8\pi^3 \omega_3^2 \sec\theta_3 \sec\theta_1 \sec\theta_2}{c^3 \sqrt{\varepsilon_I(\omega_3)\varepsilon_I(\omega_1)\varepsilon_I(\omega_2)}} |\hat{e}_3(\omega_3) \cdot \overset{\leftrightarrow}{\chi}^{(2)}_{S,eff} :
$$

$$
\times \ \hat{e}_1(\omega_1)\hat{e}_2(\omega_2)|^2 \frac{\mathrm{W}(\omega_1)\mathrm{W}(\omega_2)}{AT}
$$

$$
\sim \frac{8\pi^3 \omega_3^2}{c^3} |\hat{e}_3(\omega_3) \cdot \overset{\leftrightarrow}{\chi}^{(2)}_{S,eff} : \hat{e}_1(\omega_1)\hat{e}_2(\omega_2)|^2
$$

$$
\times \ \frac{\mathrm{W}(\omega_1)\mathrm{W}(\omega_2)}{AT}
$$

$$(2.20)$$

where $W_i = I_i AT \cos\theta_i$ and we assumed all fields have constant amplitude over area A and time T. (In SI units, the expressions on the right-hand side should be divided by $64\pi^3\varepsilon_0$). For a rough estimate, we consider the case of SF vibrational spectroscopy on a surface monolayer. The effective surface nonlinear susceptibility, $|\overset{\leftrightarrow}{\chi}^{(2)}_{S,eff}|$, in the above equation is usually of the order of 10^{-16} to 10^{-14} esu. If we assume input pulses of $\mathrm{W}(\omega_1) = \mathrm{W}(\omega_2) = 100 \ \mu\mathrm{J}=10^3$ ergs with $T = 10$ ps and $A = 10^{-4}$ cm^2, and the sum frequency is at $\omega_3 = 3\times10^{15}/\mathrm{sec}(\lambda = 600$ nm), we find $8\pi^3\omega_3^2/c^3 \sim 80$, and $\mathrm{W}(\omega_3) \approx 8 \times (10^{-10}$ to $10^{-6})$ erg $\approx 3 \times (10^2$ to $10^6)$ photons/pulse. Such a signal can be readily detected by a photomultiplier. If the input at ω_2 is changed to a 100 fs pulse with $\mathrm{W}(\omega_2) = 10 \ \mu\mathrm{J}$, then its pulse overlapping time T with the 10 ps pulse at ω_1 is 100 fs, and the effective energy of the ω_1 pulse contributing to SFG reduces to $\mathrm{W}(\omega_1) = 1 \ \mu\mathrm{J}$. Correspondingly, the SF output reduces to $\mathrm{W}(\omega_3) \approx 3 \times (10$ to $10^5)$ photons/pulse; with a pulse repetition rate of 1000 per sec, the signal strength is $3 \times (10^4$ to $10^8)$ photons/sec. Since a photodetector in the visible can easily have a sensitivity of a few photons per second, this estimate shows that SF spectroscopy should have a sensitivity to detect submonolayers of molecules and record a spectrum in a few seconds.

2.7. Summary

We summarize here the main points presented in this chapter.

- Expressions for output fields of transmitted and reflected SFG from a semi-infinite medium, resulting from coherent radiation of induced SF polarization in the medium, are provided. Measurements of transmitted and reflected SFG allow deduction of bulk and surface nonlinear susceptibilities, respectively.
- Although electric dipole approximation used to describe material response is often good enough to deal with SF spectroscopy on surfaces and interfaces, electric-quadrupole contribution should not be neglected in some practical cases.
- The tensorial form of SF nonlinear susceptibility of a material reflects the structural symmetry of the material. SFG measurements can identify two-fold and three-fold rotational symmetry axes of materials.
- SF nonlinear susceptibility of a material becomes complex when input or/and output frequencies approach resonances. The amplitude of a resonance mode in the SFG spectrum has a sign related to the polar orientation of the contributing molecular group.
- Spectra of SF nonlinear susceptibility carries molecular-level information about a material as can be seen from the microscopic expression of the nonlinear susceptibility. In particular, reflected SF spectroscopy measuring surface nonlinear susceptibility provides microscopic structure information about a surface or interface of a material.
- SF output signals from a monolayer can be estimated from theory. With ps or fs input pulses, SF spectroscopy generally has submonolayer sensitivity.

References

1. Bloembergen, N.; Pershan, P. S.: Light Waves at Boundary of Nonlinear Media. *Phys. Rev.* 1962, 128, 606–622.
2. Shen, Y. R.: *The Principles of Nonlinear Optics.* J. Wiley: New York, 1984. Chapters 1–2.

3. Shen, Y. R.: *Fundamentals of Sum-Frequency Spectroscopy*. Cambridge University Press: Cambridge, 2016. Chapter 2.
4. Shen, Y. R.: Revisiting Basic Theory of Sum-Frequency Spectroscopy. *J. Chem. Phys.* 2020, 153, 180901.
5. Jackson, J. D.: *Classical Electrodynamics*. 3rd ed. Wiley: New York, 1999.
6. Fearn, H.; James, D. F. V.; Milonni, P. W.: Microscopic Approach to Reflection, Transmission, and Ewald-Oseen Extinction Theorem. *Am. J. Phys.* 1996, 64, 986–995.
7. Cho, M.: Coherent Two-Dimensional Optical Spectroscopy. *Chem. Rev.* 2008, 108(4), 1331–1418.

Chapter 3

Experimental Considerations on Sum Frequency Spectroscopy

Compared to linear optical spectroscopy, sum frequency generation (SFG) spectroscopy with two independent input beams provides more flexibility and carries more information for material characterization. The requirement of an additional input beam in an experimental setup seems simple and straightforward, but for many researchers not used to optical techniques, it is one beam too many. We describe in this chapter some general features of experimental setups for SFG spectroscopy and post a few precautions on carrying out experiments. We shall focus on commonly adopted experimental arrangements for SFG spectroscopy and discuss possible improvements. Second harmonic generation (SHG) is to be considered as a special case of SFG.

3.1. General Considerations

A sketch of an experimental arrangement of SFG spectroscopy is shown in Fig. 3.1. Two input beams of selected frequencies and polarizations with different incident angles are overlapped on a sample. Transmitted or reflected SF output from the sample in well-defined direction with selected polarization is detected after proper spatial and spectral filtering. As discussed in Sec. 2.5, pulsed inputs are generally preferred for SFG because the signal is inversely proportional to the overlapping pulse duration and area on a sample. If the output is in the visible range, photodetectors nowadays can

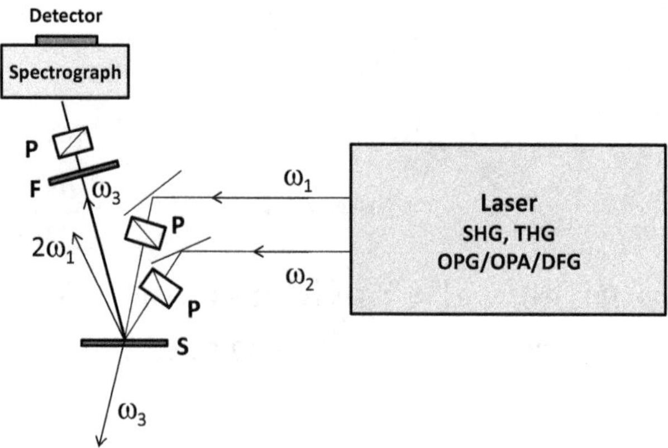

Fig. 3.1. Sketch of an SF spectroscopy setup. Two coherent light pulses, one at fixed frequency ω_1 and the other tunable at ω_2, from a laser system with associated SHG, THG, OPG/OPA, and DFG stages are directed to a sample to generate SF frequency outputs at ω_3 in the forward and reflected directions. The SF signal is recorded by a photodetector after going through spatial and spectral filters (F) and a spectrograph. S denotes the sample and P polarizers.

easily detect a few photons per pulse, making detection of sub-monolayer of molecules feasible. There is however a limit on the incoming beam intensity in order to avoid damaging a sample. Ultrashort fs pulses also may create complications because their waveform can easily be distorted by optical components in the beam path.

Consider a representative experimental arrangement sketched in Fig. 3.1. Two input pulses needed for SF spectroscopy are from a single laser system equipped with optical parametric amplifier and difference/sum frequency generation stages that can generate coherent light pulses with an overall tuning range from $16\,\mu$m in the mid IR to around $0.2\,\mu$m in the uv. Such a system with either ps or fs pulse output is commercially available. Their spatial mode quality and pulse profile however should be characterized before being incident on a sample. The two input pulses directed to a sample must be well overlapped in space and time. This is the major disadvantage of SFG spectroscopy that requires two input beams as compared to linear optical spectroscopy. (Second harmonic generation as a special case of SFG is an exception.) Thus, the input pulses ought to have a spatial profile as close to a single spatial mode as possible and a temporal profile close to transform-limited. Hot spots and patches in the field distribution and irregular spikes in the pulse profile would make overlapping of the two input pulses susceptible to change and difficult to reproduce, leading to strongly fluctuating SFG output. There are also output fluctuations due to input pulse-to-pulse variation in input beam energies and directions that should be kept low, although they can be largely suppressed by pulse-to-pulse normalization against SFG from a known reference.

For SFG spectroscopy, at least one of the input pulses is tunable in frequency and can be scanned over resonance; light absorption on resonance could heat up a sample excessively or even damage the sample. (To give some idea of how strong the pulse excitation can be, we note that a 1-mJ, 10 ps visible pulse focused on a $100\,\mu$m spot carries an energy density of $10\,\mathrm{J/cm^2}$ with a peak intensity of $\sim 1\,\mathrm{TW/cm^2}$, which is above the optical damage threshold even for many transparent solids.) Therefore, in preparation of an experiment, one must first figure out the maximum intensity permissible on the sample and arrange the beam geometry accordingly. The transmitted and reflected SF outputs from the sample are coherently radiated in well-defined directions determined by the boundary condition that input and output wave vector components parallel to the surface must be matched (i.e., $k_{1x} + k_{2x} = k_{3x}$ with x-z being the incidence plane and z the surface normal). Focusing of inputs however modifies the above boundary condition based on infinite plane wave approximation and causes the SF output to emit over a small solid angle determined by the convolution of focusing of the two inputs. The SF signal is usually more than ten orders of magnitude weaker than the input and its detection requires very effective filtering against the scattered input and background light. This can be achieved by combined spatial and spectral filtering, the former by a pinhole in the beam path after the sample and the latter in the form of narrowband color filters in series with a spectrograph. (For SHG, the use of only color filters is generally sufficient.) Finally, the filtered SF signal is detected by a photomultiplier or CCD camera with associated electronics.

SF spectroscopy with two visible input pulses is relatively easy to implement and can be used to probe electronic structure of materials. It has not yet been widely adopted presumably because it is not yet clear how much new and useful information can be extracted from SF spectra on electronic resonances. SF vibrational spectroscopy (SFVS), on the other hand, is already well established, although mainly for surface studies. It uses an IR input to excite vibrational or low-frequency resonances and a visible input to up-convert the IR excitation into SF output in the visible. The difficulty of tracing an IR beam by eye makes beam adjustment in the experiment more cumbersome. To overlap the two input beams on a sample, the IR beam is usually focused on a smaller spot on the sample than the visible beam because the latter often has more available flux. Currently, the tunable IR spectral range is limited to $\sim 16\,\mu$m, beyond which the IR pulse energy drops off rapidly. This prevents SFVS from applications to studies of low-frequency excitations in condensed matter physics and intermolecular vibrations or intersectional vibrations of large molecules in chemistry and biology.

We note that, in general, the visible input of SFVS on a sample can simultaneously generate a readily detectable, directional, second harmonic output from the sample, which may provide useful information about electronic structure of the sample. This is particularly true if the visible input is also tunable. With both IR and visible pulsed inputs tunable, double IR-visible resonances including 2D coherent transient phenomena can be interrogated.

Different versions of experimental schemes and setups have been developed for SFVS. We shall describe two commonly adopted and commercially available schemes in Secs. 3.2 and 3.3.

3.2. Narrowband Scheme

A cheaper and less complex experimental arrangement for SFVS is the narrowband scheme employing a picosecond pulsed laser system. As described in Sec. 2.5, shorter input pulses for SFG generate stronger signals. However, the spectral resolution of SFVS is limited by the linewidths of the input pulses. For most applications, a spectral resolution of a few cm^{-1} is sufficient, and the use

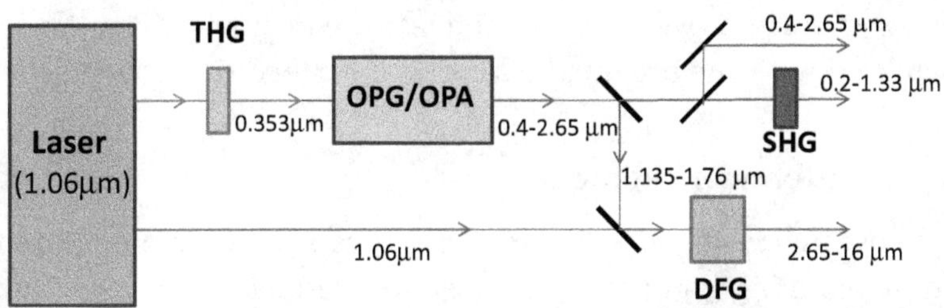

Fig. 3.2.   Schematic representation of a widely tunable coherent light source composed of a laser and a laser-pumped optical parametric generation/amplification (OPG/OPA) system supplemented with second harmonic generation (SHG), third harmonic generation (THG), and difference frequency generation (DFG) stages.

of picosecond pulses from a mode-locked laser system appears most appropriate. The block diagram in Fig. 3.2 shows that a ps mode-locked laser, incorporated with harmonic generation, optical parametric oscillator/amplifier, and sum and difference frequency stages can generate tunable ps pulses from uv ($\sim$200 nm) to mid IR ($\sim$16 μm), and at a few fixed frequencies as well. Such a widely tunable coherent light source is commercially available, and can produce $\sim$30-ps pulses with $>$0.5 mJ/pulse in the visible, $\sim$40 μJ/pulse around 10 μm, and $\sim$10 μJ/pulse near 16 μm with a linewidth $<$3 cm^{-1} and a repetition rate of 50 pulses/sec. (It would be ideal if the pulse length could be reduced to a few ps.) The system is a highly versatile and powerful tool for laser spectroscopists, and particularly suits SF spectroscopy for allowing a great deal of flexibility in experimental design.

For SFVS probing of a surface with an effective surface nonlinear susceptibility of $\sim$10^{-14} esu $[(4\pi/3) \times 10^{-20}$ m^2/V in SI units], the above system with appropriate beam geometry on a surface monolayer could generate thousands of photons per second in one frequency setting. (See Sec. 2.6.) We expect that if the tunable wavelength is scanned in 1 nm steps, a spectrum of high signal-to-noise ratio over 100nm spectral range would take less than a minute. In practice, it may take tens of minutes to record such a spectrum. The reason is that scanning (often enabled by a mechanical device) takes time; the speed of recording a spectrum is limited by how fast the input frequency can be tuned. To minimize the time for taking a spectrum, we have to eliminate frequency tuning. This can be achieved with the broadband scheme described in the next section.

3.3. Broadband Scheme

The above narrowband scheme of SF spectroscopy records a spectrum by scanning input frequency ω_2 over resonances and measuring the SF signal at $\omega_1 + \omega_2$ one at a time (i.e., a single channel measurement). Alternatively, SF spectroscopy can be carried out by mixing of a narrowband input at ω_1 with a broadband input covering a range of ω_2 frequencies (i.e., a multi-channel or multiplex measurement). This scheme can be made possible with a commercially available, high-power fs pulsed laser system and CCD camera. As sketched in Fig. 3.3, a typical fs Ti:sapphire laser system can generate 5 mJ/pulse at 800 nm with 35 fs pulse width and a repetition rate of 1 kHz. It can be used to pump optical parametric oscillator/amplifier, harmonic generation, and sum and difference frequency generation stages to provide <100 fs pulses (corresponding to a spectral full width at half maxima of >150 cm^{-1} for a Gaussian pulse) tunable over a spectral range from 16 μm to 200 nm. The achievable amount

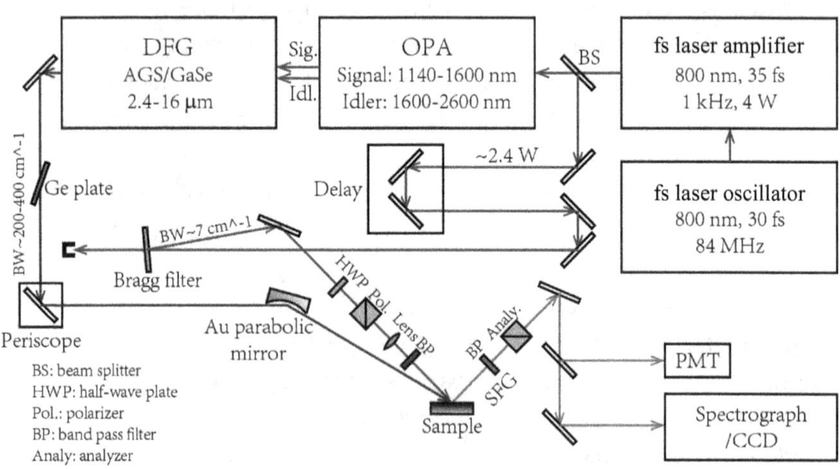

Fig. 3.3. A typical setup for multiplex broadband sum frequency spectroscopy that employs an fs Ti:sapphire laser with associated OPG/OPA and DFG stages to produce widely tunable fs pulses with ensuing beam arrangement for SFG from a sample and its detection. (Courtesy of W.T. Liu)

of output energy per pulse is $\sim 3\,\mu$J at $16\,\mu$m up to $\sim 300\,\mu$J at $2\,\mu$m. Such a laser system is more than sufficient for the broadband scheme of SF spectroscopy. On the detector side, CCD cameras have detection sensitivity of a few photons per channel, again sufficient for submonolayer detection of SF spectroscopy.

The broadband scheme for SF spectroscopy using an fs laser system was first conceived and demonstrated by Richter *et al.*[1] A typical setup is described in Fig. 3.3. The OPG/OPA and DFG/SFG stages generate the widely tunable, broadband fs pulses as the ω_2 input for SFG. The other input at ω_1 comes directly from the laser with its spectral width narrowed down by spectral filtering (the Bragg filter in Fig. 3.3) and accordingly, its pulse length stretched to ~ 1 ps. Overlapping of the ps narrowband ω_1 pulse with the fs broadband ω_2 pulse on a sample produces an fs broadband SF signal, which is spectrally dispersed through a spectrograph and detected by a CCD camera. A large part of the ω_1 pulse that does not overlap with the ω_2 pulse is clearly wasted. Fortunately, this is not a problem because there is plenty of available pulse energy at ω_1 from the laser. With such a scheme that does not require any mechanical adjustment, an SF spectrum that has its spectral range set by the bandwidth of the ω_2 pulse and spectral resolution limited by the linewidth of the ω_1 pulse may take only a few seconds. To cover a wider spectral range, discrete band-to-band frequency tuning can be used.

3.4. Spectral Analysis

The measured SF signal of both narrowband and broadband schemes discussed above is in terms of pulse energy per frequency channel, which according to Eq. (2.19) is proportional to $\left|\chi_{SD}^{(2)}(\omega_3)\right|^2$ or $\left|X_{SD}^{(2)}(\omega_3)\right|^2$ with $\overset{\leftrightarrow}{X}_{SD}^{(2)}(\omega_3) \equiv \left[\hat{e}(\omega_3)\cdot\overset{\leftrightarrow}{F}^{I-II}(\omega_3)\right]^* \cdot \overset{\leftrightarrow}{\chi}_{SD}^{(2)} : \left[\overset{\leftrightarrow}{F}^{I-II}(\omega_1)\cdot\hat{e}(\omega_1)\right]\left[\overset{\leftrightarrow}{F}^{I-II}(\omega_2)\cdot\hat{e}(\omega_2)\right]$ and $\overset{\leftrightarrow}{\chi}_{SD}^{(2)}(\omega = \omega_1 + \omega_2) = \overset{\leftrightarrow}{\chi}_{NR}^{(2)} + \overset{\leftrightarrow}{\chi}_{R}^{(2)} = \overset{\leftrightarrow}{\chi}_{NR}^{(2)} + \sum_q \frac{\overset{\leftrightarrow}{A}_q}{\omega_2 - \omega_q + i\Gamma_q}$ for discrete resonances from Eq. (2.13) [or the expression in Eq. (2.14) for continuum resonances]. While $\left|\chi_{SD}^{(2)}(\omega_3)\right|^2$ displays relevant spectral features, it does not carry the detailed information on the characteristics of resonances, and hence the microscopic information about the surface or interface, as

$\overset{\leftrightarrow(2)}{\chi}_R$ does. To extract $\chi_R^{(2)}(\omega_3)$ from $\left|\chi_{SD}^{(2)}(\omega_3)\right|^2$ is not possible without assumption and/or approximation because $\chi_{SD}^{(2)}(\omega_3)$ is complex. In the case of discrete resonances, one can use the above expression of $\chi_{SD}^{(2)}(\omega_3)$ with $\chi_{NR}^{(2)}$, A_q, ω_q, and Γ_q as adjustable parameters (or other functions to describe resonances) to fit the measured $\left|\chi_{SD}^{(2)}(\omega_3)\right|^2$ spectrum, as often reported in the literature. There is however the difficulty that the fit may not be unique, especially on the sign of A_q; an example is shown in Fig. 3.4. Furthermore, overlapping of resonance bands and strong nonresonant background from $\overset{\leftrightarrow(2)}{\chi}_{NR}(\omega_3)$

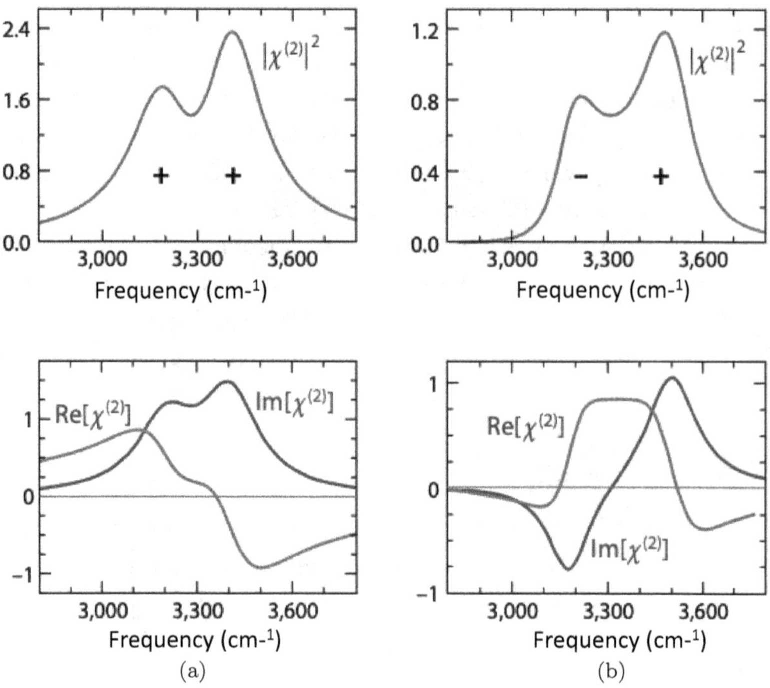

Fig. 3.4. An example showing that spectral fitting of $\left|\chi^{(2)}(\omega)\right|^2$ to deduce $\mathrm{Re}\chi^{(2)}(\omega)$, and $\mathrm{Im}\chi^{(2)}(\omega)$ is often not unique. Spectra of $\left|\chi^{(2)}(\omega)\right|^2$, $\mathrm{Re}\chi^{(2)}(\omega)$, and $\mathrm{Im}\chi^{(2)}(\omega)$ of two overlapping bands with (a) both amplitudes positive and (b) one positive Send the other negative illustrate that while the two $\left|\chi^{(2)}(\omega)\right|^2$ spectra are nearly the same, the corresponding $\mathrm{Re}\chi^{(2)}(\omega)$ and $\mathrm{Im}\chi^{(2)}(\omega)$ spectra of the two are very different. (after Ref. [2])

can drastically distort the spectral shape of $\left|\chi_{SD}^{(2)}(\omega_3)\right|^2$. In the case of strongly overlapping resonance bands or resonance continuum, the problem is more severe, although in some cases, one may assume the continuum is composed of discrete bands with specific spectral profile to perform the fitting.

The difficulty of not being able to have unique spectral fit to find $\chi_{SD}^{(2)}(\omega_3) = \left|\chi_{SD}^{(2)}(\omega_3)\right|e^{i\phi_S(\omega_3)}$ arises because we need also the phase spectrum $\phi_S(\omega_3)$ to define $\chi_{SD}^{(2)}(\omega_3)$ fully. The phase-sensitive SFVS to be discussed in the next section has been developed for this task. The spectrum of Im $\chi_{SD}^{(2)}(\omega_3)$ obtained from measurement can then directly characterize resonances.

3.5. Phase-Sensitive Sum Frequency Spectroscopy[2]

It is well known in optics that phase information can be extracted from interference measurement. To find $\phi(\omega_3)$ of SFG, we can interfere the SF output wave with an SF wave generated from a reference (or local oscillator) with $\chi_{Ref}^{(2)}(\omega_3) = \left|\chi_{Ref}^{(2)}(\omega_3)\right|e^{i\phi_{Ref}(\omega_3)}$. The interference signal has the expression

$$
\begin{aligned}
S(\omega_3, \Delta\phi) &= A\left|\chi_{SD}^{(2)} + \chi_{Ref}^{(2)}e^{-i\Delta\phi}\right|^2 + B\left|\chi_{Ref}^{(2)}\right|^2 \\
&= A\left[\left|\chi_{SD}^{(2)}\right|^2 + \left|\chi_{Ref}^{(2)}\right|^2\right] + B\left|\chi_{Ref}^{(2)}\right|^2 \\
&\quad + 2\left|A\chi_{SD}^{(2)}\chi_{Ref}^{(2)}\right|\cos(\phi_S - \phi_{Ref} - \Delta\phi)]
\end{aligned} \tag{3.1}
$$

where A and B are coefficients referring to the areas the signal and reference waves overlap and do not overlap, respectively, assuming the SF reference has a larger beam cross-section, and $\Delta\phi$ is the phase difference of the overlapping SH signal and reference waves due to their different path lengths. (More generally, $\chi_{SD}^{(2)}$ should be replaced by $\chi_{S,eff}^{(2)}$ if the electric-quadrupole contribution to the surface nonlinearity is not negligible as discussed in Sec. 2.2.) By varying $\Delta\phi$ at given ω_3, a fringe pattern of S is generated from the $\cos(\phi_S - \phi_{Ref} - \Delta\phi)$ term, from which ϕ_S can be determined if ϕ_{Ref} is known. In practice, it is simpler and more reliable if we simply keep ϕ_{Ref} fixed, replace the sample by a standard reference with known ϕ_{SR}

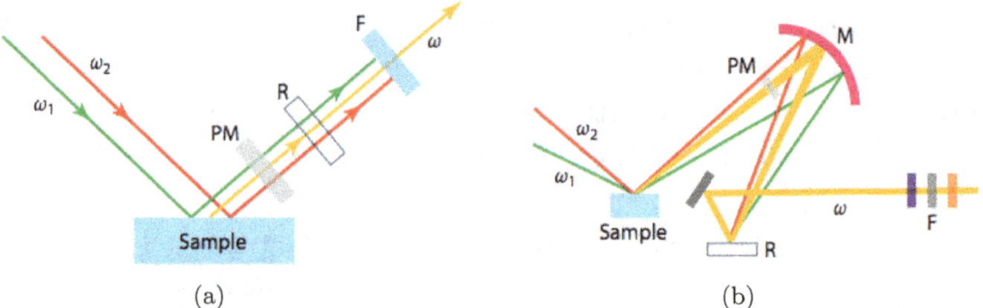

Fig. 3.5. Schematic representations of experimental arrangement for phase-sensitive sum frequency spectroscopic measurements: (a) collinear and (b) noncollinear beam geometry. F, filter; M, concave mirror; PM, silica phase modulator; R, reference (after Ref. [2]).

(for example, $\phi_{SR} = \pm 90°$ for reflected SF wave from quartz in the visible), and repeat the measurement to get $\phi_{SR} - \phi_{Ref}$. Together with the measured $\phi_S - \phi_{Ref}$, we can find $\phi_S - \phi_{SR}$ or ϕ_S with reference to ϕ_{SR}. For accurate determination of ϕ_S, it is clearly important to know the value of ϕ_{SR} accurately.

With the single-channel narrowband scheme, the phase measurement is carried out at one frequency setting at a time. Two typical experimental arrangements for phase measurement of reflected SF signal from a sample are sketched in Fig. 3.5, one with collinear beam geometry and the other noncollinear; the former is often more stable. The phase modulator (PM) in the beam path in the figures can be a silica plate, tilting of which changes $\Delta\phi$. The reference SF generator denoted by R in the beam path can situate either before or after the sample. The combined SF signal from the sample and reference is detected after the spectral filtering stage labeled by F. Figure 3.6 shows the interference patterns of (a) SFVS from the air/water interface and (b) from a quartz standard reference as examples. Accuracy of phase measurement depends on the number of data points taken for each interference fringe pattern, the more, the better. Such measurement is however very tedious and time consuming. Fortunately, it is usually sufficient to measure $\phi(\omega_3)$ at selected discrete frequencies over a spectrum because $\phi(\omega_3)$ is not expected to vary irrationally with frequency.

A less accurate, but more convenient, method to obtain spectra of both $\left|\chi_{SD}^{(2)}(\omega)\right|$ and $\phi(\omega)$ is to scan a set of four spectra: spectra

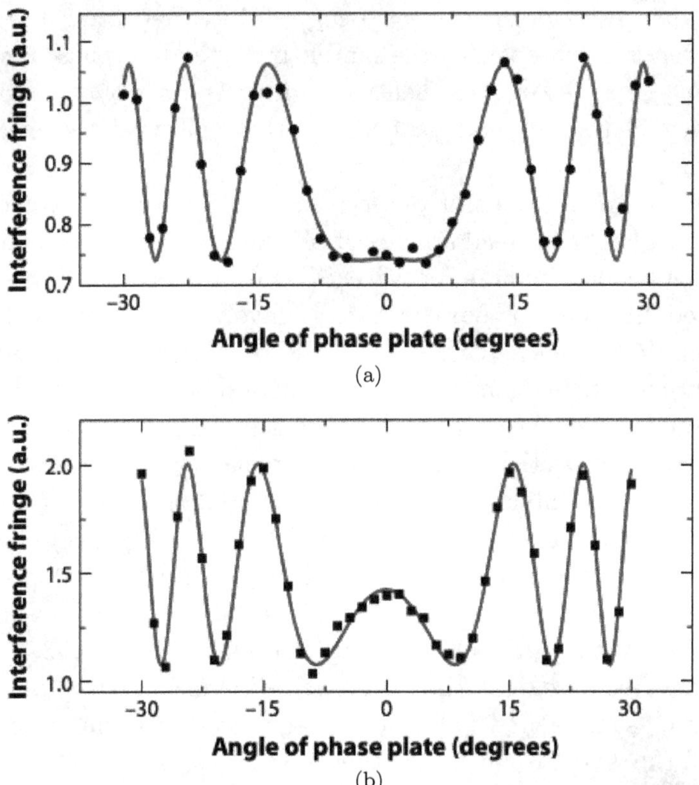

Fig. 3.6. Representative fringe patterns obtained from narrowband phase-sensitive SF vibrational spectroscopic measurement of (a) the air/water interface and (b) a z-cut quartz reference taken at the IR frequency of $3400\,\mathrm{cm}^{-1}$ (after Ref. [2]).

of $A\big|(\chi^{(2)}_{SD})_{ijk}\big|^2$ and $(A+B)\big|(\chi^{(2)}_{Ref})_{ijk}\big|^2$ measured by removing the reference (R) and the sample from the beam path, respectively, and spectra of $S(\omega_3, \Delta\phi_0)$ and $S(\omega_3, \Delta\phi_0 + \pi/2)$ measured by having the phase difference set at $\Delta\phi_0$ and $\Delta\phi_0 + \pi/2$. From Eq. (3.1), we find

$$\frac{S(\omega_3, \Delta\phi_0 + \tfrac{\pi}{2}) - \big[A\big|(\chi^{(2)}_{SD})_{ijk}\big|^2 + (A+B)\big|(\chi^{(2)}_{Ref})_{ijk}\big|^2\big]}{S(\omega_3, \Delta\phi_0) - \big[A\big|(\chi^{(2)}_{SD})_{ijk}\big|^2 + (A+B)\big|(\chi^{(2)}_{Ref})_{ijk}\big|^2\big]}$$

$$= \tan(\phi_S - \phi_{Ref} - \Delta\phi_o) \qquad (3.2)$$

from which we can deduce $\phi_S - \phi_{Ref} - \Delta\phi_o$. Repeating the set of measurements by replacing the sample by a standard phase reference yields $\phi_{SR} - \phi_R - \Delta\phi_o$, and hence ϕ_S with respect to ϕ_{SR}. Obviously, accuracy of this method depends on the quality of the measured spectra.

The above methods of phase measurement can be extended to the multiplex broadband scheme of SF spectroscopy.[3,4] (Tahara and coworkers labeled phase-sensitive SF spectroscopy as heterodyne-detected SF spectroscopy because it involves SF signal detection through interference with a local SF oscillator. This is however a misnomer. By common definition, heterodyne detection has the signal and the local oscillator at different frequencies; it is homodyne detection that has the two at the same frequency.) They are designed to extract the interference term in $S(\omega_3, \Delta\phi)$ of Eq. (3.1) and hence the phase information. It is possible to find $\phi_S(\omega_3) - \phi_R(\omega_3)$ directly from the data analysis of a single interference spectrum. For illustration, we use phase-sensitive SFVS on the water/air interface in the OH stretch range presented in Fig. 3.7 as an example. We recognize that the path difference (Z) of SF waves from the sample and reference is equivalent to the time delay (T) of the two SF pulses in time, $\Delta\phi = k_3 Z = \omega_3 T$, and the interference term of $S(\omega_3)$ (in Eq. (3.1) and Fig. 3.7(a)) can be written as $2\left|A(\omega_3)\chi_{SD}^{(2)}(\omega_3)\chi_R^{(2)}(\omega_3)\right|\cos[\phi_S(\omega_3) - \phi_R(\omega_3) - \omega_3 T]$. If the inverse of the spectral width of $S(\omega_3)$ is significantly smaller than the time delay T, it can be seen that the two parts of the above expression, $\left|A(\omega_3)[\chi_{SD}^{(2)}(\omega_3)\chi_R^{(2)}(\omega_3)\right|e^{\pm i[\phi_S(\omega_3)-\phi_R(\omega_3)-\omega_3 T]}$, after Fourier transform to the time domain, can be fairly cleanly separated in time. One can then use a filter function to select and retain only the portion of the temporal interferogram on the right $(T > 0)$, and Fourier transform it back to the spectra domain (Figs. 3.7(b) and 3.7(c)). The back and forth Fourier transform now yields a complex interferogram described by the complex function $\left|A(\omega_3)[\chi_{SD}^{(2)}(\omega_3)\chi_R^{(2)}(\omega_3)\right|e^{i[\phi_S(\omega_3)-\phi_R(\omega_3)-\omega_3 T]}$ (Fig. 3.7(c)). The real part of the interferogram is $\left|A(\omega_3)\chi_{SD}^{(2)}(\omega_3)\chi_R^{(2)}(\omega_3)\right|\cos[\phi_S(\omega_3) - \phi_R(\omega_3) - \omega_3 T]$ and the imaginary part is $\left|A(\omega_3)\chi_{SD}^{(2)}(\omega_3)\chi_R^{(2)}(\omega_3)\right|\sin[\phi_S(\omega_3) - \phi_R(\omega_3) - \omega_3 T]$. From these two

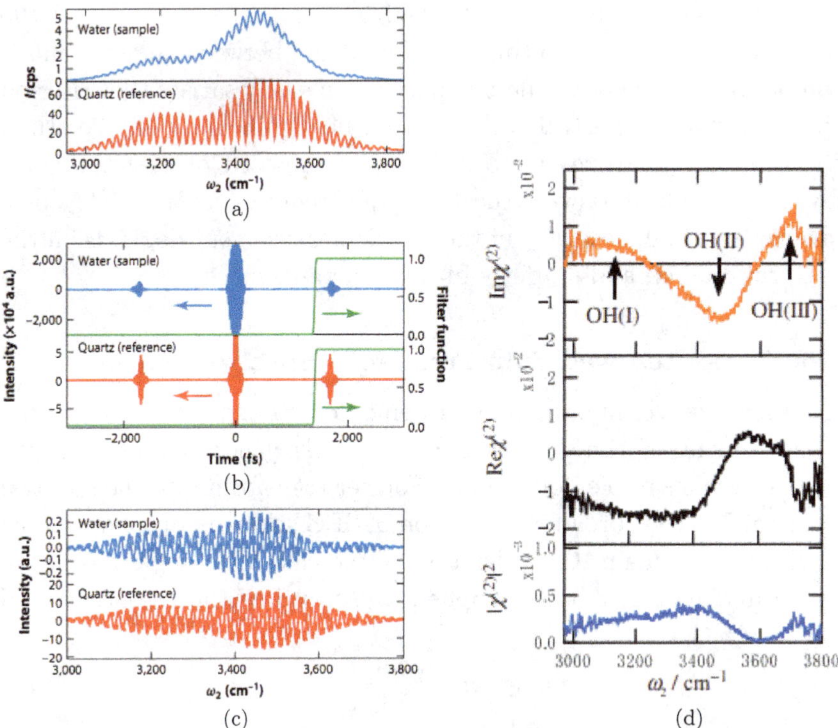

Fig. 3.7.　Three stages in the analysis of an SF spectral interferogram obtained by the multiplex broadband scheme using the setup of Fig. 3.6(b). (a) Raw spectral interferograms of SSP SF vibrational spectroscopy from the water/air interface (blue) and a z-cut quartz reference (red). (b) Time-domain interferograms obtained from Fourier transform of the interferograms in (a). The filter function (green) picks up the interferograms on the right. (c) Real (solid lines) and imaginary (dashed lines) parts of the sample and reference sample spectra obtained from an inverse Fourier transform of the filtered interferograms in (b). (d) Normalized $\left|\chi^{(2)}(\omega)\right|^2$, $\mathrm{Re}\chi^{(2)}(\omega)$, and $\mathrm{Im}\chi^{(2)}(\omega)$ spectra of the water/air interface deduced from the analysis. The positive OH(I) band in $\mathrm{Im}\chi^{(2)}(\omega)$ was later believed to be non-existent as reported by Nihonyanagi *et al.* in *J. Chem. Phys.* 143, 124707 (2015) (after Ref. [4]).

spectral interferograms, the spectra of $\left|A(\omega_3)\chi_{SD}^{(2)}(\omega_3)\chi_R^{(2)}(\omega_3)\right|$ and $\phi_S(\omega_3) - \phi_R(\omega_3) - \omega_3 T$ can be separately deduced. Again by normalizing the spectra against those of a standard reference, $\left|\chi_{SD}^{(2)}(\omega_3)\right|$ and $\phi_S(\omega_3)$ are deduced, and then $\mathrm{Re}\chi_{SD}^{(2)}(\omega_3)$ and $\mathrm{Im}\ \chi_{SD}^{(2)}(\omega_3)$ found (Fig. 3.7(d)). This method is best suited for broadband

SF spectroscopy since SF outputs from both the sample and the reference are fs pulses so that the time delay between the two pulses can be easily varied and the SF spectra can easily satisfy the criterion that the inverse spectral width should be significantly smaller than the time delay. To resolve a spectrum composed of structure of a few cm^{-1} wide, however, the technique requires a time delay of a few ps. As usual, accuracy of the phase measurement depends on the spectral quality, and stability of beam arrangement.

3.6. Time-Resolved Sum Frequency Spectroscopy

Another less commonly used scheme for SF spectroscopy is the measurement of time-resolved SFG signals generated by fs input pulses from a sample, followed by Fourier transform into the spectral domain. This is a broadband version of SFG spectroscopy carried out in the time domain. Consider the simple case of a single resonance illustrated in Fig. 3.8. An applied pulse excites the resonance and

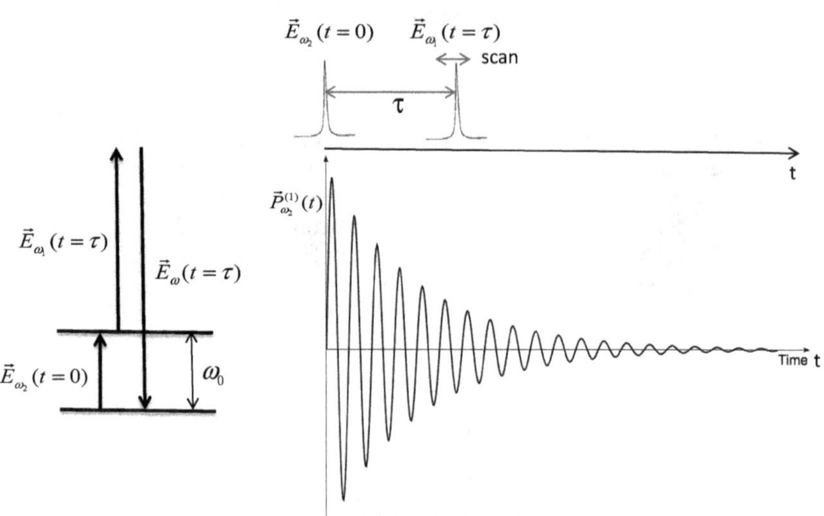

Fig. 3.8. Schematic representation describing time-dependent SFG that up-converts a free induction decay signal. The free induction decay appears after the excitation of the transition at frequency ω_0 by the ultrashort pulse $\vec{E}(\omega_2, t=0)$ at $t=0$ with $\omega_2 \sim \omega_0$. The ultrashort probe pulse $\vec{E}(\omega_1, t=\tau)$ up-converts the free induction decay at $t=\tau$ to $\vec{E}(\omega_3 = \omega_1 + \omega_0, t=\tau)$. The entire free induction decay curve is up-converted when τ is scanned.

induces a polarization that oscillates at the resonant frequency. If the pulse is significantly shorter than the relaxation time of the resonance, then after the excitation pulse is over, the polarization oscillation lasts for some time with a relaxation behavior characteristic of the resonance. This is known as free induction decay. Fourier transformation of the time-resolved free induction decay yields the resonance spectrum.

Sum frequency generation is a means to measure free induction decay in a medium. It can be regarded as a process in which the amplitude of an input field at ω_1 is modulated by the free induction decay through their interaction in the medium. One can use an ω_1 input pulse much longer than the free induction decay time and detect its modulated time-resolved output in reflection (similar to the SF pulse detection in the broadband scheme discussed in Sec. 3.4.). This can also be achieved in the time domain by using a narrow probe pulse (at ω_1), much narrower than the oscillation period, and scanning it over the temporal variation of the decaying oscillation (see Fig. 3.8.). Explicitly, the free induction decay comes from radiation emitted by an induced polarization, $P^{(1)}(t) = A\chi_R^{(1)}(t) \propto \cos(\omega_0 t)e^{-\Gamma t}$, in a medium at time t when the excitation pulse is essentially over. Interaction of an off-resonant ultrashort probe pulse, approximately described by $E(\omega_1, t) = E_1\delta(t = \tau)e^{-i\omega_1 t}$, instantaneously upconverts $P^{(1)}(t)$ at $t = \tau$ to $P^{(2)}(t) \propto E_1 e^{-i\omega_1 t}\cos(\omega_0 t)e^{-\Gamma t}$, which is a polarization at ω_1 with its amplitude modulated at frequency ω_0. (We have omitted the wave vectors in the above expressions.) If the radiated field, $E_{out}(t) \propto P^{(2)}(t)$, can be measured, then Fourier transform of $E_{out}(t)$ gives

$$E_{out}(\omega) \propto P^{(2)}(\omega) \propto \chi_R^{(2)}(\omega)$$

$$\propto \frac{1}{\omega - (\omega_1 + \omega_0) + i\Gamma} + \frac{1}{\omega - (\omega_1 - \omega_0) + i\Gamma}, \quad (3.3)$$

which is a spectrum containing two side bands at $\sim (\omega_1 \pm \omega_0)$, one appearing as SFG and the other as DFG. In case multiple resonances are simultaneously excited, $P^{(1)}(t)$ is the linear superposition of free induction decays of all excited resonances, and the spectrum of the corresponding radiation from $P^{(2)}(t)$ yields the spectrum of the resonances. We should note that amplitude modulation of a wave is a second-order process and therefore this scheme is surface-specific for media with inversion symmetry.[5] One advantage of this scheme is that it can be easily extended to two-dimensional coherent transient spectroscopy.[6]

We assumed in the above discussion that $E_{out}(t)$ can be measured, but the usual experimental arrangement only allows measurement of $\left|E_{out}(\omega)\right|^2 \propto \left|\chi_R^{(2)}(\omega)\right|^2$. To find the complex $\chi_R^{(2)}(\omega)$, one has to resort again to interference measurement of $E_{out}(t)$ with a reference $E_{Ref}(t)$ as described in Sec. 3.5. This can be achieved simply by sending an appropriate fraction of the reflected probe beam together with $E_{out}(t)$ into the photodetector.

3.7. Summary and Prospects

We summarize here the main points presented in this chapter:

- A general sketch of experimental arrangement for SF spectroscopy with pulsed inputs is given.
- Widely tunable coherent ps and fs light sources are commercially available. Precautions on setting up SF spectroscopy measurement are noted.
- The main difficulty of SF experimental arrangement lies in temporal and spatial overlapping of two input pulses. Irregular temporal and spatial intensity distributions in the input beam profiles may cause trouble.
- Input intensity on a sample is limited by optical damage or perturbation of the sample structure.
- Data fluctuations can be reduced by proper normalization against a reference.
- A narrowband scheme for SF spectroscopy using ps input pulses has good spatial and spectral resolution and is cheaper to set up, but frequency scan takes a large fraction of the measurement time.
- A broadband scheme is more expensive to set up, but without the need of frequency scan, the measurement time is greatly reduced.

- Usual SF spectroscopy measures only $\left|\chi^{(2)}(\omega)\right|^2$. Deduction of $\chi^{(2)}(\omega)$ from $\left|\chi^{(2)}(\omega)\right|^2$ by spectral fitting is possible in simple cases, but generally not reliable. Phase information on $\chi^{(2)}(\omega)$ is needed for complete determination of $\chi^{(2)}(\omega)$.
- Phase-sensitive SF spectroscopy to measure complex $\chi^{(2)}(\omega)$ can be set up for both narrowband and broadband schemes by interfering SF output from a sample with a reference.
- SF spectroscopy can also be carried out in the time domain, followed by Fourier transform, to get the spectrum.

Since the first reported SF spectroscopy measurement, the technical aspect of spectroscopy has been greatly improved following advances of laser technology. There are of course more improvements to expect. For example, if a fiber laser could be used to replace the current ps or fs laser system, the spectroscopy setup would become more compact and movable. Extension to far IR spectral range above $16\,\mu$m may be possible following development of new coherent THz sources. Stable overlapping of two pulses is difficult, but could be achieved by installation of a beam feedback loop. We note in passing that with two inputs at ω_1 and ω_2 impinging on a sample, one should expect not only output at $\omega_1+\omega_2$, but also outputs at $\omega_1-\omega_2$, $2\omega_1$ and $2\omega_2$. They could provide additional information about the sample if separately detected.

References

1. Richter, L. J.; Petralli-Mallow, T. P.; Stephenson, J. C.: Vibrationally Resolved Sum-Frequency Generation with Broad-Bandwidth Infrared Pulses. *Opt. Lett.* 1998, 23, 1594–1596.
2. Shen, Y. R.: Phase-Sensitive Sum-Frequency Spectroscopy. *Ann. Rev. Phys. Chem.* 2013, 64, 129–150.
3. Yamaguchi, S.; Tahara, T.: Heterodyne-Detected Electronic Sum Frequency Generation: "Up" versus "Down" Alignment of Interfacial Molecules. *J. Chem. Phys.* 2008, 29, 101102.
4. Nihonyanagi, S.; Yamaguchi, S.; Tahara, T.: Direct Evidence for Orientational Flip-Flop of Water Molecules at Charged Interfaces: A Heterodyne-Detected Vibrational Sum Frequency Generation Study. *J. Chem. Phys.* 2009, 130, 204704.

5. Guyot-Sionnest, P.: Coherent Processes at Surfaces — Free-Induction Decay and Photon-Echo of the Si-H Stretching Vibration for H/Si(111). *Phys. Rev. Lett.* 1991, 66, 1489–1492.
6. Cho, M.: Coherent Two-Dimensional Optical Spectroscopy. *Chem. Rev.* 2008, 108(4), 1331–1418.

More references on experimental arrangement can be found in Shen, Y.R.: Fundamentals of Sum Frequency Spectroscopy. (Cambridge University Press, Cambridge, 2016). Chapter 4.

Review Articles.

• Shen, Y. R.: Phase-Sensitive Sum-Frequency Spectroscopy. *Ann. Rev. Phys. Chem.* 2013, 64, 129–150.
• Roy, S.; Saha, S.; Mondal, J. A.: Classical and Heterodyne-Detected Vibrational Sum Frequency Generation Spectroscopy and Its Application to Soft interfaces. In Singh, D. K.; Pradhan, M.; Materny A. (eds.), *Modern Techniques of Spectroscopy: Progress in Optical Science and Photonics.* Vol. 13, pp. 87–115, Springer Nature, Singapore, 2021.

Chapter 4

Sum Frequency Spectroscopy for Bulk Characterization

Nonlinear optical spectroscopy in general and SFG/SHG spectroscopy in particular can provide more, and oftentimes unique, information on materials than linear optical spectroscopy, but they are not as popular because of complexity in both experimental arrangement and theoretical analysis. The unique features of SFG/SHG spectroscopy come mainly from symmetry consideration. We discuss here a number of cases where specific opportunities of bulk material characterization by SFG/SHG spectroscopy have been demonstrated. On probing the structural symmetry of materials, the optical frequency dependence of material response is often not so relevant, and SHG can be conveniently adopted as the tool.

4.1. Probing Bulk Structure and Phase Transition by Second Harmonic Generation

As described in Sec. 2.2, SHG/SFG is allowed under electric-dipole approximation only in media without inversion symmetry. It was noticed in the very early years of nonlinear optics that SHG could help identify the bulk structure of materials. A very weak symmetry change of a crystalline structure may be difficult for X-ray diffraction to identify, but can be readily detected by SHG. Among the 32 symmetry classes of crystals, 21 of them lack inversion symmetry. For each symmetry class, there is a nonlinear susceptibility tensor, $\overset{\leftrightarrow}{\chi}^{(2)}$, for SHG/SFG with a specific set of nonvanishing elements, which are tabulated in many books.[1] In principle, finding the set allows the identification of the crystal symmetry class. In reality, this is not practical as the measurement would take much more time and effort than X-ray diffraction. Distinguishing one symmetric class from another, however, is feasible since measurement can focus on just a few $\overset{\leftrightarrow}{\chi}{}^{(2)}$ elements. If two structural phases of a material have the same nonvanishing $\overset{\leftrightarrow}{\chi}{}^{(2)}$ element, their values are likely to be different. Accordingly, SHG can be used to probe structural phase transitions.[2] The simplest case is with ferroelectric to paraelectric phase transition; the former has a nonvanishing $\overset{\leftrightarrow}{\chi}{}^{(2)}$, but the latter has $\overset{\leftrightarrow}{\chi}{}^{(2)} = 0$. An example is shown in Fig. 4.1(a), where the spontaneous polarization P_s of $LaTaO_3$ is plotted as a function of temperature.[3] The measured values of $\chi^{(2)}_{zzz}$ (open circles) of $LaTaO_3$ fall right on the P_s curve, indicating that the nonvanishing $\chi^{(2)}_{zzz}$ is proportional to the ferroelectric order parameter, $P_s/P_{s,\mathrm{max}}$ as expected. Above the phase transition temperature T_N, both P_s and $\chi^{(2)}_{zzz}$ vanish.

To see a more complex case, we take $KNbO_3$ as an example.[4] The crystal undergoes transitions from orthorhombic to tetragonal structure at $225°C$ and from tetragonal to cubic at $435°C$. Structures

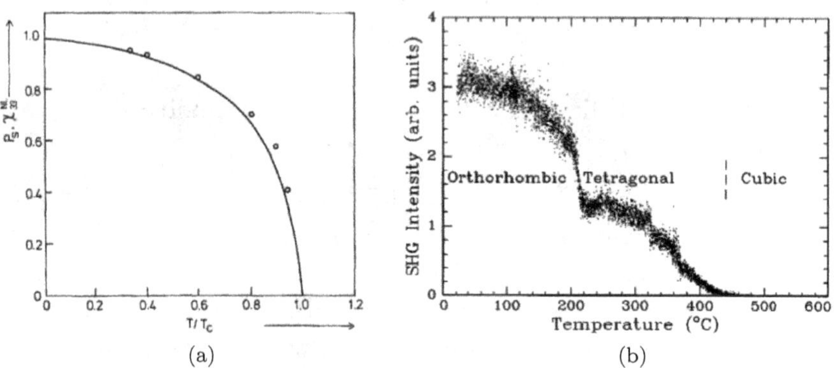

Fig. 4.1. (a) Temperature dependences of spontaneous polarization P_s (solid curve) and second-order nonlinear susceptibility (dots) of $LaTiO_3$ (after Ref. [3]). (b) Temperature dependence of SH intensity generated from $KNbO_3$ showing orthorhombic to tetrahedral phase transition at $\sim220°$ C and tetrahedral to cubic phase transition at $\sim435°$ C (after Ref. [4]).

of the first two phases have no inversion symmetry, but the cubic one has. The SHG measurement was carried out on a $KNbO_3$ film epitaxially grown on a $SrTiO_3(100)$ crystal with the beam geometry and sample orientation fixed and temperature varied. As seen in Fig. 4.1(b), variation of the SH signal with temperature shows clearly the phase transition behavior around the transition temperatures. The signal approaches zero towards the tetrahedral to cubic transition.

A ferroelectric crystal often appears with multiple domains, and SHG can be used to probe the domain structure.[5] This is because the $\overset{\leftrightarrow}{\chi}{}^{(2)}$ tensor of each domain has its symmetry axes attached to the structural axes of the domain so that in the lab coordinates, $\overset{\leftrightarrow}{\chi}{}^{(2)}$'s of different domains are different. With time-resolved SHG, dynamics of phase transitions and domain structure variation can also be probed. For instance, SHG was used earlier by Shank *et al.* to study ultrafast dynamics of laser-induced melting of silicon.[6]

We note that linear optics can also be used to probe structural changes of a medium if the relevant refractive index changes are detectable, but its capability is more limited. Since SHG (and SFG) has more independently adjustable input parameters, it can provide more information about a material. For instance, linear optics allows the identification of two-fold symmetry axes; the dielectric or refractive index tensor appears the same as isotropic for a symmetry axis beyond two-fold. On the other hand, $\overset{\leftrightarrow}{\chi}{}^{(2)}$ optics allows identification of two- and three-fold axes, but not symmetry axes four-fold or beyond. Consider the three-fold z-axis of quartz (0001) as an example. The nonlinear susceptibility $\chi_{ssp}^{(2)}$ measured by SSP SFG, with the incidence plane containing $\hat{z}$ and making an angle ϕ with the $\hat{x} - \hat{z}$ plane, can be shown to depend on ϕ with the expression $\chi_{ssp}^{(2)}(\phi) = A + B\cos 3\phi$, as we shall discuss in Sec. 4.4. In some cases, we may find $\chi_{ssp}^{(2)}(\phi) = 0$ at selected values of ϕ and the bulk SHG/SFG is strongly suppressed; as a result, SHG/SFG from a surface may become detectable and can be used to probe the surface. This will be discussed in Sec. 7.2B. More generally, an nth-rank tensor can identify up to n-fold symmetry axes. For example, $\chi_{ssp}^{(2)}(\phi)$ as a fourth-rank tensor from EQ bulk contribution can identify a bulk four-fold symmetry axis. Recently, SHG has been widely adopted to probe the structural symmetry of 2D materials.

4.2. Probing Antiferromagnetism by Second Harmonic Generation

Linear magneto-optical effects describing the change of beam characteristics upon interaction with a magnetized medium have long been used to study bulk magnetization of a material since Faraday's time. Similarly, nonlinear optical processes in a medium should also be affected by the magnetization of a medium and can be employed to probe magnetic materials. One would expect that to the lowest-order approximation, both linear and nonlinear optical effects should be proportional to the magnetization and would therefore be insensitive to antiferromagnetization. This is true for linear optics, but not for SHG/SFG, which actually can have nonvanishing electric-dipole (ED) response to antiferromagnetization. The reason is that an antiferromagetic spin arrangement may break inversion symmetry. This is most interesting because aside from SHG/SFG, only neutron scattering is known to be effective to probe antiferromagnetism.

In nonmagnetic crystals, SHG/SFG is allowed under ED approximation characterized by the nonlinear susceptibility $\overset{\leftrightarrow}{\chi}^{(2)}(i)$ if the atomic arrangement of crystal structure has no inversion symmetry, but forbidden otherwise, although it may still be detectable from the much weaker electric quadrupole–magnetic dipole (EQ/MD) contribution to $\overset{\leftrightarrow}{\chi}^{(2)}(i)$. For magnetic crystals, symmetry of spin arrangement must also be taken into account. A spin arrangement without inversion symmetry contributes an additional nonlinear susceptibility, $\overset{\leftrightarrow}{\chi}^{(2)}(c)$, to ED-allowed SHG/SFG, but because light interacts with spins only through spin-orbit coupling, $\overset{\leftrightarrow}{\chi}^{(2)}(c)$ is generally much weaker than the ED-allowed $\overset{\leftrightarrow}{\chi}^{(2)}(i)$ and comparable to the ED-forbidden $\overset{\leftrightarrow}{\chi}^{(2)}(i)$. (The EQ/MD contribution to $\overset{\leftrightarrow}{\chi}^{(2)}(c)$

is negligible.) If the spin arrangement reverses its direction, $\overset{\leftrightarrow}{\chi}^{(2)}(c)$ changes sign. The independent, nonvanishing elements of $\overset{\leftrightarrow}{\chi}^{(2)}(c)$ for crystals of different magnetic point symmetry groups are tabulated in Ref. [7]. To the lowest-order approximation, $\overset{\leftrightarrow}{\chi}^{(2)}(c)$ is proportional to the spin order parameter (or degree of spin ordering). The total nonlinear susceptibility for SHG/SFG from a magnetic crystal is $\overset{\leftrightarrow}{\chi}^{(2)} = \overset{\leftrightarrow}{\chi}^{(2)}(i) + \overset{\leftrightarrow}{\chi}^{(2)}(c)$.

We take the ferroic chromium oxide (Cr_2O_3) crystal as an example, which was first used by Froehlich and coworkers to demonstrate that SHG can be employed to probe antiferromagnetism.[8,9] The crystal structure of Cr_2O_3 is shown in Fig. 4.2(a). It belongs to the $3\bar{m}$ point group symmetry class that has inversion symmetry, and hence $\overset{\leftrightarrow}{\chi}^{(2)}(i)$ comes only from EQ/MD contribution. Above the Neel temperature $T_N = 307.5$ K, there is no spin ordering and $\overset{\leftrightarrow}{\chi}^{(2)}(c) = 0$ so that $\overset{\leftrightarrow}{\chi}^{(2)} = \overset{\leftrightarrow}{\chi}^{(2)}(i)$. Below T_N, the crystal structure remains unchanged, but the spins on Cr^{3+} are ordered as described in Fig. 4.2(b). The four spins in each unit cell are antiferromagnetically ordered. Since spins are pseudo-vectors, the inversion operation inverts the positions of the spins about the center of inversion, but

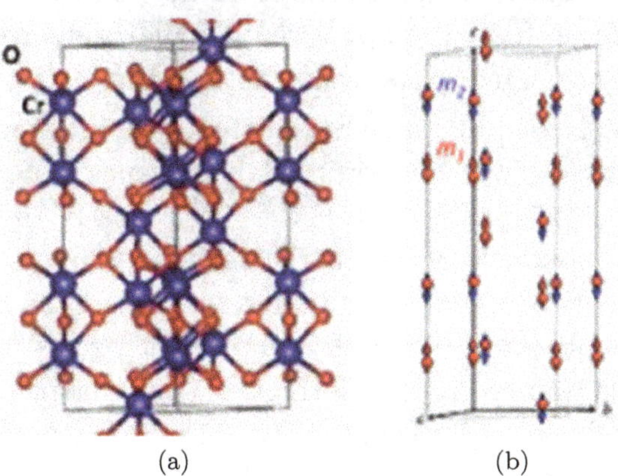

(a)   (b)

Fig. 4.2. (a) Crystal structure of Cr_2O_3. (b) Antiferromagnetic spin arrangement of Cr^{3+} in Cr_2O_3 below the Neel temperature T_N.

does not flip the spins. The four-spin arrangement of Cr_2O_3 obviously does not have inversion symmetry and $\overset{\leftrightarrow}{\chi}^{(2)}(c) \neq 0$; we then have $\overset{\leftrightarrow}{\chi}^{(2)} = \overset{\leftrightarrow}{\chi}^{(2)}(i) + \overset{\leftrightarrow}{\chi}^{(2)}(c)$. To satisfy angular momentum conservation in an SHG process, it is known that a circularly polarized input along a three-fold axis of a crystal should generate an oppositely circularly polarized SH output. A more detailed analysis taking into account the symmetry of $\overset{\leftrightarrow}{\chi}^{(2)}(i)$ and $\overset{\leftrightarrow}{\chi}^{(2)}(c)$ for Cr_2O_3 showed that the SH signal strength is given by $S \propto |i\chi_{yyy}^{(2)}(i) \pm \chi_{yyy}^{(2)}(c)|^2$,[8,9] with $+$ and $-$ referring to the input right and left circular polarization, respectively, and z is along the three-fold axis of Cr_2O_3. We should have the same SH signal for the two input circular polarizations at $T > T_N$, but different signals at $T < T_N$.

The above theoretical conclusion was experimentally verified, as depicted in Fig. 4.3. The transmitted SH spectra of d-d transitions of Cr^{3+} of Cr_2O_3 in the 1.9–2.8 eV range were taken with the input beam along the three-fold optical axis of Cr_2O_3. Two oppositely oriented antiferromagnetic domains were probed. As seen in Fig. 4.3(a) and (b), the spectra obtained at $T < T_N$ from the two circularly polarized inputs on two oppositely oriented domains are complementary to each other, i.e., the spectra from one circularly polarized input on one antiferromagnetic domain are identical to those from the oppositely circularly polarized input on the oppositely oriented antiferromagnetic domain, as expected from $S \propto |i\chi_{yyy}^{(2)}(i) \pm \chi_{yyy}^{(2)}(c)|^2$. The SH signal above T_N from $\overset{\leftrightarrow}{\chi}^{(2)}(i)$ was found to be weak. The temperature dependence of SHG measured at 2 eV is plotted in Fig. 4.3(c). It is seen clearly that the signals from the two oppositely circularly polarized inputs on a single antiferromagnetic domain are different below T_N, but they collapse into single values above T_N. Resonance features of SHG can in principle provide microscopic information about magnetic–electronic coupling associated with the optical transitions in a magnetic crystal. Separate determination of $\overset{\leftrightarrow}{\chi}^{(2)}(i)$ and $\overset{\leftrightarrow}{\chi}^{(2)}(c)$ around resonances from the measurement is possible with phase-sensitive SHG. For some magnetic crystals, such as $RMnO_3$ perovskite (R =Er, Ho, Sc, Y, Yb, etc.), the sets of nonvanishing elements of $\overset{\leftrightarrow}{\chi}^{(2)}(i)$ and $\overset{\leftrightarrow}{\chi}^{(2)}(c)$

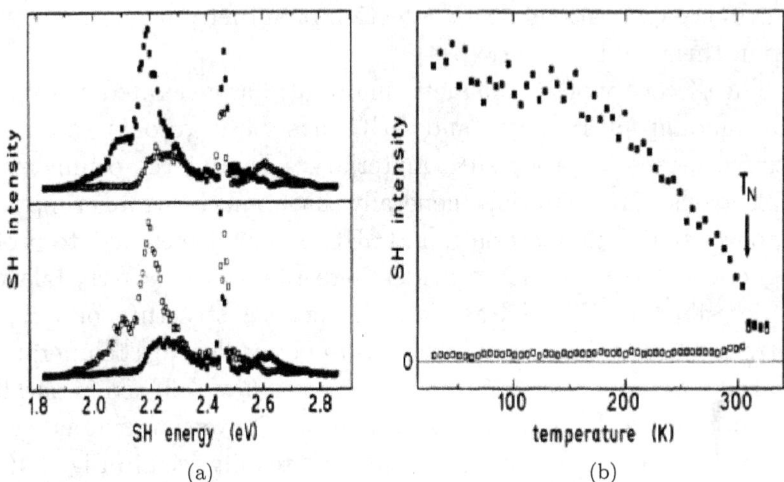

Fig. 4.3.   (a) SH spectra of two oppositely oriented antiferromagnetic domains of Cr_2O_3. Full and open squares refer to inputs with right and left circular polarizations, respectively. (b) Temperature dependence of the SH output generated from one antiferromagnetic domain of Cr_2O_3 measured at SH energy of 2 eV. The Neel temperature of Cr_2O_3 is at 307.5 K. Full and open squares refer to right and left input circular polarizations, respectively (after Ref. [8]).

are different; $\overset{\leftrightarrow}{\chi}^{(2)}(c)$ can then be separately measured using selected input–output polarization combinations.[10]

SHG/SFG is complementary to neutron or X-ray studies of antiferromagnetism. Take the case of $RMnO_3$ as an example. It was reported that neutron scattering could not distinguish two possible antiferromagnetic spin arrangements that belong to two different magnetic point groups, but SHG could from the measurement of $\overset{\leftrightarrow}{\chi}^{(2)}(c)$ because the sets of nonvanishing elements of $\overset{\leftrightarrow}{\chi}^{(2)}(c)$ for the two magnetic groups are different.[10] In the same vein, changes in antiferromagnetic spin structure can be monitored *in situ* by SHG, but would be difficult for neutron scattering. For a given antiferromagnetic spin structure, SHG can be used to measure the degree of spin ordering and antiferromagnetic phase transition as $\overset{\leftrightarrow}{\chi}^{(2)}(c)$ is proportional to its antiferromagnetic spin order parameter. Dynamics of induced spin structural changes and phase transition can also be followed by SHG. Antiferromagnets have recently been considered for spintronics applications; SHG detection of antiferromagnetism could help in their implementation.

The discovery of 2D magnetic materials has generated a tremendous amount of interest, and SHG has been recognized as an effective means to probe such materials.[11] Due to the confinement of electrons, 2D materials generally have much stronger optical responses than bulk materials, and SHG is sensitive enough to probe magnetic monolayers. We consider here the case of a CrI_3 bilayer. As described in Fig. 4.4(a)–(c), the crystal structure of CrI_3 is centrosymmetric, but the spin structure below the Neel temperature is antiferromagnetic with spins on Cr in each layer aligned in parallel but opposite in different layers. The lack of inversion symmetry of the antiferromagnetic spin arrangement is readily seen in Fig. 4.4(c); since an inversion operation does not flip spins, resulting in only an interchange of the two monolayers, the overall spin arrangement is flipped. This is contrary to the ferromagnetic spin arrangement in Fig. 4.4(d); the bilayer remains unchanged on inversion operation.

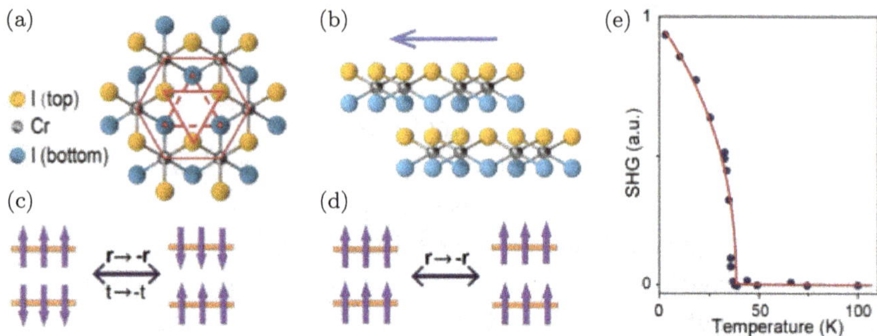

Fig. 4.4.   (a) Top view of the atomic structure of a CrI_3 monolayer. (b) Side view of the atomic structure of a CrI_3 bilayer. The arrow indicates that the two monolayers can be shifted relative to each other. (c) Spin arrangement of the antiferromagnetic state of a CrI_3 bilayer. A spatial inversion ($r \to -r$) or a time reversal ($t \to -t$) operation flips the spin arrangement and shows that the system does not have inversion symmetry. (d) The same operation on the ferromagnetic state of the CrI_3 bilayer keeps the spin arrangement unchanged and indicates that the bilayer has inversion symmetry. (e) Observed SHG signal from the bilayer as a function of temperature displaying the phase transition behavior from antiferromagnetic to nonmagnetic as temperature approaches $\sim$40 K (after Ref. [11]).

Thus, in the antiferromagnetic phase of bilayer CrI$_3$, $\overset{\leftrightarrow}{\chi}^{(2)}(i) \approx 0$, $\overset{\leftrightarrow}{\chi}^{(2)}(c) \neq 0$, and SHG from $\overset{\leftrightarrow}{\chi}^{(2)}(c)$ can be easily detected, as shown in Fig. 4.4(e). The SHG signal displays an antiferromagnetic to nonmagnetic transition behavior as the temperature approaches $T_N \sim 75$ K. A sufficiently strong applied magnetic field (~ 0.63T) along the surface normal was able to switch the antiferromagnetic spin alignment completely to ferromagnetic alignment that has an overall inversion symmetry with $\overset{\leftrightarrow}{\chi}^{(2)}(i) = \overset{\leftrightarrow}{\chi}^{(2)}(c) = 0$. The SH signal from the ferromagnetic state was indeed found to be vanishingly small. This is opposite to the result of linear magneto-optical effect, which is sensitive to ferro- and ferrimagnetization, but not to antiferromagnetization.

4.3. Detection of Charge Current, Spin Current, and Spin-Polarized Current by Second Harmonic Generation

Fields and currents in a material can change the optical response of the material. In the linear optical case, an applied dc electric field $\vec{E}$ on a medium can lead to a change in the refractive index or linear optical susceptibility tensor proportional to the field. This is the well-known Pockels effect. A similar effect can be induced by a current density $\vec{J}$ in the medium. Conversely, by measuring the change of the linear optical response, one can find the field or current distribution in the medium.

More explicitly, in the presence of $\vec{E}$ or $\vec{J}$, the linear susceptibility of a medium is given by $\overset{\leftrightarrow}{\chi}^{(1)} = \overset{\leftrightarrow}{\chi}_0^{(1)} + \Delta\overset{\leftrightarrow}{\chi}^{(1)}$. In the lowest order under the electric-dipole approximation, the induced susceptibility change is $\Delta\overset{\leftrightarrow}{\chi}^{(1)}(\vec{E}) = \overset{\leftrightarrow}{\chi}_E^{(2)} \cdot \vec{E}$ or $\Delta\overset{\leftrightarrow}{\chi}^{(1)}(\vec{J}) = \overset{\leftrightarrow}{\chi}_J^{(2)} \cdot \vec{J}$ where $\overset{\leftrightarrow}{\chi}_E^{(2)}$ and $\overset{\leftrightarrow}{\chi}_J^{(2)}$ are third-rank tensors that vanish for media with inversion symmetry. Measurement of $\Delta\overset{\leftrightarrow}{\chi}^{(1)}(\vec{E})$ or $\Delta\overset{\leftrightarrow}{\chi}^{(1)}(\vec{J})$ allows the extraction of $\vec{E}$ or $\vec{J}$. Of course, $\vec{E}$ and $\vec{J}$ can coexist in a medium, but generally, $\vec{E}$ dominates when $\vec{J}$ is small, and vice versa. Microscopically, while $\vec{E}$ perturbs the structure of a medium, $\vec{J}$ perturbs the carrier distribution in the momentum space.[12] The same description applies to nonlinear optical responses. SHG measurement of $\Delta \overleftrightarrow{\chi}^{(2)}(\vec{E}) = \overleftrightarrow{\chi}_E^{(3)} \cdot \vec{E}$ or $\Delta \overleftrightarrow{\chi}^{(2)}(\vec{J}) = \overleftrightarrow{\chi}_J^{(3)} \cdot \vec{J}$ induced by $\vec{E}$ or $\vec{J}$ can be used to determine $\vec{E}$ or $\vec{J}$. Since more parameters can be extracted from SHG measurement than from the linear optical case, the detection of not only charge current but also spin current and spin-polarized current from SHG is possible. In specific cases, $\vec{E}$ or $\vec{J}$ can selectively alter the symmetry and induces a $\Delta \chi_{ijk}^{(2)}$ in the otherwise vanishing $\chi_{ijk}^{(2)}$. The first experiment on charge current induced SHG was reported by Aktsipetrov *et al.*[13]

Consider, as an example, the case of a GaAs(001) wafer with its surface normal along the crystal axis [001], taken as $\hat{z}$, and a charge current flowing parallel to the wafer along $\hat{x}$.[14] Initially, without current, GaAs belongs to the $\bar{4}3m$ symmetry class with $\chi_{ijk}^{(2)}(i \neq j \neq k)$ being the only nonvanishing $\overleftrightarrow{\chi}^{(2)}$ elements, but the current changes the symmetry to $4mm$ and induces nonvanishing $\Delta \chi_{ijk}^{(2)}$ that has one or all subindices being x. If an x-polarized input beam impinges on the wafer along $\hat{z}$, one expects a current-induced SHG signal from the bulk GaAs proportional to $|\Delta \chi_{xxx}^{(2)}|^2 \propto |J_x|^2$. In practice, one can set the input or output polarizer slightly off $\hat{x}$ and create a residual SHG in the absence of a current much stronger than the current-induced SHG to serve as a local oscillator for homodyne detection of the latter. The change of the SHG signal induced by the current is then, to a good approximation, proportional to $\Delta \chi_{xxx}^{(2)} \propto J_x$. This was the case in the experiment of Ruzicka *et al.* shown in Fig. 4.5(a), where the sample geometry is given in the inset and the measured current-induced SHG, labeled as ΔP, is plotted as a function of current density J. The expected linear dependence of ΔP on J is seen. The image of the current-induced SHG is displayed in Fig. 4.5(b).

Spin-polarized current can be detected by linear magneto-optical effect, but can also be detected by current-induced SHG.[15] The nonvanishing elements of $\Delta \overleftrightarrow{\chi}^{(2)}(\vec{J})$ induced by the spin-polarized current are different from those induced by charge current and can again be found from symmetry consideration: a symmetry operation on the lab coordinates that keeps $\vec{J}$ unchanged, but changes $\Delta \chi_{ijk}^{(2)}$

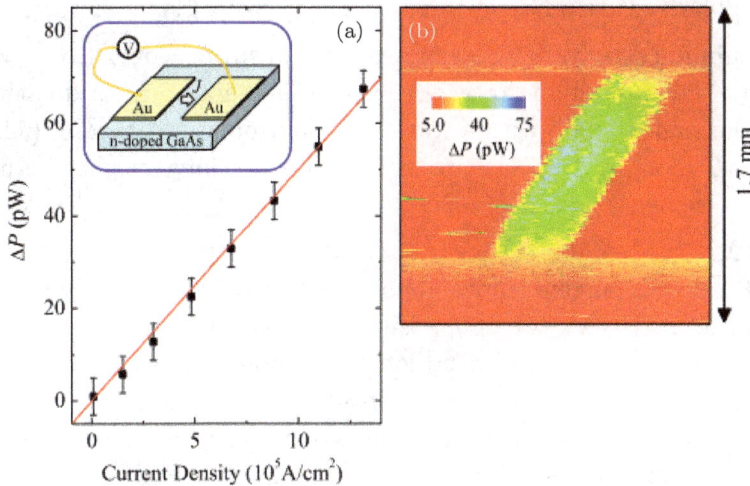

Fig. 4.5. (a) Current-induced SH signal (ΔP) from GaAs(100) as a function of current density. The inset describes the device geometry. (b) SH image of the current density distribution is obtained by scanning SH microcopy. The green strip shows that more intense SH is generated from the gap area between the gold electrodes (after Ref. [14]).

to $-\Delta\chi_{ijk}^{(2)}$, or vice versa, should lead to $\Delta\chi_{ijk}^{(2)} = 0$. A spin-polarized current density is not a simple vector, but is the product of a vector and a pseudo-vector, $\vec{J}_{sp} = \overleftrightarrow{J}_c\hat{s}$, with $\vec{J}_c$ describing the charge current density in which the spins of all carriers are aligned along $\hat{s}$. Being a pseudovector, $\hat{s}$ changes to $-\hat{s}$ upon mirror reflection if $\hat{s}$ is parallel to the mirror plane, but remains unchanged if $\hat{s}$ is perpendicular to the mirror plane. Consider a longitudinal spin-polarized current with $\hat{s} \parallel \vec{J}_c \parallel \hat{z}$ in a system with a mirror symmetry plane in $\hat{y} - \hat{z}$. Under the reflection operation about the mirror plane, $\vec{J}_{sp} \parallel \hat{z}$ changes to $-\vec{J}_{sp}$, but $\Delta\chi_{ijk}^{(2)}$ remains unchanged if the subindices contain no x or two x. On the other hand, since $\Delta\chi_{ijk}^{(2)} \propto \vec{J}_{sp}$, we must have $\Delta\chi_{ijk}^{(2)}$ changed to $-\Delta\chi_{ijk}^{(2)}$ upon reflection. This shows that $\Delta\chi_{ijk}^{(2)}$ with two or no x in the subindices should vanish. The same discussion can be extended to mirror symmetry planes $\hat{x} - \hat{y}$ and $\hat{z} - \hat{x}$. Thus, for a system with the three orthogonal mirror planes and a longitudinal spin-polarized current, $\vec{J}_{sp} \parallel \hat{z}$, all $\Delta\chi^{(2)}_{ijk}$ with x, y, z appearing in the subindices zero or two times must vanish, i.e., only $\Delta\chi^{(2)}_{ijk}$ with $i \neq j \neq k$ can be induced by the longitudinal spin-polarized current. For the same system with a transverse spin-polarized current ($\hat{s} \perp \vec{J}_c$) with $\vec{J}_c$ along $\hat{z}$ and $\hat{s}$ along $\hat{x}$, the nonvanishing current-induced $\Delta\chi^{(2)}_{ijk}$ elements are $\Delta\chi^{(2)}_{yzz}$, $\Delta\chi^{(2)}_{zyz}$, $\Delta\chi^{(2)}_{zzy}$, $\Delta\chi^{(2)}_{yxx}$, $\Delta\chi^{(2)}_{xyx}$, $\Delta\chi^{(2)}_{xxy}$, and $\Delta\chi^{(2)}_{yyy}$.

Pure spin current appears when two equal but oppositely spin-polarized currents move in opposite directions in a medium, resulting in zero charge current and no local spin polarization, but a flow of aligned spins in one direction. Presumably, the ohmic loss would be suppressed if the charge current in circuits could be replaced by the pure spin current. Because charge flow and magnetization are absent, it is difficult to probe flowing pure spin current by usual electric–magnetic means. Detection by SHG is however possible knowing that each of the two oppositely spin-polarized currents, $\vec{J}_c\hat{s}$ and $(-\vec{J}_c)(-\hat{s})$, induces the same $\Delta\overset{\leftrightarrow}{\chi}^{(2)}$, and the two combined induce a total $2\Delta\overset{\leftrightarrow}{\chi}^{(2)}$ for SHG. Experimental demonstration of SHG detection of pure spin current has been reported,[16] but the use of SHG to study spin-polarized and pure spin currents and their dynamics has not yet been well explored.

We should mention that the inverse process of current-induced SHG is the injection of current by mixing fundamental and SH waves in a medium.[17,18] With appropriate fundamental and SH polarization combination, charge, spin-polarized, or pure spin current can be selectively injected in the medium. Such currents generated without electrodes can be detected by SHG without electrodes. We will not dwell more on this topic here, but refer the readers to Ref. [19].

4.4. Bulk Characterization by Second Harmonic and Sum Frequency Spectroscopy

As mentioned in Sec. 2.3, because there are more independent input parameters, SHG/SFG carries more information about a material than linear optics. This is particularly true with spectroscopic measurements. SHG/SFG can probe double resonant transitions, but linear spectroscopy cannot. Even on single resonant transitions, SHG/SFG has different selection rules than linear absorption spectroscopy (and Raman spectroscopy): under electric-dipole approximation, it can only access transitions that are both one-photon and two-photon active. Yet with different input–output polarization combinations, it can extract more independent transition matrix elements. For illustration, we describe here the case of infrared-visible SF spectroscopy on bulk phonons of crystalline quartz.[20]

Crystalline α-quartz has a rhombohedral structure with $D_3(32)$ point symmetry. It has a three-fold symmetry axis $(\hat{c})$ and three two-fold symmetry axes perpendicular to $\hat{c}$. The three orthogonal principal axes of the crystal are $\hat{a}$, $\hat{b}$ and $\hat{c}$, with $\hat{a} \perp \hat{b} \perp \hat{c}$, and $\hat{a}$ is along a two-fold axis. The corresponding nonvanishing $\overleftrightarrow{\chi}^{(2)}$ elements are $\chi_{aaa}^{(2)} = -\chi_{abb}^{(2)} = -\chi_{bab}^{(2)} = -\chi_{bba}^{(2)}$, $\chi_{abc}^{(2)} = -\chi_{bac}^{(2)}$, $\chi_{cab}^{(2)} = -\chi_{cba}^{(2)}$ and $\chi_{bca}^{(2)} = -\chi_{acb}^{(2)}$. If we choose a lab coordinate system with $\hat{z} \parallel \hat{c}$ and $\hat{x}$ separated from $\hat{a}$ by angle ϕ, the nonvanishing $\overleftrightarrow{\chi}^{(2)}$ elements in the lab coordinates in terms of those in the crystal coordinates are

$$\chi_{xxx}^{(2)} = -\chi_{xyy}^{(2)} = -\chi_{yxy}^{(2)} = -\chi_{yyx}^{(2)} = \chi_{aaa}^{(2)} \cos 3\phi$$

$$-\chi_{yyy}^{(2)} = \chi_{yxx}^{(2)} = \chi_{xyx}^{(2)} = \chi_{xxy}^{(2)} = \chi_{aaa}^{(2)} \sin 3\phi,$$

$$\chi_{xyz}^{(2)} = -\chi_{yxz}^{(2)} = \chi_{abc}^{(2)}, \tag{4.1}$$

$$\chi_{xzy}^{(2)} = -\chi_{yzx}^{(2)} = \chi_{acb}^{(2)},$$

$$\chi_{zxy}^{(2)} = -\chi_{zyx}^{(2)} = \chi_{cab}^{(2)}.$$

The nonlinear susceptibility $\chi_{\hat{e}_3 \hat{e}_1 \hat{e}_2}^{(2)}$ measured by SFG with input–output polarization combination $(\hat{e}_3, \hat{e}_1, \hat{e}_2)$ is related to $\chi_{ijk}^{(2)}$ in the lab coordinates by $\chi_{\hat{e}_3 \hat{e}_1 \hat{e}_2}^{(2)} = \sum_{i,j,k} (\hat{e}_3 \cdot \hat{i}) \chi_{ijk}^{(2)} (\hat{j} \cdot \hat{e}_1)(\hat{k} \cdot \hat{e}_2)$, from which the three polarization combinations that can access the nonvanishing $\overset{\leftrightarrow}{\chi}^{(2)}$ elements are

$$\chi^{(2)}_{SSS} = a\chi^{(2)}_{yyy} = \alpha\chi^{(2)}_{aaa}\sin 3\phi,$$

$$\chi^{(2)}_{SPP} = b_1\chi^{(2)}_{yxx} + b_2\chi^{(2)}_{yxz} + b_3\chi^{(2)}_{yzx}$$

$$= \beta_1\chi^{(2)}_{aaa}\sin 3\phi + \beta_2\chi^{(2)}_{abc} + \beta_3\chi^{(2)}_{acb}, \tag{4.2}$$

$$\chi^{(2)}_{PSP} = c_1\chi^{(2)}_{syx} + c_2\chi^{(2)}_{xyz} + c_3\chi^{(2)}_{zyx}$$

$$= \gamma_1\chi^{(2)}_{aaa}\sin 3\phi + \gamma_2\chi^{(2)}_{abc} + \gamma_3\chi^{(2)}_{cba}.$$

When the input IR frequency is close to phonon resonances, we have, following Eq. (2.13),

$$\chi^{(2)}_{\hat{e}_3\hat{e}_1\hat{e}_2} = \chi^{(2)}_{NR,\hat{e}_3\hat{e}_1\hat{e}_2} + \sum_q \frac{A_{q,\hat{e}_3\hat{e}_1\hat{e}_2}}{(\omega_2 - \omega_q + i\Gamma_q)} \tag{4.3}$$

and a similar equation for $\chi^{(2)}_{ijk}$ in the lab coordinates and $\chi^{(2)}_{lmn}$ in the crystal coordinates. The previous relations among $\chi^{(2)}_{\hat{e}_3\hat{e}_1\hat{e}_2}$, $\chi^{(2)}_{ijk}$ and $\chi^{(2)}_{ijk}$ are also held by $A_{q,\hat{e}_3\hat{e}_1\hat{e}_2}$, $A_{q,ijk}$ and $A_{q,lmn}$. The measured SF output intensity is proportional to $|\chi^{(2)}_{\hat{e}_3\hat{e}_1\hat{e}_2}|^2$.

As mentioned earlier, SFG can only probe resonances that are both one- and two-photon active under electric-dipole approximation, or in the phonon case, only modes that are both IR and Raman active. For α-quartz, there are only three such phonon modes in the spectral range between 750 and 1300 cm^{-1}, which are seen in the SSS SF vibrational spectrum presented in Fig. 4.6(a) taken from reflection from a z-cut quartz plate.[20] For comparison, the Raman spectrum of α-quartz is shown in Fig. 4.6(b); five phonon modes appear in the same spectral range.[21] Due to strong absorption, the IR absorption spectrum of α-quartz consists of broad bands and is more difficult to analyze.[22,23] Reflected SFG from the z-cut quartz with $\hat{z}$ being a three-fold axis is expected to show a three-fold symmetry with respect to sample rotation about $\hat{z}$. This can be seen explicitly from the expression of the SF signal, $S \propto |\chi^{(2)}_{\hat{e}_3\hat{e}_1\hat{e}_2}|^2$ with $\chi^{(2)}_{\hat{e}_3\hat{e}_1\hat{e}_2}$ given in Eq. (4.2) for all three polarization combinations, i.e., S has the

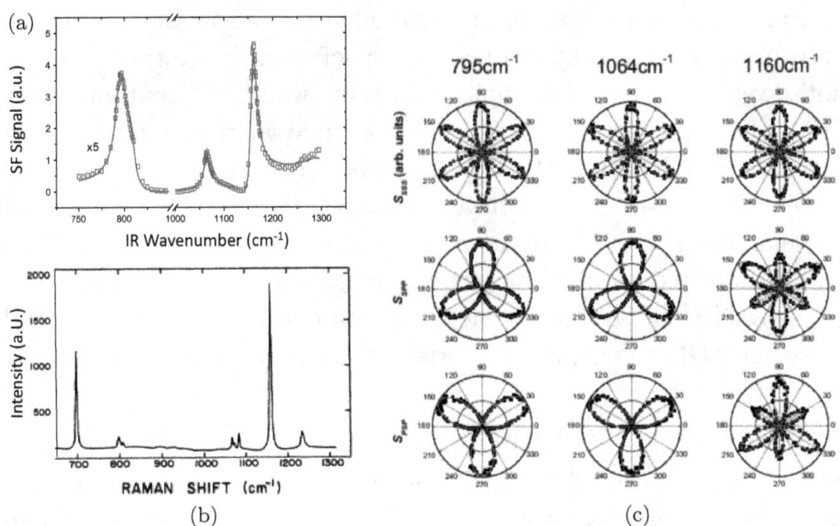

Fig. 4.6. (a) SF phonon spectrum of α-quartz (0001) obtained with SSS polarization combination (after Ref. [20]). (b) Raman spectrum of α-quartz. (c) SF signal versus azimuthal rotation of an a-quartz(0001) at phonon resonances of 795, 1064 and 1160 cm^{-1} obtained with SSS, SPP and PSP polarization combinations. The solid lines are theoretical fits (after Ref. [20]).

form $S \propto |A + B\sin 3\phi|^2$. Experimental observation is plotted in Fig. 4.6(c).[20] Note that if A = 0, S has six-fold rotational symmetry and if $|A| \sim |B|$, S has three-fold symmetry with three clear lobes. Phonon spectra of α-quartz over a spectral range extended to 300 cm^{-1} have been measured by SHG using a tunable IR free electron laser.[24] The observed modes in the 750$-$1300 cm^{-1} range agree well with those detected by SFG.

SHG/SFG can also be used to extract information on the electronic structure of crystals without inversion symmetry. In the early days of nonlinear optics, SHG spectroscopy was already used to probe interband transitions of III$-$V compounds, revealing resonant enhancement at singularity points.[25$-$27] Even for media with inversion symmetry, such as liquids, transmitted SHG/SFG can still be used to probe their electric-quadrupole transitions. However, in recent years, interest in detailed spectroscopic studies of bulk materials has waned. There is not much incentive to glean information about a material beyond what can be obtained by linear absorption, Raman, and fluorescence spectroscopy. We therefore will not dwell more on this topic, but only want to point out that SHG/SFG spectroscopy could provide unique, useful information about materials, especially new materials. For example, theoretical studies of SHG in crystals have discovered that the contribution of intraband transitions to resonant SHG is related to Berry connections and is expected to have further enhancement at Weyl points,[28–30] but how SHG can be used to help characterize Weyl semimetals, and more generally topological materials, has not yet been explored.

4.5. Summary and Prospects

SHG/SFG as a tool for the characterization of bulk materials has some distinct merits over linear optics. These were introduced in this chapter:

- The sets of independent, nonvanishing elements of nonlinear susceptibilities for SHG/SFG are different for different symmetry classes of materials. Their measurement can help distinguish structural phases and track phase transitions of materials.
- SHG can be used to probe antiferromagnetic structure, or magnetic structure in general, if the spin arrangement has no inversion symmetry, irrespective of whether the lattice structure has inversion symmetry or not.
- SHG with appropriate input–output polarization combinations allows the detection of charge current, spin-polarized current and pure spin current.
- SF phonon spectroscopy detects phonons that are both IR and Raman active; this selection rule leads to a simpler phonon spectra observed by SFG and helps identify phonon modes of materials.
- With more independently adjustable input parameters, SHG/SFG spectroscopy can provide more information about structures of materials than linear optics. It can probe electric quadrupole/magnetic dipole transitions in media with inversion symmetry.

SF spectroscopy has not yet been well explored for characterization of bulk materials. Coherent 2D spectroscopy to interrogate electron−electron and electron−phonon coupling is an example. It will be interesting to learn if SF spectroscopy can be an effective means to study the properties of novel materials not accessible by other spectroscopic techniques. The range of applicability of SF spectroscopy is currently limited by available IR sources at low frequencies.

References

1. Butcher, P. N.: *Nonlinear Optical Phenomena*. Engineering Experiment Station: Columbus, 1965.
2. Vogt, H.: Study of Structural Phase-Transitions by Techniques of Nonlinear Optics. *Appl. Phys.* 1974, 5, 85−96.
3. Glass, A. M.: Dielectric Thermal and Pyroelectric Properties of Ferroelectric $LiTaO_3$. *Phys. Rev.* 1968, 172, 564−571.
4. Copalan, V.; Raj, R.: Domain Structure and Phase Transitions in Epitaxial $KNBO_3$ Thin Films Studied by *In Situ* Second Harmonic Generation Measurements. *Appl. Phys. Lett.* 1996, 68, 1323.
5. Uesu, Y.; Kurimura, S.; Yamamoto, Y.: Optical Second-Harmonic Images of 90° Domain-Structure in $BaTiO_3$, and Periodically Inverted Anti-parallel Domains in $LiTaO_3$. *Appl. Phys. Lett.* 1995, 66, 2165−2167.
6. Shank, C. V.; Yen, R.; Hirlimann, C.: Femtosecond-Time-Resolved Surface Structural Dynamics of Optically-Excited Silicon. *Phys. Rev. Lett.* 1983, 51, 900−902.
7. Birss, R. R.: Symmetry and Magnetism. North-Holland Pub. Co., Interscience Publishers: Amsterdam, New York, 1964.
8. Fiebig, M.; Frohlich, D.; Krichevtsov, B. B.; Pisarev, R. V.: Second-Harmonic Generation and Magnetic-Dipole-Electric-Dipole Interference in Antiferromagnetic Cr_2O_3. *Phys. Rev. Lett.* 1994, 73, 2127−2130.
9. Pisarev, R. V.; Fiebig, M.; Frohlich, D.: Nonlinear Optical Spectroscopy of Magnetoelectric and Piezomagnetic Crystals. *Ferroelectrics* 1997, 204, 1−21.
10. Fiebig, M.; Frohlich, D.; Kohn, K.; Leute, S.; Lottermoser, T.; Pavlov, V. V.; Pisarev, R. V.: Determination of the Magnetic Symmetry of Hexagonal Manganites by Second Harmonic Generation. *Phys. Rev. Lett.* 2000, 84, 5620−5623.

11. Sun, Z.; Yi, Y.; Song, T.; Clark, G.; Huang, B.; Shan, Y.; Wu, S.; Huang, D.; Gao, C.; Chen, Z.; McGuire, M.; Cao, T.; Xiao, D.; Liu, W.T.; Yao, W.; Xu, X.; We, S.W.: Giant Nonreciprocal Second Harmonic Generation from Antiferromagnetic Bilayer CrI_3. *Nature* 2019, 572, 497−501.

12. Khurgin, J. B.: Current-Induced Second-Harmonic Generation in Semiconductors. *Appl. Phys. Lett.* 1995, 67, 1113−1115.

13. Aktsipetrov, O.; Bessonov, Y.; Fedyanin, A.; Valdner, V.: DC-Induced Generation of the Reflected Second Harmonic in Silicon. *JETP Lett.* 2009, 89, 58−62.

14. Ruzicka, B. A.; Werake, L. K.; Xu, G. W.; Khurgin, J. B.; Sherman, E. Y.; Wu, J. Z.; Zhao, H.: Second-Harmonic Generation Induced by Electric Currents in GaAs. *Phys. Rev. Lett.* 2012, 108, 077403.

15. Wang, J.; Zhu, B. F.; Liu, R. B.: Second-Order Nonlinear Optical Effects of Spin Currents. *Phys. Rev. Lett.* 2010, 104, 256601.

16. Werake, L. K.; Zhao, H.: Observation of Second-Harmonic Generation Induced by Pure Spin Currents. *Nat. Phys.* 2010, 6, 875−878.

17. Dupont, E.; Corkum, P. B.; Liu, H. C.; Buchanan, M.; Wasilewski, Z. R.: Phase-Controlled Currents in Semiconductors. *Phys. Rev. Lett.* 1995, 74, 3596−3599.

18. Atanasov, R.; Hache, A.; Hughes, J. L. P.; vanDriel, H. M.; Sipe, J. E.: Coherent Control of Photocurrent Generation in Bulk Semiconductors. *Phys. Rev. Lett.* 1996, 76, 1703−1706.

19. Ruzicka, B. A.; Zhao, H.: Optical Studies of Ballistic Currents in Semiconductors. *J. Opt. Soc. Am. B* 2012, 29, A43−A54.

20. Liu, W. T.; Shen, Y. R.: Sum-Frequency Phonon Spectroscopy on Alpha-Quartz. *Phys. Rev. B* 2008, 78, 024302.

21. Tekippe, V. J.; Ramdas, A. K.; Rodrigue.S: Piezospectroscopic Study of Raman-Spectrum of Alpha-Quartz. *Phys. Rev. B* 1973, 8, 706−717.

22. Spitzer, W.G.; Kleinman, D.A.: Infrared Lattice Bands of Quartz. *Phys. Rev.* 121, 1324−1335.

23. He, M,; Yan, W.; Chang, Y. Liu, K.; Liu, X.: Fundamental Infrared Absorption Features of Alpha-Quartz: An Unpolarized Single-Crystal Absorption Infrared Spectroscopic Study. *Vibra. Spectr.* 2019, 101, 52−63.

24. Winta, C.J.; Gewinner, S.; Schöllkopf, W.; Wolf, M.; Paarmann, A.: Second Harmonic Phonon Spectroscopy of Alpha-Quartz. *Phys. Rev. B* 2018, 97, 094108.

25. Parsons. F.G.; Chang, R.K.: Measurement of the Nonlinear Susceptibility Dispersion by Dye Lasers. *Opt. Comm.* 1971, 3, 173−176.

26. Lotem, H.; Koren, G.; Yacoby, Y.: Dispersion of Nonlinear Optical Susceptibility in GaAs and GaSb. *Phys. Rev. B* 1974, 9, 3532−3540.

27. Bergfeld, S.; Daum, W.: Second-Harmonic Generation in GaAs: Experiment versus Theoretical Predictions of $\chi^{(2)}_{xyz}$. *Phys. Rev. Lett.* 2003, 90, 036801.
28. Aversa, C.; Sipe, J.E.: Nonlinear Optical Susceptibilities of Semiconductors: Results with a Length–Gauge analysis. *Phys. Rev. B* 1995, 52, 14636–14645.
29. Morimoto, T.; Nagaosa, N.: Topological Nature of Nonlinear Optical Effects in Solids. *Sci. Adv.* 2016, 2, e1501524.
30. Parker, D.E.; Morimoto, T.; Orenstein, J.; Moore, J.E.: Diagrammatic Approach to Nonlinear Optical Response with Application to Weyl Semimetals. *Phys. Rev. B* 2019, 99, 045121.

Review Articles.

- Fiebig, M.; Pavlov, V. V.; Pisarev, R. V.: Second-Harmonic Generation As A Tool for Studying Electronic and Magnetic Structures of Crystals: Review. *J Opt Soc Am B-Optical Physics* 2005, 22, 96–118.
- Ruzicka, B. A.; Zhao, H.: Optical Studies of Ballistic Currents in Semiconductors. *J Opt Soc Am B* 2012, 29, A43–A54.
- Wang, Y.; Xiao, J.; Yang, S.; Wang, y.; Zhang, X.: Second Harmonic Generation Spectroscopy on Two-Dimensional Materials. Opt Mat Express 2019, 1136–1149.

Chapter 5

Sum Frequency Chiral Spectroscopy

Chirality originates from the Greek word for "hand" and was first used by William Thomson (Lord Kelvin) to describe a three-dimensional geometric shape or form that has a distinguishable mirror image such as a hand: "I call any geometric figure or group of points, 'Chiral', and say that it has chirality if its image in a plane mirror, ideally realized, cannot be brought to coincide with itself."[1] (See Fig. 5.1 for illustration.) It appears ubiquitously in nature over a wide spectrum from elementary particles, nucleus, atoms, molecules, and crystals to shells, plants, planets, and even the universe (Fig. 5.2(a)). Molecular chirality in particular is intimately related to our life. Why more than 90% of biological molecules are homochiral is still a big mystery connected to the origin of life. Enantiomers (a word of Greek origin referring to pairs of molecules that are mirror images of each other) interacting differently with cells can have very different effects on living bodies. For example, right-handed lemone smells like orange, but left-handed lemone smells like lemon; only the right-handed morphine and cocaine are physiologically active. Identifying chirality and chiral structure of molecules has been most important for organic chemistry and biology.[2]

There are not many techniques available for probing chirality. The conventional one is circular dichroism (CD) spectroscopy (and the associated optical rotatory dispersion (ORD) spectroscopy). Most biology labs these days have a CD spectrometer for characterization of chirality of molecules, but the sensitivity

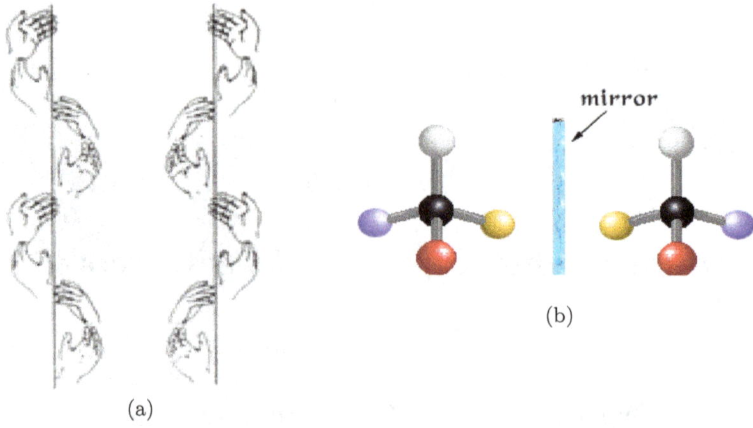

Fig. 5.1. Illustration of chirality. (a) Chirality is the Greek word for "hand". According to Lord kelvin, an object "has chirality if its image in a plane mirror, ideally realized, cannot be brought to coincide with itself." (b) A simple molecule with chiral structure.

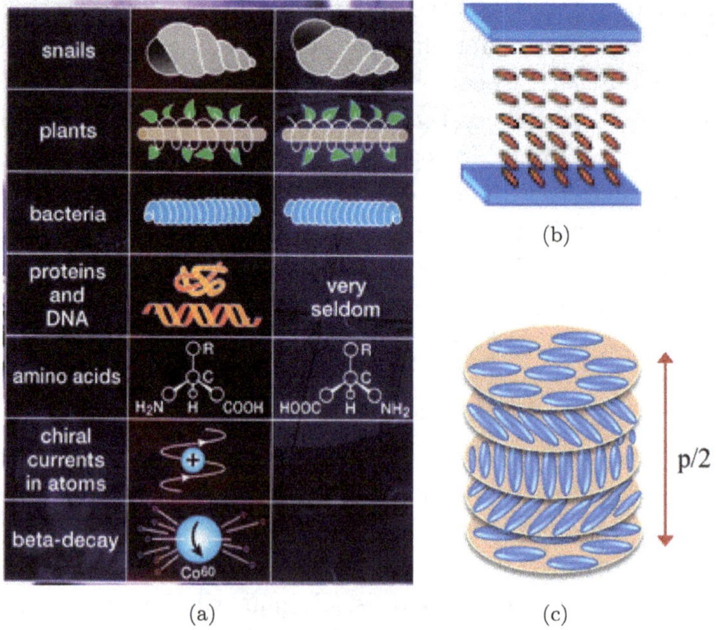

Fig. 5.2. (a) Chiral structures in nature, from very small to very large. (Curtesy of K.-H. Ernst) (b) A twisted nematic liquid crystal film. (c) A cholesteric liquid crystal film.

is limited. Two-photon and Raman spectroscopy can also probe chirality; yet their sensitivity is even worse.

We discuss in this chapter how SFG can be a powerful tool for probing chirality.[3,4] It is known that chiral materials have no inversion symmetry even in the liquid phase and therefore SFG is electric-dipole-allowed in such materials. Because SFG can have submonolayer sensitivity, we expect SFG spectroscopy to be more sensitive than CD spectroscopy. The higher sensitivity leads to many advantages of using SF spectroscopy to probe chirality as we shall see in this chpter. Our discussion will focus on studies of chiral liquids since they are most relevant to practical interest.

5.1. Circular Dichroism Spectroscopy versus Sum Frequency Spectroscopy

We begin our discussion on the question why linear CD/ORD has poor detection sensitivity. In a chiral medium, ORD/CD appears because the refractive indices, n_+ and n_-, for right and left circularly polarized light, respectively, are different: $n_\pm = n_0 \pm \Delta n = (n' + in'') \pm (\Delta n' + i\Delta n'')$ with $\Delta n'$ and $\Delta n''$ responsible for ORD and CD. On passing through a medium of thickness d, opposite circularly polarized beams of wavelength λ experience a phase and absorbance difference described by $2\Delta kd = (4\pi/\lambda)(\Delta n' + i\Delta n'')d$, the real part of which leads to ORD (rotation of linear polarization of the input) and the imaginary part to CD (difference in attenuations of right and left circularly polarized beams). The sensitivity of ORD/CD measurement is commonly limited to $(2\pi/\lambda)\Delta nd \sim 10^{-4}$ in the visible and $\sim 10^{-3}$ in the IR. For chiral molecular liquids, we usually have $|\Delta n| \sim 10^{-3}$ or smaller for electronic transitions and $|\Delta n| \sim 10^{-4}$ or smaller for vibrational transitions. In order to detect $|\Delta n| = 10^{-4}$, we need to have d $> 0.1\,\mu$m if $\lambda = 0.6\,\mu$m and d $> 1\,\mu$m if $\lambda = 6\,\mu$m. The very weak CD/ORD signal limits its ability to probe chirality *in situ* and to study chiral functions and dynamics of practical chiral systems.

Why is $|\Delta n|$ so small? This is because under the electric-dipole (ED) approximation, CD/ORD is forbidden, and becomes allowed only when electric-quadrupole (EQ, including magnetic-dipole) contribution to the refractive index of a medium is taken into account.[5] The reason is as follows. Since chirality is a property that exists only in 3D structural form, it must be represented by a rank-3 tensor, but Δn is a rank-2 tensor under ED approximation. In order to describe CD/ORD, we need to go beyond the ED approximation by incorporating the wave-vector ($\vec{k}$) dependence in n; to the first order in $\vec{k}$ (i.e., EQ contribution), we can write $n_{ij} = n_{0,ij} + \Delta n_{ij} = n_{0,ij} + \sum_k G_{ijk} k_k$ in the Cartesian tensorial notation, and the third-rank tensor G_{ijk} with $i \neq j \neq k$ describes chirality. As an EQ material response, $\Delta n_{ij} = \sum_k G_{ijk} k_k$ is expected to be smaller than n_0 by $\sim 10^{-4}$ to 10^{-3}.

Compared to Δn_{ij} in CD/ORD, the SF response coefficient $\chi_{ijk}^{(2)}$ is intrinsically a third-rank tensor, and SFG is ED-allowed even in isotropic chiral media. Generally, for a medium that lacks inversion symmetry only because of its chiral structure, we have $\chi_{ijk}^{(2)}(i \neq j \neq k)$ ED-allowed and serve as a chiral response coefficient for the chiral medium, but the achiral part of $\chi_{ijk}^{(2)}$ still vanishes under ED approximation. Contrary to CD/ORD that measures ED-forbidden Δn_{ij} over an ED-allowed background of $n_{0,ij}$, chiral SFG measures ED-allowed $\chi_{ijk}^{(2)}(i \neq j \neq k)$ over an ED-forbidden achiral background. As we shall see later, the achiral response in SFG can be further suppressed by selected input/output beam polarization combinations. Thus, the signal-to-background ratio is $|\chi_{chiral}^{(2)}/\chi_{achiral}^{(2)}| \gg 1$ for chiral SFG measurement in comparison with $|\Delta n/n_0| \ll 1$ for CD/ORD. The chiral detection sensitivity of SFG is no longer limited by the background, but limited by the detectable value of $\chi_{ijk}^{(2)}(i \neq j \neq k)$, indicating that SFG could be a sensitive technique to probe material chirality. The chiral nonlinearity coming from chiral molecular structure is usually weaker by more than one order of magnitude than nonlinearity induced by non-chiral, but non-centrosymmetric, molecular structure and/or arrangement. Even so, the sensitivity of SFG to detect chirality through $\chi_{ijk}^{(2)}(i \neq j \neq k)$ has been found to be significantly higher than CD/ORD. We now give a more detailed description of the theory of SF chiral spectroscopy in what follows.

There are two types of chirality associated with chiral molecular materials. One is from chiral bonding structure of molecules or molecular units; a simple example is shown in Fig. 5.1(b). The other is from chiral distribution of molecular units, illustrated in Figs. 5.2(b) and 5.2(c); the molecular units can be achiral as in the case of a twisted nematic liquid crystal film (Fig. 5.2(b)) or chiral as in the cases of cholesteric liquid crystals (Fig. 5.2(c)) and DNA. In anisotropic achiral media, it is possible to have $\chi_{ijk}^{(2)}(i \neq j \neq k) \neq 0$, (for example, in III-V semiconductors with cubic zincblende structure), but it has nothing to do with chirality. In the following, we shall restrict our discussion to isotropic chiral media with chirality from bonding structure and only touch upon chirality from molecular distribution in connection with chiral surfaces (Sec. 5.7), which are of particular importance for biointerfaces (Sec. 10.2).

5.2. Basic Theory of Chiral Sum Frequency Spectroscopy

We give in this section a sketch of the basic theory underlying chiral SF spectroscopy. Details can be found elsewhere.[3] We first need to learn more about $\chi_{ijk}^{(2)}(\omega_3 = \omega_1 + \omega_2)$ that SFG spectroscopy measures. For chiral media, the bulk nonlinear susceptibility is composed of chiral and achiral parts, $\chi_{ijk}^{(2)} = (\chi_{chiral}^{(2)})_{ijk} + (\chi_{achiral}^{(2)})_{ijk}$. For an isotropic chiral medium, we have $(\chi_{achiral}^{(2)})_{ijk} = 0$ under ED approximation as well as $(\chi_{chiral}^{(2)})_{iij} = (\chi_{chiral}^{(2)})_{iji} = (\chi_{chiral}^{(2)})_{jii} = 0$ because an isotropic chiral medium has rotational symmetry about $\hat{x}, \hat{y}$, and $\hat{z}$. Its lack of inversion symmetry due to chiral structure however results in $(\chi_{chiral}^{(2)})_{ijk} = \chi_{ijk}^{(2)}(i \neq j \neq k) \neq 0$, and its invariance upon $90°$ rotation about $\hat{i}$ followed by $90°$ rotation about $\hat{j}$ or $\hat{k}$ requires $\chi_{xyz}^{(2)} = \chi_{yzx}^{(2)} = \chi_{zxy}^{(2)} = -\chi_{yxz}^{(2)} = -\chi_{xzy}^{(2)} = -\chi_{zyx}^{(2)} \equiv \chi_C^{(2)}$. For the two enantiomers of right-handed (RH) and left-handed (LH) chirality, we have $\chi_C^{(2)}(\text{RH}) = -\chi_C^{(2)}(\text{LH})$. If we go beyond the ED approximation, then $(\chi_{achiral}^{(2)})_{ijk}$ becomes nonvanishing as the EQ contribution chips in. We have, following Eq. (2.9), $(\chi_{achiral}^{(2)})_{ijk}(\vec{k}_3 \neq \vec{k}_1 + \vec{k}_2) = -i\vec{k}_3 \cdot \vec{\chi}_{q3,ijk}^{(2)} + \vec{\chi}_{q1,ijk}^{(2)} \cdot i\vec{k}_1 + \vec{\chi}_{q2,ijk}^{(2)} \cdot i\vec{k}_2$. We note that $\vec{\chi}_{q\alpha,ijk}^{(2)}$ is a fourth-rank tensor with $\vec{k}_3 \cdot \vec{\chi}_{q3,ijk}^{(2)} \equiv \sum_{\kappa} k_{3,\kappa} \chi_{q3,(\bar{\kappa}i)jk}^{(2)}$, $\vec{\chi}_{q1,ijk}^{(2)} \cdot \vec{k}_1 \equiv \sum_{\kappa} \chi_{q1,i(\bar{\kappa}j)k}^{(2)} k_{1,\kappa}$, and $\vec{\chi}_{q2,ijk}^{(2)} \cdot \vec{k}_2 \equiv \sum_{\kappa} \chi_{q2,ij(\bar{\kappa}k)}^{(2)} k_{2,\kappa}$, where a bracketed subindice $(\bar{\kappa}\xi)$ denotes the field that provides EQ contribution through $\nabla\vec{E}$ with the spatial derivative along κ and the field component along ξ. The only nonvanishing elements of $\vec{\chi}_{q\alpha}^{(2)}$ for a medium of isotropic symmetry are $\chi_{q\alpha,llmm}^{(2)}$, $\chi_{q\alpha,lmml}^{(2)}$, and $\chi_{q\alpha,lmlm}^{(2)}$, and $\chi_{q\alpha,llll}^{(2)} = \chi_{q\alpha,llmm}^{(2)} + \chi_{q\alpha,lmml}^{(2)} + \chi_{q\alpha,lmlm}^{(2)}$.

To access the ED-allowed $(\chi_{chiral}^{(2)})_{ijk}(i \neq j \neq k)$, we need to have three orthogonal polarization components from the three respective input/output fields; this means that the input/output polarization combination must be either SPP or PSP, or PPS. To access $(\chi_{achiral}^{(2)})_{ijk}$ that has nonvanishing elements $\chi_{q\alpha,llmm}^{(2)}$, $\chi_{q\alpha,lmml}^{(2)}$, and $\chi_{q\alpha,lmlm}^{(2)}$, the polarization combination must be SSP, SPS, PSS, or PPP. These two sets of polarization combinations are mutually exclusive, allowing separate measurements of $(\chi_{chiral}^{(2)})_{ijk}$ and $(\chi_{achiral}^{(2)})_{ijk}$ with no interference from each other.

We can also see from the symmetry argument that $(\chi_{chiral}^{(2)})_{ijk}$ for isotropic chiral media should vanish for SHG. This is because SHG remains unchanged if the two input fields are interchanged and hence, $\chi_{ijk}^{(2)}(2\omega = \omega + \omega) = \chi_{ikj}^{(2)}(2\omega = \omega + \omega)$, but $(\chi_{chiral}^{(2)})_{ijk}$ has the additional symmetry relation $[\chi_{chiral}^{(2)}(2\omega = \omega + \omega)]_{ijk} = -[\chi_{chiral}^{(2)}(2\omega = \omega + \omega)]_{ikj}$. In order to satisfy both relations, $[\chi_{chiral}^{(2)}(2\omega = \omega + \omega)]_{ijk}$ must vanish. Thus, SHG spectroscopy is incapable of probing chirality of bulk isotropic chiral media. However, the above symmetry argument is not valid for anisotropic chiral media, and chiral SHG does not vanish in such media, including surfaces/interfaces, as we shall discuss in Sec. 5.6.

Another important feature of $(\chi_{chiral}^{(2)})_{ijk}$ for isotropic chiral media is that $(\chi_{chiral}^{(2)})_{ijk}$ would vanish if it had no frequency dispersion, meaning that if the input/output frequencies are far away from resonances, $(\chi_{chiral}^{(2)})_{ijk}$ should be very weak. This can be seen readily from the microscopic expression for $[\chi_{chiral}^{(2)}(\omega_3 = \omega_1 + \omega_2)]_{ijk}$ following Eq. (2.15). The bulk chiral nonlinearity $(\chi_{chiral}^{(2)})_{ijk}(i \neq j \neq k)$ is generally related to the molecular polarizability $\alpha_{\xi\eta\zeta}^{(2)}(\xi \neq \eta \neq \zeta)$ by

$$(\chi_{chiral}^{(2)})_{ijk} = N \sum_{\xi,\eta,\zeta} \alpha_{\xi\eta\zeta}^{(2)} \langle (\hat{i} \cdot \hat{\xi})(\hat{j} \cdot \hat{\eta})(\hat{k} \cdot \hat{\zeta}) \rangle \qquad (5.1)$$

Here, as already mentioned in Sec. 2.4, the effects of local field and intermolecular interactions are included in the definition of $\alpha_{\xi\eta\zeta}^{(2)}$, and the angular brackets denote average over molecular orientations. For isotropic chiral media, it can be shown that $\langle (\hat{i} \cdot \hat{\xi})(\hat{j} \cdot \hat{\eta})(\hat{k} \cdot \hat{\zeta}) \rangle = 1/6$ for all sets of $(\hat{i}, \hat{j}, \hat{k})$ and $(\hat{\zeta}, \hat{\eta}, \hat{\zeta})$ with $i \neq j \neq k$ and $\xi \neq \eta \neq \zeta$, and we have

$$\chi_C^{(2)} = (\chi_{chiral}^{(2)})_{ijk}(i \neq j \neq k)$$

$$= \frac{1}{6}N \left[(\alpha_{chiral}^{(2)})_{x'y'z'} - (\alpha_{chiral}^{(2)})_{x'z'y'} + (\alpha_{chiral}^{(2)})_{y'z'x'} \right. \qquad (5.2)$$

$$\left. - (\alpha_{chiral}^{(2)})_{y'x'z'} + (\alpha_{chiral}^{(2)})_{z'x'y'} - (\alpha_{chiral}^{(2)})_{z'y'x'} \right]$$

where (x', y', z') denotes the molecular coordinate system. From the microscopic expression for $\alpha_{\xi\eta\zeta}^{(2)}$ in Eq. (2.17), we see that if the frequency dependence is neglected (equivalent to taking the frequency denominators out of the summation sign), $(\alpha_{chiral}^{(2)})_{\xi\eta\zeta}$ is reduced to the form $(\alpha_{chiral}^{(2)})_{\xi\eta\zeta} = C \sum_g \rho_{gg}^0 \langle g|r_\xi r_\eta r_\zeta|g \rangle$. Substitution of this expression into the above equation leads to $\chi_C^{(2)} = NC \sum_g \rho_{gg}^0 \langle g|\vec{r} \cdot (\vec{r} \times \vec{r})|g \rangle = 0$. We therefore expect non-resonant SFG from isotropic chiral media to be very weak. Fortunately, the main interest of SF chiral spectroscopy is exactly in probing the resonant structure of chiral media. Electronically resonant $(\chi^{(2)}_{chiral})_{ijk}$ of a monolayer of chiral molecules can be readily detected as we shall describe in Sec. 5.4. Through rearranging the microscopic expression of $(\chi^{(2)}_{chiral})_{ijk}$, we can find $\chi^{(2)}_{C} \propto (\omega_1 - \omega_2)$ for isotropic chiral media and explicitly $(\chi^{(2)}_{chiral})_{ijk} = 0$ for SHG with $\omega_1 = \omega_2$, but this is not true for anisotropic chiral media.

We shall provide a few examples of using SF chiral spectroscopy to probe chirality in electronic and vibrational transitions of chiral media in Secs. 5.4 and 5.5 with further discussion on how $(\alpha^{(2)}_{chiral})_{\xi\eta\zeta}$ and $(\chi^{(2)}_{chiral})_{ijk}$ reflect molecular chirality.

5.3. Experimental Considerations

The experimental arrangement of chiral SF spectroscopy is the same as the one described in Fig. 3.1 and both electronic and vibrational transitions of a medium can be probed as sketched in Fig. 2.2(b). Experimental procedure generally follows the description in Chapter 3. Both transmitted and reflected SFG, in principle, can be used to find bulk $\chi^{(2)}_{chiral}$ and $\chi^{(2)}_{achiral}$. However, as discussed in Secs. 2.1 and 2.2, the SF output field, proportional to $\chi^{(2)}_{chiral}/\Delta k_z$ or $\chi^{(2)}_{achiral}/\Delta k_z$, is much weaker in reflection than in transmission because the former has a much larger phase mismatch $|\Delta k_{rz}|$ ($\sim 10^5 \, \text{cm}^{-1}$) than the latter ($|\Delta k_{tz}| \sim 10^3 \, \text{cm}^{-1}$). The transmitted SFG signal is limited by the effective sample thickness, $1/|\Delta k_{tz}|$, which is of the order of $10 \, \mu\text{m}$. Because of this limitation, it is more suitable for chiral SFG to probe thin chiral samples. According to the estimate in Sec. 2.5, we should be able to detect $\sim 10^4$ photons/sec of SFG from $|\chi^{(2)}/\Delta k_z| = 10^{-16}$esu ($4 \times 10^{-22} \, \text{m}^2/\text{V}$ in SI units). This means that we can detect $\sim 10^4$ photons/sec of chiral SFG from a sample with $|\chi^{(2)}_{chiral}| = 10^{-13}$esu ($4 \times 10^{-17} \, \text{m/V}$) and $10 \, \mu\text{m}$ thick while a pure chiral molecular material usually has $|\chi^{(2)}_{chiral}| = 10^{-10} - 10^{-12}$esu($4 \times 10^{-14} - 4 \times 10^{-16} \, \text{m/V}$). Assuming a minimum detection sensitivity of few photons per second, chiral SFG would be able to detect chirality in a $10 \, \mu\text{m}$ sample with 0.01% to

0.1% of chiral molecular concentration. We also note that to access the $(\chi^{(2)}_{chiral})_{ijk}(i \neq j \neq k)$ elements by polarization combinations SPP. PSP, and PPS, we must have one of the P-polarized beams propagating in the sample at an oblique angle. For example, with $\hat{z}$ along the surface normal, S along y, and P in the $x - z$ plane, the PSP combination of SFG can access $(\chi^{(2)}_{chiral})_{xyz}$ and $(\chi^{(2)}_{chiral})_{zyx}$, and SPP can access $(\chi^{(2)}_{chiral})_{yxz}$ and $(\chi^{(2)}_{chiral})_{yzx}$.

Ordinary SF spectroscopy measures only the spectrum of $|\chi^{(2)}|^2$ (Sec. 2.1). Since $\chi^{(2)}_C(\text{RH}) = -\chi^{(2)}_C(\text{LH})$, it is not capable of distinguishing the two enantiomers. As we shall see in later sections, left-hand(S) and right-hand(R) enantiomers always have identical $|\chi^{(2)}|^2$ spectra, while the racemic mixture of the two has no detectable chiral SF spectrum. To distinguish the enantiomers, we need to resort to interference measurement to extract the phase or sign of $\chi^{(2)}_{chiral}$. One simple way is to arrange for one of the beams to have a mixed S$\pm$ P polarization so that both $\chi^{(2)}_{chiral}$ and $\chi^{(2)}_{achiral}$ contribute to the SF output, $S_{S\pm P} \propto |\chi^{(2)}_{chiral} \pm f\chi^{(2)}_{achiral}|^2$ with f being a constant.[6] It is seen that $S_{S+P} - S_{S-P} \propto 2Re[f^*\chi^{(2)}_{chiral}\chi^{(2)*}_{achiral}]$ has opposite signs for the two enantiomers from $\chi^{(2)}_{chiral} = \chi^{(2)}_C$ and $-\chi^{(2)}_C$, respectively. Another way is to carry out phase-sensitive SF spectroscopy (Sec. 3.5) by interfering the chiral SF signal with a reference (with the known $\chi^{(2)}_{Ref}$ of the reference replacing $\chi^{(2)}_{achiral}$ in the above equations). This allows deduction of the spectrum of complex $\chi^{(2)}_{chiral}$.[7] We shall provide more details with examples in Secs. 5.4 and 5.5.

5.4. Chiral Sum Frequency Electronic Spectroscopy

It was first suggested in 1965 that SFG could be observed in chiral liquids.[8] Experimental confirmation was reported in the ensuing year,[9] but the result could not be repeated by others. Not until 2001 was it realized that SFG off resonance in the early studies was perhaps too weak to be detected, but near resonance, SFG spectra from chiral liquids could be easily obtained.[6,10]

We consider here chirality from two types of structures. One has chirality arising directly from chiral structure of molecules and the other has chirality induced in achiral moieties of molecules by interaction with neighboring chiral structure. Binapthanol (BN) and amino acids molecules can be taken as representatives of the two types, respectively. We discuss first chiral SF spectroscopy on electronic transitions using BN as an example.

The molecular structure of BN is described in Fig. 5.3(a). The molecule is a twisted dimer composed of two coupled identical monomers. The chiral structures of the two enantiomers, denoted by R and S for right-hand and left-hand, respectively, come from clockwise and anti-clockwise twists of the monomers. Each monomer has a set of electronic states. If there were no coupling between the two monomers, the dimer would have pairs of degenerate states. With

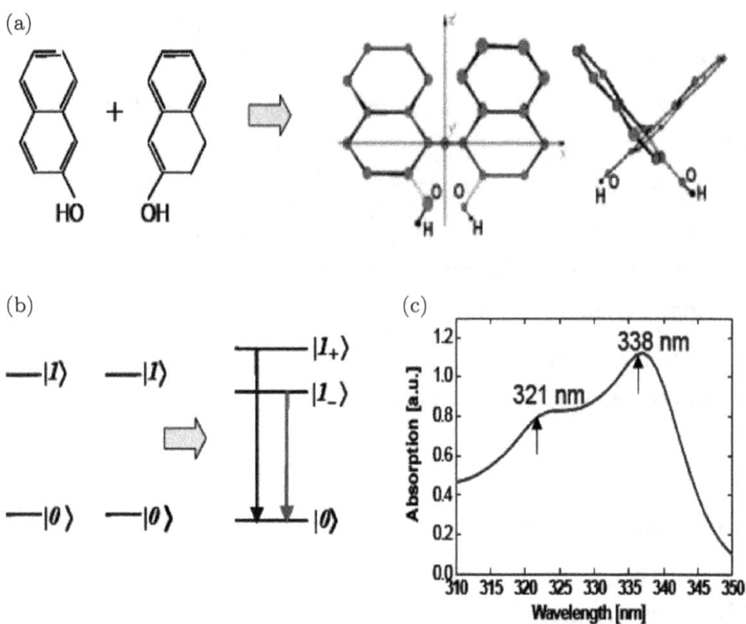

Fig. 5.3.   (a) Structure of a 1,1′-bi-2-naphthol (BN) chiral molecule seen as a twisted dimer composed of two monomers linked together by a C-C bond. (b) Energy level diagrams of BN showing formation of exciton-split states from lifted degeneracy of the lowest pair of excited monomer states by coupling of the monomers, and optical transitions between the ground state and the exciton-split states, and (c) the corresponding absorption spectrum of BN in tetrahydrofuran (after Ref. [10]).

coupling, the degeneracy is lifted and each pair of excited states is split into two with symmetric and anti-symmetric wave functions, respectively. The energy level diagram of BN depicting the ground state and the first pair of electronic excited states is presented in Fig. 5.3(b). Transitions from the ground state to the pair of excited states show up as two peaks with different chirality in the absorption spectrum of BN displayed in Fig. 5.3(c). We expect that chiral SF spectroscopy on BN should also reveal the two peaks with opposite chirality. This was confirmed in a transmitted SF spectroscopic measurement on 0.7M BN solutions in tetrahydrofuran using ps input pulses, one fixed at 9,400 cm^{-1} and another tunable from 19,600 to 22,200 cm^{-1}.[6] As presented in Figs. 5.4(a) and 5.4(b), the SPP and PSP $|\chi^{(2)}_{chiral}|^2$ spectra of BN indeed exhibit two peaks corresponding to the two absorption peaks in Fig. 5.3(c). The spectra for R- and S-BN solutions are the same as they should be, and the spectrum vanishes for the racemic mixture. To distinguish the two enantiomers, phase-sensitive SF spectroscopy was carried out. The BN solution

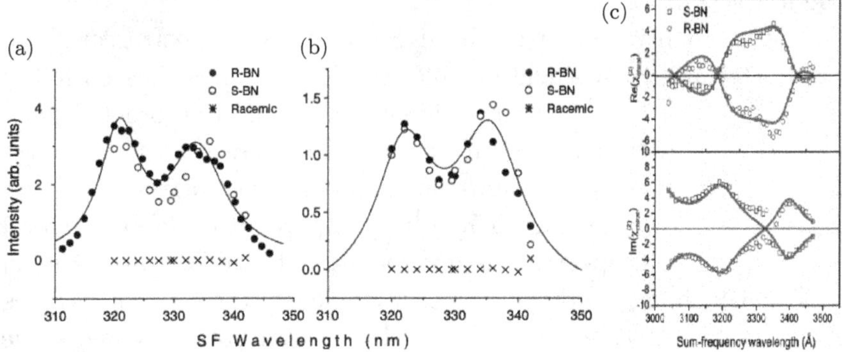

Fig. 5.4. Chiral SF electronic spectra, $|\chi_C|^2$ versus ω_{SF}, of the lowest pair of exciton-split transitions of BN for R-, S-, and racemic mixture of 0.7 M BN solutions in tetrahydrofuran, with (a) SPP and (b) PSP polarization combinations. The spectra for R- and S-enantiomers are identical within experimental uncertainties and the spectrum for racemic mixture vanishes. The peak value of $|\chi_C/N|$ at 338 nm was found to be $\sim 1.5 \times 10^{-40}$ m^4/V with N $= 4.2 \times 10^{26}$ molecules/m^3 for pure BN (after Ref. [10]). (c) SPP Im χ_C spectra of R- and S-BN obtained from phase-sensitive SF spectroscopic measurement showing the two are opposite in sign. The solid curves in (a)-(c) are the theoretical fits (after Ref. [7]).

was in contact with a y-cut quartz plate properly oriented at two positions to contribute $\pm\chi_Q^{(2)}$ to the transmitted SFG such that the SF output signal from the quartz-BN assembly was given by $S_\pm \propto |\chi_{chiral}^{(2)} \pm \chi_Q^{(2)}|^2$. Since $\chi_Q^{(2)}$ for quartz is real and both $\chi_Q^{(2)}$ and $|\chi_{chiral}^{(2)}|$ could be separately measured, $\mathrm{Re}\chi_{chiral}^{(2)}$ and $\mathrm{Im}\,\chi_{chiral}^{(2)}$ could be deduced from $S_+ - S_- \propto 2\mathrm{Re}(\chi_{chiral}^{(2)}\chi_Q^{(2)*})$ and $|\chi_{chiral}^{(2)}|$. The spectra so obtained are presented in Fig. 5.4(c). Because $\chi_{chiral}^{(2)}$ has opposite signs for R and S enantiomers, the spectra for the two enantiomers are opposite to each other.

Both CD and chiral SFG can identify chirality of BN, but their spectra actually describe different aspects of the molecular chiral structure. Since CD comes from EQ response to the gradient of electric field, it is nonlocal and probes nonlocal chiral structure. In the BN case, it depends on the vector describing the relative position of the two monomers. The ED-allowed chiral SFG is a local response, depending only on the chiral structure of electric dipole transition moments of BN induced by the twisted (chiral) coupling between the monomers.[11]

Being highly sensitive, chiral SF electronic spectroscopy can probe chirality of a single monolayer. The nonlinear susceptibility is expected to be a few times larger for orientation-ordered than randomly oriented molecules at a surface. The observed chiral spectra for the two lowest electronic transitions of a monolayer of BN on water are shown in Fig. 5.5(d).[12] Compared to the bulk chiral spectra, the PSP spectrum is now barely detectable, and the band at $\lambda = 320\,\mathrm{nm}$ is greatly suppressed. Obviously the difference comes from the fact that adsorbed BN molecules are polar-oriented while BN molecules in liquid are randomly oriented. Indeed, the spectra can be fit well by theoretical calculation assuming BN molecules in the monolayer are well orientated with their symmetric axis along the surface normal and the OH terminals immersed in water, as described in the inset of Fig. 5.5(d).[12] The achiral spectra of the oriented surface BN monolayer being ED-allowed can also be readily observed, but requires a different set of polarization combinations, and the spectra for R-BN, S-BN, and racemic mixture are identical.

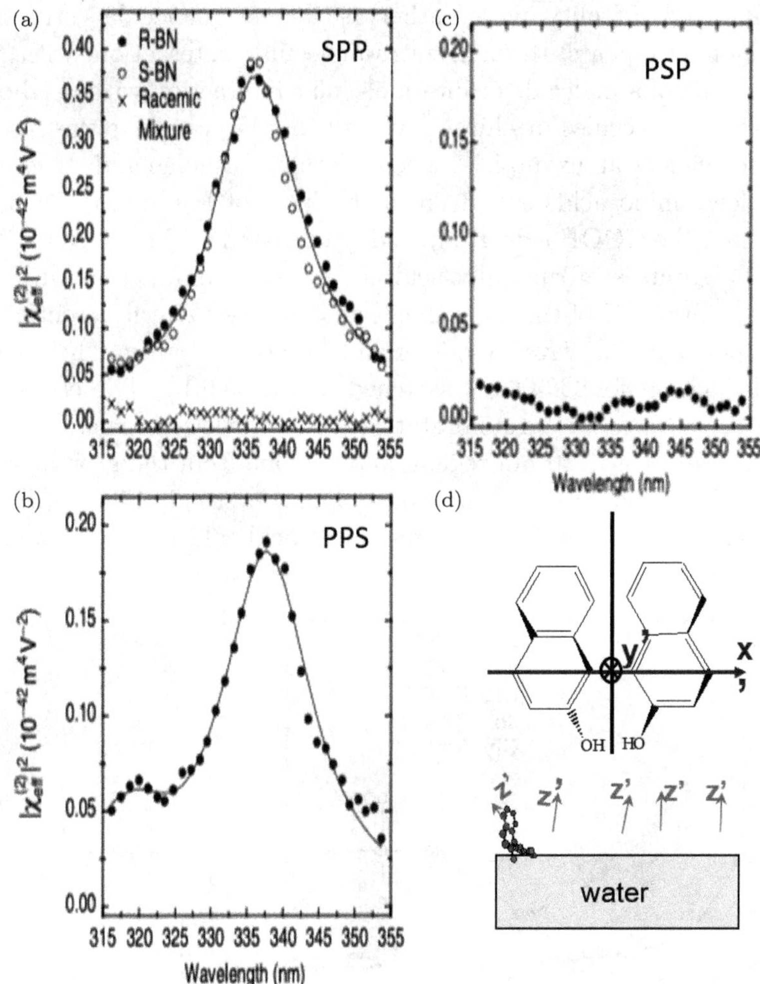

Fig. 5.5.   Chiral SF electronic spectra for monolayers of R-, S-, and racemic BN monolayers on water obtained with polarization combinations (a) SPP, (b) PPS, and (c) PSP. The solid curves are theoretical fits. The spectra are very different from the corresponding bulk spectra in Fig. 5.4 because the monolayers are well ordered in orientation as sketched in (d). The peak value of $|\chi_C/N|$ in (a) is $\sim 8 \times 10^{-40}\, m^4/\text{V}$ (after Ref. [12]).

Induced chirality in an otherwise achiral molecular structure commonly appears through intra-molecular interaction with the chiral part of a molecule or intermolecular interaction with neighboring chiral molecules or chiral environment. We consider the case of amino acids as an example.[13] The molecular formulae and structures of a few amino acids are given in the inset of Fig. 5.6(a). In basic solution, the COOH side group is deprotonated to form COO^-. The COO^- group is intrinsically achiral, but upon interaction with the chiral center "C" of the molecule, it experiences a chiral potential V_C that perturbs its wave functions and induces chirality. The energy level diagram for COO^- is sketched in Fig. 5.6(b). The electronic transitions from the ground state to the two lowest excited states are in the 180 to 210 nm region, and the one from the ground state to the third excited state is around 169 nm. As seen in Fig. 5.6(a), the SPP chiral SF signal from several amino acid solutions is resonantly

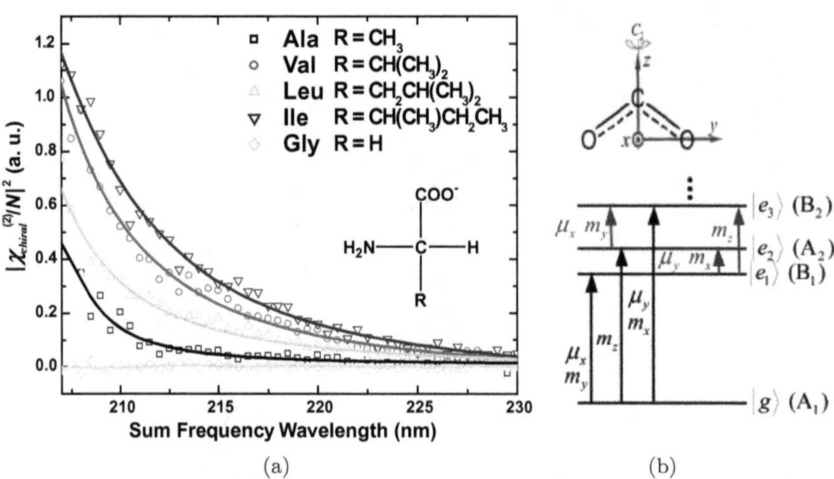

(a)   (b)

Fig. 5.6. (a) Chiral SPP SF electronic spectra for various left-hand (S)-amino acids in 4M NaOH solution with their molecular formulae and structure shown in the inset. It is seen that for all amino acids, $|\chi_C/N|^2$ is resonantly enhanced toward electronic transitions in the uv. The solid curves are guides for eyes. (b) The structure and the four lowest electronic energy levels of a COO^- group are used in a dynamic coupling model (Ref. [14]) to evaluate the frequency dependence and the ranking of chiral strength of the five amino acids. Arrows indicate the allowed electric-dipole and magnetic-dipole transitions (after Ref. [13]).

enhanced as the sum frequency increases toward these transitions, in agreement with the CD result. The theory for CD of amino acids was extended to chiral SFG and was able to explain the relative magnitude of $|\chi^{(2)}_{chiral}|$ for different amino acids in Fig. 5.6(a).[14] The same theoretical approach should be applicable to induced change of chirality in chiral and achiral molecules in chiral or achiral environments if appropriate perturbing potential describing the environmental effect is used.

5.5. Chiral Sum Frequency Vibrational Spectroscopy

Vibrational spectroscopy generally provides more information about structure of materials than electronic spectroscopy. Vibrational transitions are however much weaker than electronic transitions. It is possible to use CD and Raman optical activity (ROA) to obtain chiral vibrational spectra, but the measurement is quite challenging, especially for thin samples. Transmitted SF spectroscopy has been found to have sufficient sensitivity to probe chiral vibrational transitions although the signal is significantly weaker than that from chiral electronic transitions.[6]

The chiral SF vibrational spectra of three liquids, carvone, limone, and menthal acetate, are presented in Fig. 5.7 as examples.[6] We focus here on limone. Both chiral and achiral SF vibrational spectra of limone in the CH_x stretch range are shown in Fig. 5.8 labeled with polarization combinations SPP and PSP for $|\chi^{(2)}_{chiral}|^2$ chiral spectra in Figs. 5.8(a) and 5.8(b) and PPP for $|\chi^{(2)}_{achiral}|^2$ achiral spectra in Fig. 5.8(c). The two prominent peaks at 2839 and 2879 cm^{-1} in the $|\chi^{(2)}_{chiral}|^2$ spectra belong to symmetric CH_2 and CH_3 stretches, respectively, but it is interesting to note that they are rather weak in the PPP $|\chi^{(2)}_{achiral}|^2$ spectra, which instead have the anti-symmetric CH_3 stretch at 2936 cm^{-1} being the most prominent. This manifests the fact that different optical transitions generally have different degrees of chirality. The chiral intensity spectra are identical for the two enantiomers, and the spectrum vanishes for racemic mixtures. As mentioned in Sec. 5.3, in order to distinguish the two enantiomers, we can let one of the input beams be S±P

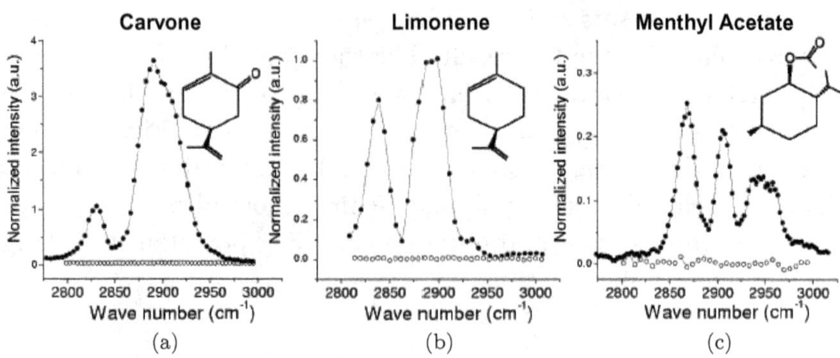

Fig. 5.7. Chiral SPP SF vibrational spectra for right-hand (R) and racemic carvone, limone, and menthyl acetate liquids in the CH stretching region. Molecular structures are given in the insets (after Ref. [6]).

polarized so that $\chi^{(2)}_{achiral}$ interferes with $\chi^{(2)}_{chiral}$ in the SF output. Figures 5.8(d) and 5.8(e) depict the spectra with P(S +P)P and P(S-P)P polarization combinations for the R and S enantiomers of limone that can be expressed by $S_{S\pm P} \propto |\chi^{(2)}_{chiral} \pm f\chi^{(2)}_{achiral}|^2$. The corresponding difference spectra of $S_{S+P} - S_{S-P} \propto 2\mathrm{Re}[f^*\chi^{(2)}_{chiral}\chi^{(2)*}_{achiral}]$ displayed in Fig. 5.8(f) are opposite for the two enantiomers as expected.

The spectrum of $|\chi^{(2)}_{chiral}|^2$ extracted from spectral analysis and normalized against a quartz reference is given in Fig. 5.8(g). The peak value of $|\chi^{(2)}_{chiral}|$ for limone is seen to be $\sim 10^{-14}\,\mathrm{m/V}$. Compared to the value of ED-allowed $|\chi^{(2)}_{achiral}|$ from polar-ordered CH_x stretches, it is about two orders of magnitude smaller. Knowing that $|\chi^{(2)}_{chiral}|$ is also ED-allowed, this is surprising; one would expect that $|\chi^{(2)}_{chiral}|$ induced by chiral bonding structure could be weak and comparable to $|\chi^{(2)}_{achiral}|$, but a difference of two orders of magnitude was not foreseen. It turns out that there is a deeper reason for the reduction.[15] As described in Sec. 2.4, the SF process probing a vibrational transition can be considered as an IR excitation followed by an anti-Stokes Raman transition and we have, at a vibrational resonance, $\chi^{(2)}_{ijk} \propto M_{ij}\mu_k$ (or $\alpha^{(2)}_{\xi\eta\zeta} \propto M_{\xi\eta}\mu_\zeta$ in molecular coordinates), where μ and M denote IR and Raman matrix elements for the transition, respectively. For $(\chi^{(2)}_{chiral})_{ijk}(i \neq j \neq k)$, we

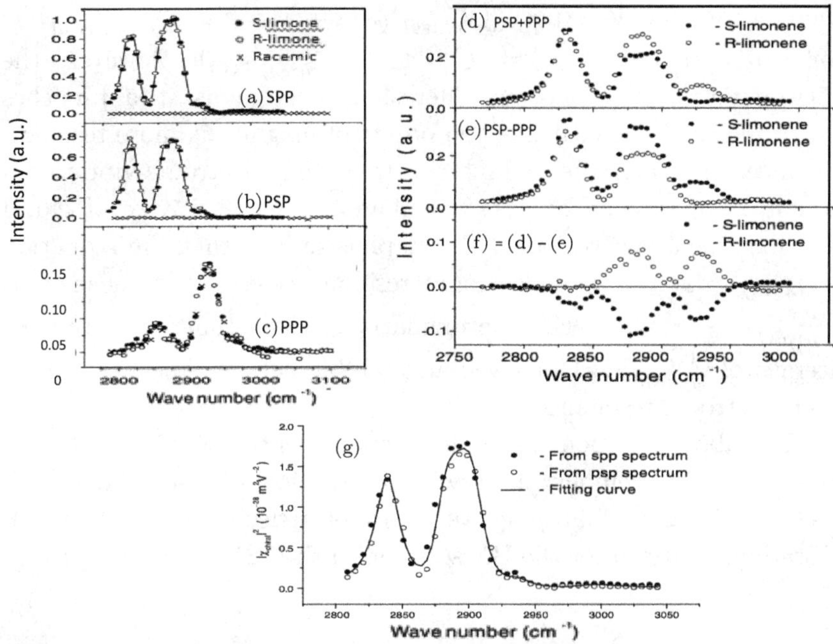

Fig. 5.8. SF vibrational spectra of S (filled dots), R (open circles), and racemic (crosses) limone liquid versus input IR frequency in the CH stretching range: (a) and (b) display SPP and PSP chiral spectra, respectively, and (c) PPP, achiral spectrum. The solid lines are guides for eyes. The difference between chiral and achiral spectra shows that chiral strength is very different for different vibrational modes. (d) and (e) display mixed chiral and achiral SF vibrational spectra obtained from S(S +P)P and S(S-P)P polarization combinations that appear different for the two enantiomers.The difference spectra of (d) and (e) for R and S limones are given in (f). They are the same but opposite in sign for the two enantiomers. (g) Normalized spectrum of $|\chi^{(2)}_{chiral}|^2$ (against quartz) deduced from analysis of the SPP (filled dots) and PSP (open circles) chiral spectra. The peak value of $|\chi^{(2)}_{chiral}/N|$ at $\sim 2875\,\mathrm{cm}^{-1}$ is $\sim 3 \times 10^{-42}\,\mathrm{m^4/V}$ with $N = 3.7 \times 10^{26}\,/\mathrm{m^3}$ for pure limone. The solid curve is a theoretical fit (after Ref. [6]).

must have $M_{ij}(= -M_{ji})$ with $i \neq j$. It is known in molecular physics that the antisymmetric $M_{ij}(= -M_{ji})$ away from electronic resonances is much weaker than the symmetric one. Theory shows that if the frequency dispersion in antisymmetric M_{ij} were neglected, $M_{ij}(= -M_{ji})$ would vanish. With dispersion, it is still much smaller than the symmetric M_{ij} away from electronic resonances,[15] and therefore we should expect $|(\chi^{(2)}_{chiral})_{ijk}(i \neq j \neq k)| \ll |(\chi^{(2)}_{achiral})_{iik}|$ (or $|(\alpha^{(2)}_{chiral})_{\xi\eta\zeta}(\xi \neq \eta \neq \zeta)| \ll |(\alpha^{(2)}_{achiral})_{\xi\xi\zeta}|$). However, the antisymmetric M_{ij} becomes increasingly large as it approaches electronic resonances; it has two orders of magnitude more resonant enhancement than the symmetric M_{ij}. This electronic-vibrational double resonance effect greatly enhances the sensitivity of chiral SF vibrational spectroscopy. We emphasize here that the reduction of $|\chi^{(2)}_{chiral}|$ away from electronic resonances does not mean that $\chi^{(2)}_{chiral}$ vanishes under ED approximation as sometimes stated in the literature; $\chi^{(2)}_{chiral}$ is ED-allowed and can be readily detected in SFVS near electronic resonances.

The above-mentioned double resonance effect of chiral SFG was experimentally demonstrated with a solution of 0.46M R-BN in acetone.[16] Figure 5.9(a) shows a set of reflected chiral SPP SF vibrational spectra for the BN solution in the 1250-1550 cm^{-1} range

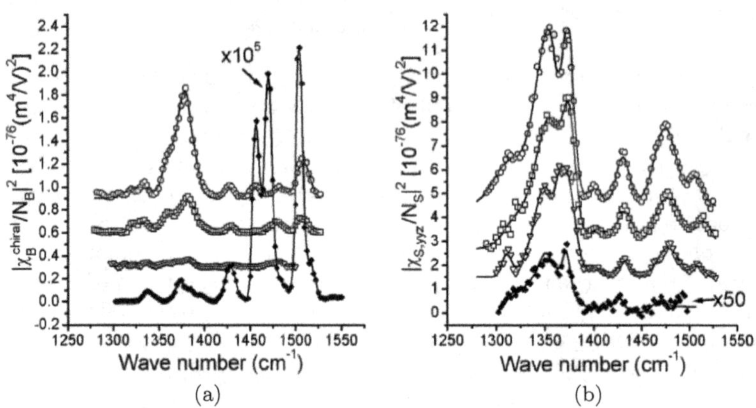

Fig. 5.9. Sets of SF vibrational spectra of R-BN exhibiting double-resonance enhancement. (a) SPP chiral SF vibrational spectra of a 0.46M solution of R-BN in acetone, and (b) achiral spectra of an R-BN monolayer on water, with the sum frequency set at 3.71 eV (open circles), 3.65 eV (open squares), 3.60 eV (open down triangles), and 2.48 eV (solid diamonds). The 2.48 eV spectra in (a) and (b) are enhanced by 10^5 and 50, respectively. Vertical shifts are made to separate the spectra in the figures and the lines are guides for eyes. It is particularly obvious in the chiral spectra in (a) that different vibrational modes are enhanced differently because of different chiral electron-vibraton couplings (after Ref. [16]).

taken with the sum frequency set at 2.48, 3.60, 3.65, and 3.71 eV. The two lowest electronic transitions of BN are at 3.67 and 3.89 eV. It is seen that the spectral intensity increases very rapidly as the sum frequency approaches the first electronic transition. The spectrum taken at 3.71 eV is $\sim 10^5$ times stronger than the one at 2.48 eV. Resonant enhancements of different vibrational modes are very different, providing structural information on how the excited electronic states couple with various vibrational modes. The observation that the mode at 1375 cm^{-1} has an enhancement 100 times more than the modes at 1455, 1470, and 1505 cm^{-1} is particularly revealing. To see that only chiral vibrational modes exhibit the exceptionally strong doubly resonant enhancement, we display in Fig. 5.9(b) the corresponding set of achiral SF SSP vibrational spectra for the BN solution for comparison. The observed resonant enhancement toward the electronic transition is only ~ 100. Close to the electronic transitions, the peak values of $|\alpha^{(2)}_{chiral}|$ approach those of $|\alpha^{(2)}_{achiral}|$ within a factor of 2 or 3.

Because $|\chi^{(2)}_{chiral}|$ of BN is even much weaker than ED-forbidden $|\chi^{(2)}_{achiral}|$, SF chiral spectroscopy on BN monolayers away from electronic resonances would be difficult. However, the extraordinarily strong double-resonance enhancement makes it possible. Figure 5.10 shows the chiral spectrum of a monolayer of R-BN molecules on water with the sum frequency set at 3.71 eV. The resonantly enhanced vibrational modes are clearly seen in the spectrum, which is different from the spectrum of BN solution in Fig. 5.9 presumably because BN molecules in the monolayer are well orientation-ordered (Sec. 5.4). No spectrum can be discerned for a racemic BN monolayer as expected.

5.6. Second Harmonic Generation as a Chiral Probe

Under the ED approximation, SHG is forbidden in isotropic chiral media (Sec. 5.2), but should still be allowed if the media are anisotropic. Consider, for example, a uniaxial chiral medium with azimuthal isotropy. Let the uniaxis be along z. Symmetry argument

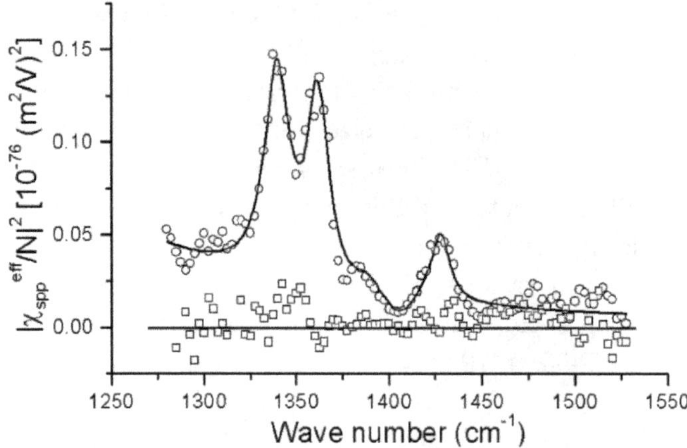

Fig. 5.10. SPP SF chiral vibrational spectra of monolayers of R-BN (circles) and BN racemic mixture (square) on water with the sum frequency set at 3.71 eV, demonstrating that with double-resonance enhancement, SF chiral vibration spectroscopy has monolayer sensitivity (after Ref. [16]).

shows that under ED approximation $\chi^{(2)}_{achiral}$ should still vanish and the only nonvanishing elements of $[\chi^{(2)}(2\omega = \omega + \omega)]_{ijk}$ are the chiral elements $\chi^{(2)}_{xyz} = \chi^{(2)}_{xzy} = -\chi^{(2)}_{yxz} = -\chi^{(2)}_{yzx}$. Chiral SHG from such azimuthally isotropic chiral materials can be observed even away from any resonance because with anisotropic refractive indices, phase matching ($\Delta k = 0$) of SHG is possible, greatly increasing the transmitted SH output. Chiral liquid crystals can be used as an example.[17] The observed S-polarized SH output from a chiral smectic liquid crystal film versus the incident angle of the P-polarized input at 1.06 μm is plotted in Fig. 5.11 and is seen to peak sharply at the phase matching angle. Also shown in Fig. 5.11 is the complete suppression of SHG in a racemic LC film.

Polar-oriented chiral molecules adsorbed on a substrate appear as an anisotropic layer. In addition to chirality, the inversion symmetry of the layer is also broken by polar orientation. Both chiral and achiral $[\chi^{(2)}(2\omega = \omega + \omega)]_{ijk}$ are now ED-allowed. Azimuthal anisotropy

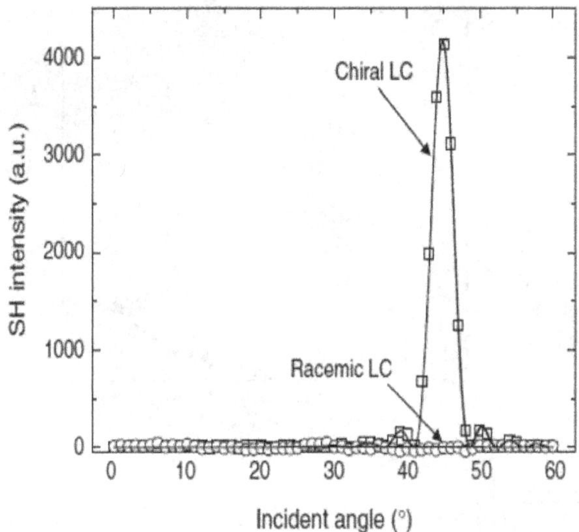

Fig. 5.11.   Second harmonic signal from a homeotropically aligned chiral smectic-A liquid crystal film as a function of the incident angle obtained with the fundamental input at 1.06 μm and a P-in, S-out polarization combination, showing a phase-matching peak at $\sim$45°. The signal from a racemic mixture is hardly discernible as expected (after Ref. [17]).

leads to the nonvanishing achiral elements $\chi^{(2)}_{zzz}, \chi^{(2)}_{zxx} = \chi^{(2)}_{zyy}$, and $\chi^{(2)}_{xzx} = \chi^{(2)}_{yzy} = \chi^{(2)}_{xxz} = \chi^{(2)}_{yyz}$, with z along the surface normal, and nonvanishing chiral elements $\chi^{(2)}_{xyz} = -\chi^{(2)}_{yxz} = \chi^{(2)}_{xzy} = -\chi^{(2)}_{yzx}$. They can be measured, respectively, by SHG with (S-in, P-out) and (P-in, S-out) polarization combinations. Hicks and coworkers first demonstrated chiral SH spectroscopy on monolayers of BN on water.[18] The P-in, S-out SH spectrum for a BN(R) layer is given in Fig. 5.12(a), showing resonant behavior as the SH frequency varies over the range from the lowest electronic transition up to the higher and stronger transitions. Using oppositely circularly polarized inputs, they observed opposite SH spectra for R-BN and S-BN as shown in Fig. 5.12(b).[19] They and others later extended chiral SF spectroscopy to biological molecules at interfaces.

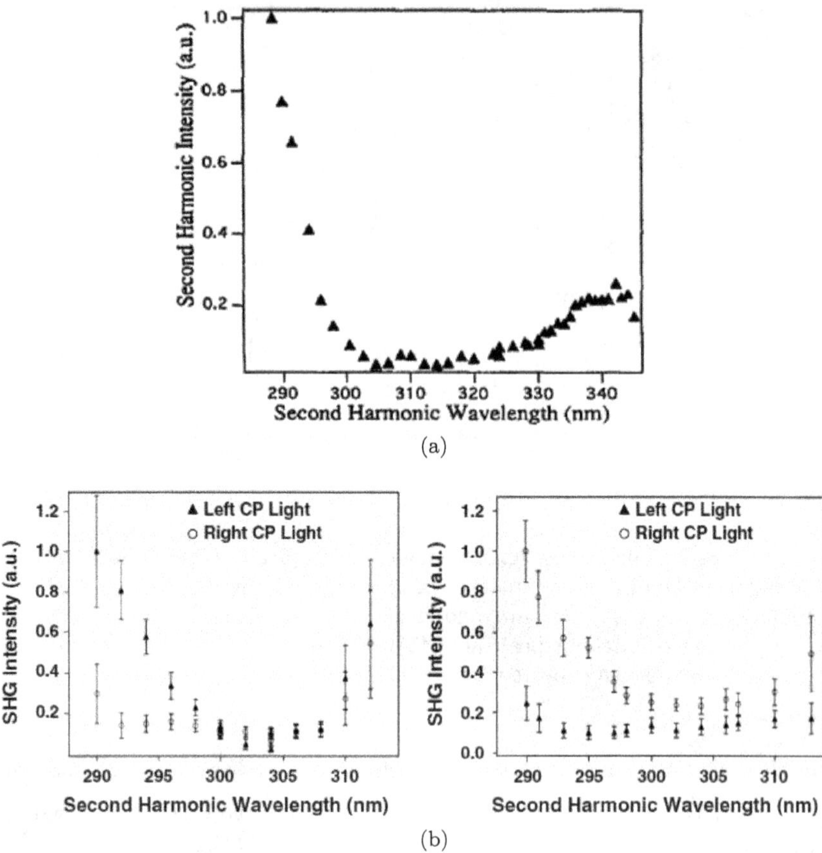

Fig. 5.12. S-polarized second harmonic (SH) intensity spectra of monolayers of (a)R-BN on water obtained with P-in, S-out polarization combination (after Ref. [19]), and (b)R-BN and (c)S-BN on water generated by left and right circularly polarized input (after Ref. [18]).

5.7. Sum Frequency Vibrational Spectroscopy for Surface Chirality

Because chiral SFVS is ED-allowed, it is in principle not surface specific. In fact, if the chirality comes from molecular bonding structure, it would be difficult to detect surface chirality away from electronic transitions because as we discussed in Sec. 5.5, $|\chi^{(2)}_{chiral}|$ is two orders of magnitude smaller than $|\chi^{(2)}_{achiral}|$(or $|\alpha^{(2)}_{chiral}| \sim 10^{-2}|\alpha^{(2)}_{achiral}|$). This was illustrated by the case of monolayer BN in Sec. 5.5. The situation however changes if the surface chirality comes from chiral distribution of molecular units. Consider molecular units with achiral polarizability $(\alpha^{(2)}_{achiral})_{\xi\xi\zeta}$ arranged spatially in a chiral (twisted) form. The system should have a nonvanishing spatial average of $\langle\alpha^{(2)}_{ijk}\rangle_{chiral}$ in the lab coordinates and its magnitude should be close to $(\alpha^{(2)}_{achiral})_{\xi\xi\zeta}$ with some reduction due to angular average, $|\langle\alpha^{(2)}_{ijk}\rangle_{chiral}| \sim |(\alpha^{(2)}_{achiral})_{\xi\xi\zeta}|$, and accordingly, the chiral surface nonlinear susceptibility is comparable to its achiral counterpart, $|\chi^{(2)}_{S,chiral}| \sim |\chi^{(2)}_{S,achiral}|$, which is much larger than $|\chi^{(2)}_{S,chiral}|$ from chiral bonding structure. The explicit expressions relating $\chi^{(2)}_{ijk}$ to $\alpha^{(2)}_{\xi\eta\zeta}$ can be obtained from Eq. (5.1) via coordinate transformation, and have been worked out by Moad *et al.* with molecular orientation specified by the Euler angles;[20] they explicitly show the dependence of $(\chi^{(2)}_{S,chiral})_{ijk}$ on both chiral and achiral $\alpha^{(2)}_{\xi\eta\zeta}$. In the special case of isotropic chiral bulk, they reproduce the result that $(\chi^{(2)}_{B,chiral})_{ijk}$ depends only on $(\alpha^{(2)}_{chiral})_{\xi\eta\zeta}$, as described in Eq. (5.2). Surface chirality of biomolecules adsorbed at interfaces probed by SFVS is usually dominated by chiral distribution of their molecular units as we shall see later in Sec. 10.2.

There is the question whether chiral SFVS can be surface specific if surface chirality is dominated by chiral distribution of molecules. We refer back to the description in Sec. 2.2 and learn that the chiral SF output field should be proportional to $\chi^{(2)}_{S,chiral} - \chi^{(2)}_{B,chiral}/i\Delta k_z$. With $\chi^{(2)}_{S,chiral}$ dominated by chiral molecular distribution, we have $|\chi^{(2)}_{S,chiral}| = N_I d|\langle\alpha^{(2)}\rangle_{chiral}| \sim N_I d|\alpha^{(2)}_{achiral}|$ and $|(\chi^{(2)}_{B,chiral})_{ijk}| \sim N_B|\alpha^{(2)}_{chiral}|$ where N_I and N_B are the chiral molecular densities at the surface and in the bulk, respectively, and d, the surface layer thickness. Since $|\alpha^{(2)}_{chiral}| \sim 10^{-2}|\alpha^{(2)}_{achiral}|$, we find $|\chi^{(2)}_{S,chiral}|/|\chi^{(2)}_{B,chiral}/i\Delta k_z| \sim |(d\Delta k_z)(\alpha^{(2)}_{achiral}/\alpha^{(2)}_{chiral})| \sim 10^2(d|\Delta k_z|)$. For reflected chiral SF output with $|\Delta k_z| \sim 2k_z(\omega_{vis})$ taken to be $\sim 2.5 \times 10^5 \, \text{cm}^{-1}$ for a visible input and $d \sim 1$ nm, the ratio is $|\chi^{(2)}_{S,chiral}|/|\chi^{(2)}_{B,chiral}/i\Delta k_z| \sim 2.5$, indicating that the surface and bulk contributions to reflected chiral SFVS are of the same order of magnitude; chiral SFVS is not surface-specific in general. The surface contribution to reflected chiral SFVS can dominate over the bulk if the bulk is a film with thickness significantly smaller than $1/|\Delta k_z|$, or if the bulk chiral molecular concentration is sufficiently dilute (i.e., $N_I >> N_B$). We note that to probe surface chirality from chiral molecular bonding structure with $|\chi^{(2)}_{S,chiral}| = N_I d |\alpha^{(2)}_{chiral}| \sim 10^{-2} N_I d |\alpha^{(2)}_{achiral}|$ is difficult; we have to resort to electronic-vibrational double resonance enhancement as described in Sec. 5.5.

5.8. Chiral Sum Frequency Microscopy

Optical microscopy that allows distinction of chiral versus achiral objects could be of interest in chemistry and bioscience. Chiral optical microscopy based on CD/ORD suffers from weak chiral imaging against strong achiral background. Obviously, chiral SF imaging benefiting from weak achiral background could be appealing, particularly if it could identify molecular species without labeling through their vibrational spectra. Over the past decades, various nonlinear optical microscopies have been developed to serve bioscience. Among them, two-photon fluorescence microscopy has been most successful. SHG and coherent anti-Stokes Raman scattering (CARS) microscopies have also received a great deal of attention. SHG has been exploited to image chiral surfaces,[21] but no other nonlinear optical techniques deal with chiral microscopy.

Chiral SF microscopy was first demonstrated in Ref. [22]. Chiral SFG with the help of an ordinary optical microscope (Fig. 5.13(A)) was used to image a 100x100 μm^2 thin film composed of two patches of R-BN and racemic BN separated by a 55 μm-wide glass stripe. The two input frequencies were fixed in the visible with the sum frequency at 4.48 eV in resonance with the electronic transition band of BN. Both fluorescence and chiral SF images were taken, as presented in Figs. 5.13(Ba) and 5.13(Bb).[22] The fluorescence image cannot distinguish the different BN species, but the chiral SF image

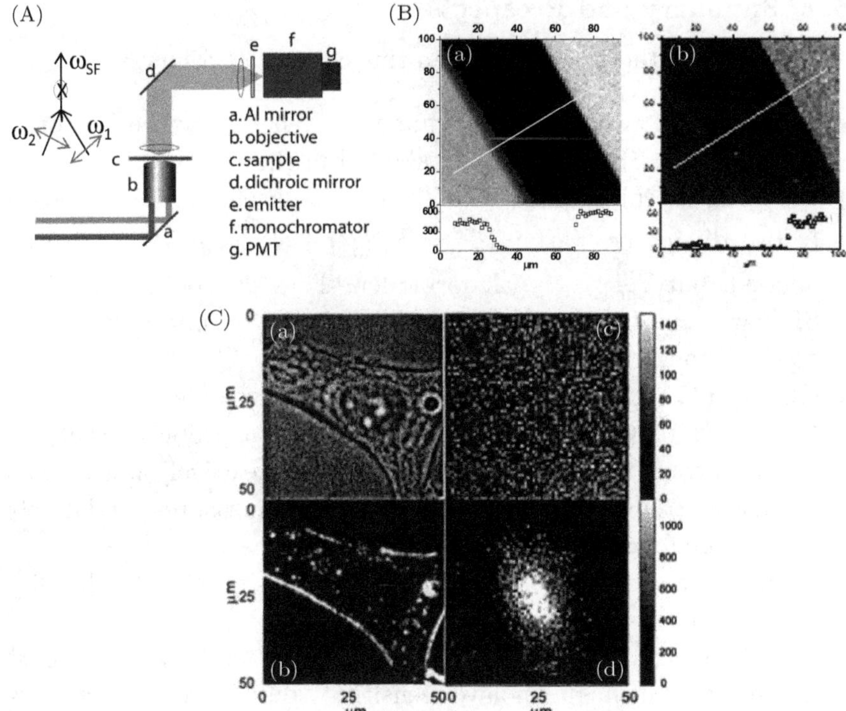

Fig. 5.13. (A) Experimental arrangement for transmitted chiral SF microscopy. (B) Fluorescence and chiral SF microscopic images (on the left-hand and right-hand side, respectively) of a racemic BN solution and an R-BN solution separated by a 55 μm glass spacer. The fluorescence and SF intensities versus position along the white lines in the images are plotted below the respective images (after Ref. [22]). (C) Bright, dark, fluorescence, and chiral SFG microscopic images of a HeLa cell in frames (a), (b), (c), and (d), respectively (after Ref. [23]).

can since SFG occurs only in R-BN. Chiral SF microscopy has also been tried on imaging cells in solution.[23] Figure 5.13(C) displays the ordinary microscopic images of the nuclear region of a HeLa cell (fixed by 4% formaldehyde in PBS buffer) together with the fluorescence and chiral SFG images. The chiral SF image has clear contrast in the nuclear region, but the fluorescence image does not. Chiral SF microscopy has not yet been well explored for applications. The microscopic technique could be further improved if needed.

5.9. Summary and Prospects

We summarize the major points in this chapter as follows:

- Chiral media lack inversion symmetry even if they are isotropic, and therefore SFG is generally allowed in chiral media.
- For isotropic chiral media, the SF response coefficient is composed of two distinctive parts: $\overleftrightarrow{\chi}^{(2)} = \overleftrightarrow{\chi}^{(2)}_{chiral} + \overleftrightarrow{\chi}^{(2)}_{achiral}$; $\overleftrightarrow{\chi}^{(2)}_{chiral}$ is ED-allowed, but $\overleftrightarrow{\chi}^{(2)}_{achiral}$ is ED-forbidden. They can be measured by SFG with mutually exclusive sets of input/output polarization combinations.
- Chiral SFG rapidly diminishes away from electronic resonance.
- Compared to CD, SFG is more sensitive to probe chirality of resonant transitions. The former measures a relatively weak signal against a strong background, but the latter measures a relatively strong signal against a weak background.
- Chiral SF spectroscopy has monolayer sensitivity to probe chirality of electronic transitions.
- Chiral SF spectroscopy can probe chirality of bulk vibrational transitions, but has monolayer sensitivity only close to electronic-vibrational double resonances. The double resonance of $|\chi^{(2)}_{chiral}|^2$ can have several orders-of-magnitude resonance enhancement.
- Chiral SHG is ED-forbidden in isotropic bulk chiral media but can be used to probe chirality of anisotropic chiral media or oriented molecular layers.
- Chirality from chiral arrangement of molecules is orders of magnitude larger than chirality from bonding structures of chiral molecules. Surface chirality from chiral distribution of surface molecules can be probed by chiral SFVS.
- Chiral SF microscopy is capable of resolving chiral and achiral objects. The potential of chiral SF microscopy has not yet been well recognized.

Chiral SF spectroscopy appears to have many advantages over other techniques to probe chirality, but its possible applications to chemistry and bioscience are yet to be explored. Its sensitivity is limited by the effective sample thickness that is usually restricted by phase mismatch of SFG in a chiral liquid at ~ 10 μm. Thus, for routine check of right or left chirality of molecules dissolvable in solution, chiral SF spectroscopy is not likely to be able to replace CD spectroscopy that has no limit on sample length. However, the high sensitivity of the technique together with inherent ultrafast time resolution provides many opportunities for chiral studies. In situ measurements of chirality and chiral functions are possible with chiral SF spectroscopy. Information about chiral structure and its dependence on environment can be obtained through spectral responses from vibrational resonances and electronic-vibrational double resonances. Induced chirality and chiral transformation by external perturbation, as well as their dynamics, can be monitored. Chiral imaging of biological systems in real time could also be of interest. At present, chiral SF spectroscopy is probably the only viable means to probe chirality of surfaces, monolayers, and nm thin films. Advance in this area would rely strongly on theoretical success in relating chiral spectroscopic information to chiral structure and properties of a material.

References

1. Kelvin, W.: *Baltimore Lectures on Molecular Dynamics and Wave Theory of Light.* C.J. Clay & Sons: London, 1904.
2. Barron, L. D.: *Molecular Light Scattering and Optical Activity.* 2nd ed. Cambridge University Press: Cambridge, UK; New York, 2004.
3. Belkin, M. A.; Shen, Y. R.: Nonlinear Optical Spectroscopy as a Novel Probe for Molecular Chirality. *Int. Rev. Phys. Chem.* 2005, 24, 257–299.
4. Ji, N.; Shen, Y. R.: A Novel Spectroscopy Tool for Molecular Chemistry. *Chrality* 2006, 18, 146–158.
5. Rosenfeld, L.: Quantenmechanische Theorie der Natiirlichen optischen Aktivitat von Fliissigkeiten und Gasen. *Z. Phys.* 1928, 52, 161–174.
6. Belkin, M. A.; Kulakov, T. A.; Ernst, K. H.; Yan, L.; Shen, Y. R.: Sum-Frequency Vibrational Spectroscopy on Chiral Liquids: A Novel Technique to Probe Molecular Chirality. *Phys. Rev. Lett.* 2000, 85, 4474–4477.
7. Ji, N.; Ostroverkhov, V.; Belkin, M.; Shiu, Y. J.; Shen, Y. R.: Toward Chiral Sum-Frequency Spectroscopy. *J. Am. Chem. Soc.* 2006, 128, 8845–8848.

8. Giordmaine, J. A.: Nonlinear Optical Properties of Liquids. *Phys. Rev.* 1965, 138, 1599–1606.

9. Rentzepis, P. M.; Giordmaine, J. A.; Wecht, K. W.: Coherent Optical Mixing in Optically Active Liquids. *Phys. Rev. Lett.* 1966, 16, 792–794.

10. Belkin, M. A.; Han, S. H.; Wei, X.; Shen, Y. R.: Sum-Frequency Generation in Chiral Liquids near Electronic Resonance. *Phys. Rev. Lett.* 2001, 87, 113001.

11. Belkin, M. A.; Shen, Y. R.; Flytzanis, C.: Coupled-Oscillator Model for Nonlinear Optical Activity. *Chem. Phys. Lett.* 2002, 363, 479–485.

12. Han, S. H.; Ji, N.; Belkin, M. A.; Shen, Y. R.: Sum-Frequency Spectroscopy of Electronic Resonances on A chiral Surface Monolayer of Bi-naphthol. *Phys. Rev. B* 2002, 66, 165415.

13. Ji, N.; Shen, Y. R.: Optically Active Sum Frequency Generation from Molecules with a Chiral Center: Amino Acids as Model Systems. *J. Am. Chem. Soc.* 2004, 126, 15008–15009.

14. Ji, N.; Shen, Y. R.: A Dynamic Coupling Model for Sum Frequency Chiral Response from Liquids Composed of Molecules with a Chiral Side Chain and an Achiral Chromophore. *J. Am. Chem. Soc.* 2005, 127, 12933–12942.

15. Belkin, M. A.; Shen, Y. R.; Harris, R. A.: Sum-Frequency Vibrational Spectroscopy of Chiral Liquids off and close to Electronic Resonance and the Antisymmetric Raman Tensor. *J. Chem. Phys.* 2004, 120, 10118–10126.

16. Belkin, M. A.; Shen, Y. R.: Doubly Resonant IR-UV Sum-Frequency Vibrational Spectroscopy on Molecular Chirality. *Phys. Rev. Lett.* 2003, 91, 213907.

17. Han, S. H.; Belkin, M. A.; Shen, Y. R.: Optically Active Second-Harmonic Generation from a Uniaxial Fluid Medium. *Opt. Lett.* 2004, 29, 1527–1529.

18. Byers, J. D.; Yee, H. I.; Hicks, J. M.: A Second-Harmonic Generation Analog of Optical-Rotatory Dispersion for the Study of Chiral Monolayers. *J. Chem. Phys.* 1994, 101, 6233–6241.

19. Petralli-Mallow, T.; Wong, T. M.; Byers, J. D.; Yee, H. I.; Hicks, J. M.: Circular-Dichroism Spectroscopy at Interfaces – a Surface Second Harmonic Generation Study. *J. Phys. Chem.-Us* 1993, 97, 1383–1388.

20. Moad, A. J.; Simpson, G. J.: A Unified Treatment of Selection Rules and Symetry Relations for Sum-Frequency and Second harmonic Spectroscopies. *J. Phys. Chem. B* 2004, 108, 3548–3562.

21. Kriech, M. A.; Conboy, J. C.: Label-Free Chiral Detection of Melittin Binding to a Membrane. *J. Am. Chem. Soc.* 2003, 125, 1148–1149.

22. Ji, N.; Zhang, K.; Yang, H.; Shen, Y. R.: Three-Dimensional Chiral Imaging by Sum-Frequency Generation. *J. Am. Chem. Soc.* 2006, 128, 3482–3483.
23. Zhang, K.; Ji, N.; Shen, Y. R.; Yang, H.: Optically Active Sum Frequency Generation Microscopy for Cellular Imaging. *Springer Series Chem.* 2007, 88, 825–827.

Review Articles.

- Ji, N.; Shen, Y. R.: A Novel Spectroscopic Probe for Molecular chirality. *Chirality* 2006, 18, 146–158.
- Yan, E. C. Y.; Fu, L.; Wang, Z.; Liu, W.: Biological Macromolecules at Interfaces Probed by Chiral Vibrational Sum Frequency Generation Spectroscopy. *Chem. Rev.* 2014, 114, 8471–8498.
- Ishibashi, T.; Okuno, M.: Heterodyne-Detected Chiral Vibrational Sum Frequency Generation Spectroscopy of Bulk and Interfacial Samples. *Mol. Laser Spectrosc.* 2020, 2, 315–348.

Chapter 6

Molecular Adsorption at Surfaces and Interfaces

A large part of surface science and technology deals with atomic and molecular adsorption at interfaces. In one aspect, surface contamination by molecular adsorption may deteriorate functional properties of materials. Corrosion is an example. In another aspect, molecular adsorption by design can modify surface functional properties and initiate or impede reactions to suit needs. Coating and catalysis are examples. In other cases, adsorbed monolayers such as Langmuir–Blodgett films can find important applications in their own right. Microscopic information on how adsorbates interact with substrates, how they orient, distribute, and behave at surfaces or interfaces, and how they influence interfacial properties is of great interest.

Experimental techniques to investigate atomic and molecular adsorption have been extensively developed. Many of them require samples in ultrahigh vacuum with focus on probing adsorbed species, adsorption sites, and adsorption kinetics.[1] For studies of electronic and vibrational structures of adsorbates, optical spectroscopic techniques are probably most viable. While simple reflection-absorption spectroscopy is capable of probing submonolayers of adsorbates, it has severe intrinsic limitation; in comparison, SFG spectroscopy generally can provide more information.

6.1. General Description on Reflection–Absorption Spectroscopy and SFG Spectroscopy

Reflection–absorption spectroscopy (RAS) measures the induced change of the reflection spectrum of an interface, $\Delta R(\omega)$ of $R_0(\omega)$, when an adsorbed layer appears at the interface. It is from the spectral change that the absorption spectrum of the adsorbed layer can be deduced. The underlying theory for RAS given in the literature is usually based on a three-layer model.[2-4] Instead, we describe here a less model-dependent, more physical, approach.

It is well known that the reflected field from an interface is given by $\vec{E}_{R0} = \overset{\leftrightarrow}{F}_R \cdot \vec{E}_I$, where $\overset{\leftrightarrow}{F}_R$ is the Fresnel reflection coefficient, and $\vec{E}_I$ is the incoming field. In the presence of a layer of adsorbates at the interface, the fields, $\vec{E}_I$ and $\vec{E}_{R0}$, induce a sheet of oscillating dipoles in the layer that also radiates and contributes to the reflected field. In terms of induced dipoles per unit volume, $\vec{P}^{(1)} = \overset{\leftrightarrow}{\chi}_{ad}^{(1)} \cdot (\vec{E}_I + \vec{E}_{R0})$, the field of radiation in the reflected direction from a thin sheet of thickness d is given by[5]

$$\Delta \vec{E}_R = \frac{i2\pi\omega^2}{c^2 k_z} \overset{\leftrightarrow}{f} \cdot i\hat{k} \times [i\hat{k} \times \vec{P}^{(1)}]d$$

$$= \frac{i2\pi\omega^2}{c^2 k_z} \overset{\leftrightarrow}{f} \cdot i\hat{k} \times [i\hat{k} \times \overset{\leftrightarrow}{\chi}_{ad}^{(1)} \cdot (\vec{E}_I + \vec{E}_{R0})]d, \tag{6.1}$$

where $\overset{\leftrightarrow}{f} = 1$ if the effect of the substrate on the radiation is neglected, $\vec{k}$ denotes the wave vector of the reflected field, and z is along the surface normal. The total reflected field is $\vec{E}_R = \vec{E}_{R0} + \Delta \vec{E}_R$ with $|\Delta \vec{E}_R| << |\vec{E}_R|$, and the total reflectivity is given by

$$R = |\vec{E}_R|^2 / |\vec{E}_I|^2 \approx [|\vec{E}_{R0}|^2 + 2\text{Re}(\vec{E}_{R0}^* \cdot \Delta \vec{E}_R)]/|\vec{E}_I|^2 = R_0 + \Delta R$$

with $R_0 = |\vec{E}_{R0}|^2 / |\vec{E}_I|^2$,

$$\Delta R = \text{Im}\{(\overset{\leftrightarrow}{F}_R \cdot \vec{E}_I)^* \tfrac{4\pi\omega^2}{c^2 k_z} \overset{\leftrightarrow}{f} \cdot \hat{k} \times [\hat{k} \times \overset{\leftrightarrow}{\chi}_{ad}^{(1)} \cdot (\vec{E}_I + \overset{\leftrightarrow}{F}_R \cdot \vec{E}_I)]d/|\vec{E}_I|^2\}. \tag{6.2}$$

We see that ΔR or $\Delta R/R_0$ is linearly proportional to the linear susceptibility of the adsorbate layer, $\overset{\leftrightarrow}{\chi}_{ad}^{(1)}$, and the proportional coefficient can be evaluated if the optical constants of the substrate are known. Thus, we can deduce $\overset{\leftrightarrow}{\chi}_{ad}^{(1)}$ from the measured ΔR.

For illustation, we assume an isotropic adsorbate layer on a transparent substrate and an S-polarized input beam incident on it. From Eq. (6.2), we find

$$\Delta R/R_0 = (2\mathrm{Re}\Delta E_R)/E_{R0} = [4\pi\omega^2 df(1+F_R)/F_R c^2 k_z]\mathrm{Im}\ \chi_{ad}^{(1)}.$$
$$(6.3)$$

Since $\mathrm{Im}\ \chi_{ad}^{(1)}$ is proportional to the absorption coefficient of the adsorbates and the quantity in the square brackets is real, the spectrum of $\Delta R/R_0$ directly displays the absorption spectrum of the adsorbates. For P-polarized input and/or anisotropic adsorbed layers, $\Delta R/R_0 \propto \mathrm{Im}\ \chi_{ad}^{(1)}$ is still true as long as the substrate is transparent, although the square-bracketed proportional constant in Eq. (6.3) is different. For absorbing substrates, however, analysis of RAS spectra can be quite challenging because $|\Delta R| << R_0$ and $\overset{\leftrightarrow}{F}_R$ is complex so that $\Delta R/R_0$ is no longer proportional to $\mathrm{Im}\ \chi_{ad}^{(1)}$ and cannot directly reflect the absorption spectrum of the adsorbates. In order to extract the spectrum of $\mathrm{Im}\ \chi_{ad}^{(1)}$ from $\Delta R/R_0$, we need to know not only the optical constants of the substrate, but also the functional forms of $\mathrm{Re}\ \chi_{ad}^{(1)}$ and $\mathrm{Im}\ \chi_{ad}^{(1)}$. If the absorption spectrum of the substrate is smooth and broad, then sharp resonance modes of adsorbates can still be identified from the spectrum of $\Delta R/R_0$. For substrates of very large optical dielectric constants, such as metals in the infrared regime, we still have approximately $\Delta R/R_0 \propto \mathrm{Im}\ \chi_{ad}^{(1)}$ if a P-polarized input at a grazing incident angle is used.[3,4]

RAS can have submonolayer sensitivity and has been generally adopted to study adsorbates at interfaces, often with focus on vibrational spectra of adsorbates. More recently, however, it has been widely used to probe electronic transitions of 2D materials on substrates. The main difficulty that limits the sensitivity comes from $|\Delta R| << R_0$ because usually the measurement sensitivity is limited to $|\Delta R/R_0| \sim 10^{-4}$ so that random fluctuations of R_0 in the measurement must be kept low. If recording of a spectrum is only of interest, the polarization-modulation (PM) scheme of RAS can help significantly since the modulation yields a difference signal from S- and P-polarized reflection that is no longer sensitive to fluctuations in the beam path. The sensitivity of PM-RAS is also better if the adsorbed layer is highly anisotropic along the surface normal, e.g., adsorbates with induced dipoles closely aligned with the surface normal. It can be further enhanced by using appropriately mixed polarization modulation to make the difference signal from the bare substrate nearly vanish.

In comparison, SFG spectroscopy for adsorbates has several advantages over RAS because of its much larger signal-to-background ratio. Similar to RAS, an adsorbate layer generates an SF field in addition to the SF field generated from a substrate. However, unlike RAS, SFG from a substrate with inversion symmetry is weak and comparable to that from the adsorbate layer, allowing much easier extraction of the latter from measurement. Even if the substrate is a crystal without inversion symmetry, it is possible to use appropriate input/output polarization combinations to strongly suppress reflected SFG from the substrate.

To be more quantitative, we follow the description in Sec. 2.1 and write the reflected SF output field from an adsorbate-covered substrate as

$$\vec{E}(\omega_3) = \vec{E}_{sub} + \Delta E_{ad}$$

where $\vec{E}_{sub}$ is from the substrate and ΔE_{ad} from the adsorbed layer with $|E_{sub}| \sim |\Delta \vec{E}_{ad}|$. The latter has the same expression as Eq. (6.1) except that the induced polarization, $\vec{P}^{(1)} = \overset{\leftrightarrow}{\chi}_{ad}^{(1)} \cdot (\vec{E}_I + \vec{E}_{R0})$, is replaced by $\vec{P}^{(2)}(\omega_3) = \overset{\leftrightarrow}{\chi}_{ad}^{(2)} \cdot [\vec{E}_{1I}(\omega_1) + \vec{E}_{1R0}(\omega_1)][\vec{E}_{2I}(\omega_2) + \vec{E}_{2R0}(\omega_2)]$. From Eq. (2.4) with $\vec{E}_{Sr}^{I}(\omega_3, z)$

replaced by ΔE_{ad}, we have

$$\vec{E}(\omega_3) = \vec{E}_{sub} + \Delta\vec{E}_{ad}$$

$$= \frac{i2\pi\omega_3^2}{c^2 k_z}\{\overset{\leftrightarrow}{F}_T(\omega_3) : (-\frac{\overset{\leftrightarrow}{\chi}_{sub}^{(2)}}{i\Delta k_{sub,z}}) : \overset{\leftrightarrow}{F}_T(\omega_1)\overset{\leftrightarrow}{F}_T(\omega_2)$$

$$+ \overset{\leftrightarrow}{f} \cdot i\hat{k} \times (i\hat{k} \times \overset{\leftrightarrow}{\chi}_{ad}^{(2)} d \cdot [1 + \overset{\leftrightarrow}{F}_R(\omega_1)]$$

$$[1 + \overset{\leftrightarrow}{F}_R(\omega_2)])\} : |\vec{E}_1\vec{E}_2|e^{i(-k_z z - \omega_3 t)}$$

$$(6.4)$$

We are interested in obtaining the spectrum of Im $\overset{\leftrightarrow}{\chi}_{ad}^{(2)}$ through measurement of $\vec{E}(\omega_3)$ or $|\vec{E}(\omega_3)|^2$. In contrast to RAS, the signal-to-background ratio for SFG is

$$|\Delta S|/S_0 \equiv (S - S_0)/S_0 = (|\vec{E}_{sub} + \Delta\vec{E}_{ad}|^2 - |\vec{E}_{sub}|^2)/|\vec{E}_{sub}|^2,$$

$$(6.5)$$

and if $|\frac{\overset{\leftrightarrow}{\chi}_{sub}^{(2)}}{\Delta k_{sub,z}}| \sim |\overset{\leftrightarrow}{\chi}_{ad}^{(2)} d|$, the ratio is of the order of 1. For reflected SFG from transparent substrates with inversion symmetry, we often have $|\frac{\overset{\leftrightarrow}{\chi}_{sub}^{(2)}}{\Delta k_{sub,z}}| << |\overset{\leftrightarrow}{\chi}_{ad}^{(2)} d|$ and accordingly, $S \equiv |\vec{E}_{sub} + \Delta\vec{E}_{ad}|^2 \approx |\Delta\vec{E}_{ad}|^2 \propto |\overset{\leftrightarrow}{\chi}_{ad}^{(2)}|^2$. For absorbing substrates, extraction of $\overset{\leftrightarrow}{\chi}_{ad}^{(2)}$ from S or ΔS is again more difficult, although sharp resonant modes can still be identified from sharp features in the spectrum of S if the spectrum of the substrate is smooth and broad. It is however possible to use phase-sensitive SFG spectroscopy to deduce $\overset{\leftrightarrow}{\chi}_{ad}^{(2)}$ directly from measured $\vec{E}(\omega_3)$. We can express Eq. (6.4) as $\vec{E}(\omega_3) = \frac{i2\pi\omega_3^2}{c^2 k_z}\{\overset{\leftrightarrow}{X}_{sub}^{(2)} + \overset{\leftrightarrow}{X}_{ad}^{(2)}\} : |\vec{E}_1\vec{E}_2|e^{i(-k_z z - \omega_3 t)}$ with $\overset{\leftrightarrow}{X}_{sub}^{(2)} \propto \overset{\leftrightarrow}{\chi}_{sub}^{(2)}$ and $\overset{\leftrightarrow}{X}_{ad}^{(2)} \propto \overset{\leftrightarrow}{\chi}_{ad}^{(2)}$. Phase-sensitive measurements on a substrate with and without adsorbates yield, respectively, Im $\overset{\leftrightarrow}{X}_{sub}^{(2)}$ and Im($\overset{\leftrightarrow}{X}_{sub}^{(2)} + \overset{\leftrightarrow}{X}_{ad}^{(2)}$). Subtraction of one from the other gives Im $\overset{\leftrightarrow}{X}_{ad}^{(2)}$, from which Im $\overset{\leftrightarrow}{\chi}_{ad}^{(2)}$ can be obtained if the optical constants of the substrate are known.

As a second-order optical process, SFG spectroscopy for studies of molecular adsorption is generally more flexible and carries more information than RAS. It is better suited for probing orientation, conformation, and arrangement of adsorbates, investigating adsorption kinetics and dynamics, monitoring surface reactions including catalysis in real atmosphere and real time, and tracking ultrafast surface dynamics. Some of these will be described in the following sections. The main disadvantage is that the experimental arrangement is more complex than RAS.

6.2. Spectroscopic Detection of Adsorbates

To establish a tool for surface studies, one often needs to demonstrate that it can detect adsorbates at an interface. This is the case with second harmonic generation (SHG). It was found in 1981 that similar to surface enhanced Raman scattering (SERS), SHG on a roughened silver surface could be enhanced by local field enhancement over four orders of magnitude.[6] Submonolayer adsorbates could be readily detected on such a surface.[7] It was then realized from estimate that SHG should have enough sensitivity to detect a monolayer of adsorbates even on smooth surfaces.[8] Independent, nonvanishing elements of $\overleftrightarrow{\chi}_{ad}^{(2)}$ for an adsorbed layer at a smooth interface could be measured by SHG with selected input/output polarizations, from which the average orientation of adsorbed molecules could be extracted.[9] (More details will be given in Sec. 6.3.) To persuade the surface science community that SHG could be a viable surface probe, it was used to monitor alkali adsorption on Rh(111) and oxidation of a Si(111) (7x7) surface in ultrahigh vacuum (UHV).[10,11] SHG spectroscopy on electronic transitions of adsorbates was also demonstrated with a tunable input beam,[12] as shown in Fig. 6.1. However, SHG is not suited for vibration spectroscopy of adsorbates because commonly available IR detectors are not sensitive enough to detect the weak SHG signal in the IR. It is obvious that if we use SFG instead of SHG, we can easily extend spectroscopic studies of adsorbates to a broad range from IR to UV.

Sum frequency vibrational spectroscopy (SFVS) allows identification of adsorbed species from observed spectra. As an example,

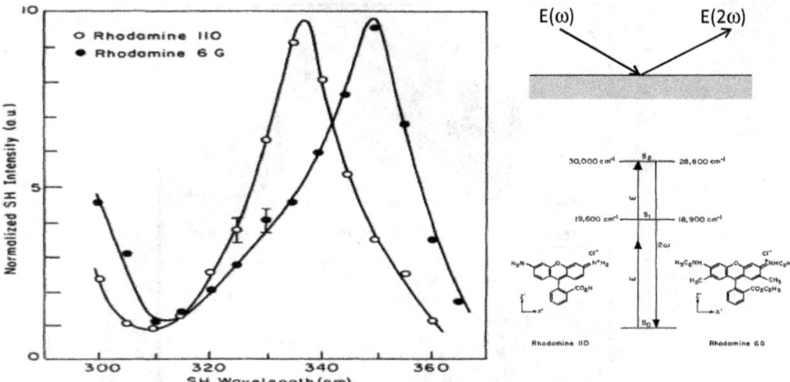

Fig. 6.1. SH spectra of S_0 to S_2 electronic transition of rhodamine 6G and rhodamine 110 monolayers adsorbed on fused silica. The SH beam geometry, structures, and energy level diagram of the molecules, and the SH resonant transition are described on the right (after Ref. [12]).

we present in Fig. 6.2 the vibrational spectra of CO adsorbed on Pd(111) in UHV taken with PPP-SFVS and PM-RAS, respectively.[13] Two peaks associated with stretch vibrations of CO adsorbed at top and hollow sites of Pd(111) are seen. This example shows that like RAS, SFG spectroscopy can be employed to study adsorbates on well-characterized metal and semiconductor surfaces in UHV. Since UHV is not required for SFG measurement, SF spectroscopy can be readily adopted to probe adsorbates *in situ* under ambient gas, and monitor their changes with changes of gas composition, pressure and temperature, as well as with time, providing possibility to study surface reactions such as catalysis in real atmosphere and real time, which will be discussed in Sec. 6.7.

Because RAS and SFG spectroscopy follow different selection rules, their spectra on adsorbates generally are not the same. The case of adsorbed monolayers with long alkyl chains is a striking example. Figure 6.3 presents the CH stretch spectra of a palmitic acid monolayer on water taken by PM-RAS and SSP SFVS.[14] The RAS spectra for three different surface pressures display three modes at 2854, 2919, and 2960 cm^{-1} attributed to methylene symmetric, methylene asymmetric, and methyl asymmetric stretches of the alkyl chain, respectively. (The modes appear as dips because reflectance

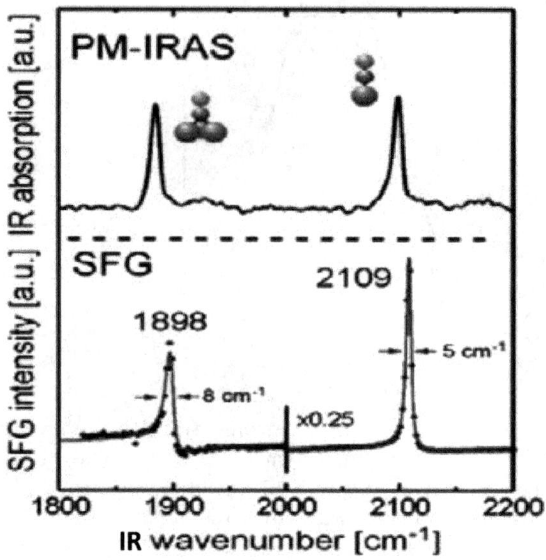

Fig. 6.2. (a) PM-IRRAS and (b)SFVS spectra of CO adsorbed on Pd(111) at CO pressures of 170 and 100 mbar, respectively, and temperature of 190K. Both spectra show two CO stretch modes: CO at top sites at ~ 2109 cm^{-1} and CO at bridge sites at ~ 1898 cm^{-1} (after Ref. [13]).

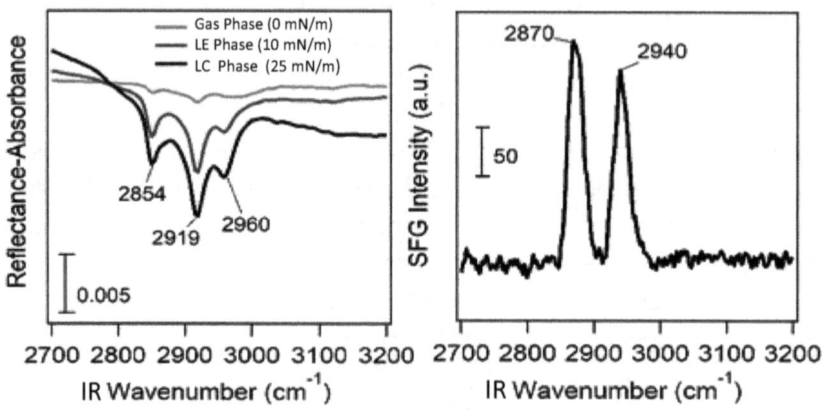

Fig. 6.3. CH stretch vibrational spectra of palmitic acid [CH$_3$(CH$_2$)$_{14}$COOH] monolayers at the air/water interface. (a)IRRAS spectra of monolayers in gas, liquid expanded, and liquid condensed phases specified by their surface tensions, and (b)SSP SF vibrational spectra for a monolayer in the liquid condensed phase. Because of the different selection rules, the CH$_2$ stretch modes dominate the IRRAS spectra, while the CH$_3$ symmetric mode and its Fermi resonance dominate the SF spectra (after Ref. [14]).

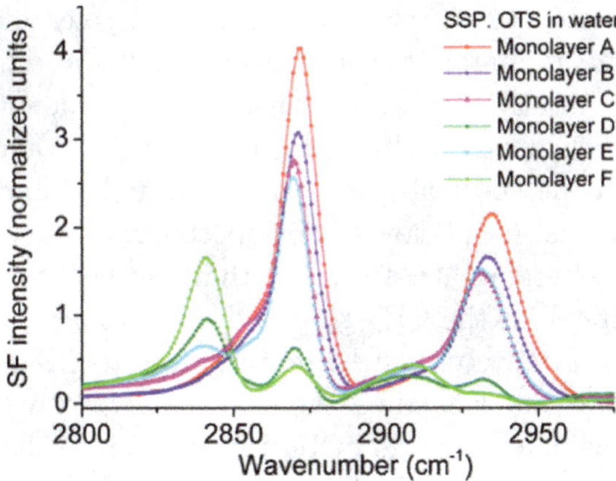

Fig. 6.4. CH stretch spectra, taken by SSP SFVS, of OTS monolayers of different qualities deposisted on silica and immersed in water. Monolayears A and B have the best quality; only CH_3 stretch modes are present, indicating that their alkyl chains are all-trans. The quality of monolayers C to F progressively decreases as evidenced by the decrease of the CH_3 modes and increase of the CH_2 stretch modes at 2840 and 2920 cm^{-1} resulting from increased amount of gauche defects in the alkyl chains (after Ref. [15]).

from the substrate was not subtracted.) The SFVS spectrum, however, only reveals vibrational modes of the terminal methyl group; the two modes at 2870 and 2940 cm^{-1} belong, respectively, to the methyl symmetric stretch and the Fermi resonance between the methyl symmetric stretch and the overtone of the bending mode. The difference between RAS and SFVS spectra arises from symmetry requirement of the two techniques. Because the monolayer is highly packed and the alkyl chains are all-trans (with the monolayer in the liquid condensed (LC) phase), the chain of methylene (CH_2) groups wrapping around the chain axis has near inversion symmetry. As a result, while all CH_2 stretch modes are prominent in the RAS spectra, they are almost completely suppressed in the SF vibrational spectrum.

Information about chain conformation can also be gleaned by SFVS. Gauche defects on a chain break the inversion symmetry of CH_2 groups along the otherwise all-trans chain and result in the appearance of CH_2 stretch modes in the spectrum. This is illustrated in Fig. 6.4,[15] where the SSP CH stretch spectra of a set of OTS monolayers on fused silica in contact with water are presented. These monolayers had different degrees of disorder in their alkyl chains, but it was not easy to tell the difference from contact angle measurement. Figure 6.4, however, shows clearly their spectra are very different. Only CH_3 modes at 2870 and 2940 cm^{-1} are present in the spectra of monolayers A and B, indicating that their alkyl chains are all-trans. The spectrum of monolayer C has a weak band emerging at 2854 cm^{-1} that can be attributed to the symmetric stretch of the CH_2 group adjacent to the terminal CH_3, suggesting appearance of gauche defects close to CH_3. Increasingly more gauche defects existed on the chains of monolayers D, E, and F, resulting in much stronger CH_2 modes and much weaker CH_3 modes from disoriented CH_3. The CH_2 symmetric stretch mode is now dominated by a band attributed to symmetric stretches of CH_2 groups away from the terminal CH_3. This is because as chains become more disordered, more gauche defects appear away from the methyl terminal. The average number of gauche defects per chain is limited by the available free space for the chains in a monolayer of given chain density. Presumably one could deduce information about the average number of defects on a monolayer of alkyl chains from its CH stretch spectrum. Unfortunately, to our knowledge, this has not yet been worked out theoretically. Polarization-dependent SF vibrational spectra in general carry information about orientation and conformation of adsorbed molecules, but for complex molecular adsorbates, theoretical help is often needed to extract such information from spectral features.

Molecular adsorption on surfaces exposed to a real atmosphere is hardly avoidable, and SFG spectroscopy is capable of detecting and analyzing contaminants. In fact, the first surface SF vibrational spectrum was taken accidentally from a contaminated fused silica surface in air. Harry Tom, a physics graduate student at Berkeley at the time, built a nanosecond optical parametric oscillator tunable in the IR, which was meant to be used in a collaborative project with Y.T. Lee's group on vibrational excitation of molecular beams. He took an opportunity to use it to test out SFVS. To his surprise, the SF spectrum taken from a silica surface showed sharp features in the CH stretch range although nothing was intentionally deposited on the silica surface. Later, it was realized that because the measurement was performed in Lee's molecular beam lab, the silica surface must have been contaminated by oil molecules in air released from pumping the molecular beam chamber. The result was unpublished due to uncertainty in interpretation.[16] We now know surface contamination by molecular adsorption from air occurs ubiquitously and unexpectedly. For example, carboxylic acids typically present only in parts per billion in atmosphere were found to overwhelmingly adsorb on TiO_2 despite the presence of much higher alcohol concentration in the atmosphere.[17] For such studies, SFVS obviously can be an effective tool to probe and identify contaminants on substrates *in situ.*[18]

Molecular adsorption at buried interfaces can also be studied by SFVS. Adsorption of solutes from solution to liquid/solid interfaces is probably most important because of its pertinence to many disciplines. Among many well-known problems are adsorptions of ions and molecules on oxides from water solution relevant to geoscience, on metals relevant to electrochemistry, and on biological membrane relevant to bioscience. We shall go through these topics in later sections and chapters.

Another often useful piece of information about adsorbates is their orientation that arises from how they interact with a surface and can modify the surface properties. The information is embedded in the polarization dependence of SFVS, and is the topic to be discussed in the next section.

6.3. Determination of Molecular Orientation at an Interface

As described in Sec. 2.3, the nonlinear susceptibility, $\overset{\leftrightarrow}{\chi}^{(2)}$, of a material is the sum of nonlinear polarizability, $\overset{\leftrightarrow}{\alpha}^{(2)}$, of individual molecules in a unit volume with their orientations specified by an orientation distribution. Surface nonlinear susceptibility for a layer of adsorbates can be expressed as $\overset{\leftrightarrow}{\chi}^{(2)}_{ad} = N_S < \overset{\leftrightarrow}{\alpha}^{(2)}_{ad} > = N_S \int f(\Omega)\overset{\leftrightarrow}{\alpha}^{(2)}_{ad}(\Omega)d\Omega$, with N_S denoting the surface density of adsorbates and $f(\Omega)$ describing the orientation distribution function. More specifically, such a relation refers to resonant $\overset{\leftrightarrow}{\chi}^{(2)}_{ad}$ and $\overset{\leftrightarrow}{\alpha}^{(2)}_{ad}$ for a vibrational mode of a subgroup of the adsorbate molecule with $f(\Omega)$ being the orientation distribution of the subgroup. (The same relation holds with the resonant amplitude $\overset{\leftrightarrow}{A}^{(2)}$ of the mode replacing $\overset{\leftrightarrow}{\chi}^{(2)}_{ad}$ and the corresponding amplitude $\overset{\leftrightarrow}{a}^{(2)}$ of the mode in $\overset{\leftrightarrow}{\alpha}^{(2)}_{ad}$ replacing $\overset{\leftrightarrow}{\alpha}^{(2)}_{ad}$.) If M independent nonvanishing elements of $\overset{\leftrightarrow}{\chi}^{(2)}_{ad}$ can be measured, then we should be able to use the above relation between $\overset{\leftrightarrow}{\chi}^{(2)}_{ad}$ and $\overset{\leftrightarrow}{\alpha}^{(2)}_{ad}$ (or between $\overset{\leftrightarrow}{A}^{(2)}$ and $\overset{\leftrightarrow}{a}^{(2)}$) to determine m unknown independent elements of $\overset{\leftrightarrow}{\alpha}^{(2)}_{ad}$ (or $\overset{\leftrightarrow}{a}^{(2)}$) and M-m parameters (assuming $M > m$) that specify $f(\Omega)$ to yield approximate information about the orientation distribution. We provide a more in-depth description on the formalism in what follows.

Following the discussion in Sec. 2.3 and Eq. (2.15), we can write the relation between $\overset{\leftrightarrow}{\chi}^{(2)}_{ad}$ and $\overset{\leftrightarrow}{\alpha}^{(2)}_{ad}$ explicitly as

$$
\begin{aligned}
\chi^{(2)}_{ad,ijk} &= N_S \sum_{\xi,\eta,\zeta} \langle (\hat{i}\cdot\hat{\xi})(\hat{j}\cdot\hat{\eta})(\hat{k}\cdot\hat{\zeta})\rangle \alpha^{(2)}_{\xi\eta\zeta} \\
&= N_S \sum_{\xi,\eta} \int (\hat{i}\cdot\hat{\xi})(\hat{j}\cdot\hat{\eta})(\hat{k}\cdot\hat{\zeta})\alpha^{(2)}_{\xi\eta\zeta} f(\theta,\phi)\sin\theta\, d\theta d\phi
\end{aligned}
\tag{6.6}
$$

Here, (i,j,k) refer to the lab coordinates (x,y,z) and (ξ,η,ζ) to the molecular coordinates (x',y',z'); the two coordinate systems are related by the usual coordinate transformation with θ being the polar angle between $\hat{z}'$ and $\hat{z}$, and ϕ being the azimuthal angle of $\hat{z}'$ with respect to $\hat{x}$ in the $x-y$ plane; $f(\theta,\phi)$ is the orientation distribution function in θ and ϕ. Measurement of M independent $\chi^{(2)}_{ad,ijk}$ elements (specifically, the amplitude $A^{(2)}_{ijk}$ of a selected resonant mode) allows determination of M unknown parameters in the integral on the right-hand side of Eq. (6.6) (with reference to the selected resonant mode). Since $(\hat{i}\cdot\hat{\xi})(\hat{j}\cdot\hat{\eta})(\hat{k}\cdot\hat{\zeta})$ as a function of θ and ϕ is known, the parameters to be determined are $N_S a^{(2)}_{\xi\eta\zeta}$ of the resonant mode and those specifying $f(\theta,\phi)$. If all nonvanishing elements of $N_S \overset{\leftrightarrow}{a}^{(2)}$ for the mode are known, the M measured $\chi^{(2)}_{ad,ijk}$ can allow determination of

M parameters in $f(\theta, \phi)$. Unfortunately, this is usually not the case. Then, if the number of unknown $N_S a^{(2)}_{\xi\eta\zeta}$ is m, the M measurements can only determine M-m parameters in $f(\theta, \phi)$, although they are usually chosen to be the dominant parameters that define $f(\theta, \phi)$. This limits information on the molecular orientation that can be extracted from SFVS measurement.

The simplest, but fairly common, case deals with a vibrational mode of molecules or molecular subgroups having a three-fold or higher rotational symmetry axis (along $\hat{z}'$). The set of nonvanishing elements of $\overleftrightarrow{\alpha}^{(2)}$ (or $\overleftrightarrow{a}^{(2)}$) are then $\alpha^{(2)}_{z'z'z'}$ and $\alpha^{(2)}_{x'x'z'} = \alpha^{(2)}_{y'y'z'} = r\alpha^{(2)}_{z'z'z'}$. For a polar oriented, azimuthally isotropic, layer of such molecules or molecular groups, we can take 3 independent SFVS measurements with input/output polarization combinations SSP, SPS, and PPP to find $\chi^{(2)}_{ad,xxz} = \chi^{(2)}_{ad,yyz}$, $\chi^{(2)}_{ad,xzx} = \chi^{(2)}_{ad,yzy}$, and $\chi^{(2)}_{ad,zzz}$, with $\hat{z}$ along the layer normal. (Because the frequencies of visible input and SF output are close, PSS measurement yields nearly the same spectrum as SPS.) They allow determination of $\alpha^{(2)}_{z'z'z'}$ (or $a^{(2)}_{z'z'z'}$) and r plus only one parameter for $f(\theta, \phi) = f(\theta)$. From Eq. (6.6), we find, for such cases,

$$\chi^{(2)}_{ad,zzz} = N_S \alpha^{(2)}_{z'z'z'} [r <\cos\theta> + (1-r) <\cos^3\theta>]$$

$$\chi^{(2)}_{ad,yyz} = \tfrac{1}{2} N_S \alpha^{(2)}_{z'z'z'} [(1+r) <\cos\theta> + r <\cos^3\theta>]$$

$$\chi^{(2)}_{ad,yzy} = \chi^{(2)}_{ad,zyy} = \tfrac{1}{2} N_S \alpha^{(2)}_{z'z'z'} (1-r)(<\cos\theta> - <\cos^3\theta>)$$

We can assume $f(\theta, \phi) = \delta(\cos\theta - \cos\theta_0)$, or $f(\theta, \phi) = $ constant for $0 \leq \theta \leq \theta_0$ and 0 otherwise, with θ_0 to be determined. In practice, to avoid absolute measurement on $\chi^{(2)}_{ad,ijk}$, ratios of $\chi^{(2)}_{ad,ijk}$ are taken, so that the common factor, $N_S \alpha^{(2)}_{\zeta\zeta\zeta}$, in the above equations is eliminated. The two ratios then permit determination of r and θ_0.

We use the case of an azimuthally isotropic layer of 4"-n-phenyl-4-cyano-p-terphenyl (5CT) on water as an example.[19] The molecular structure of 5CT, sketched in Fig. 6.5(a), is composed of three sections: a CN headgroup, a para-terphenyl chromophore, and a

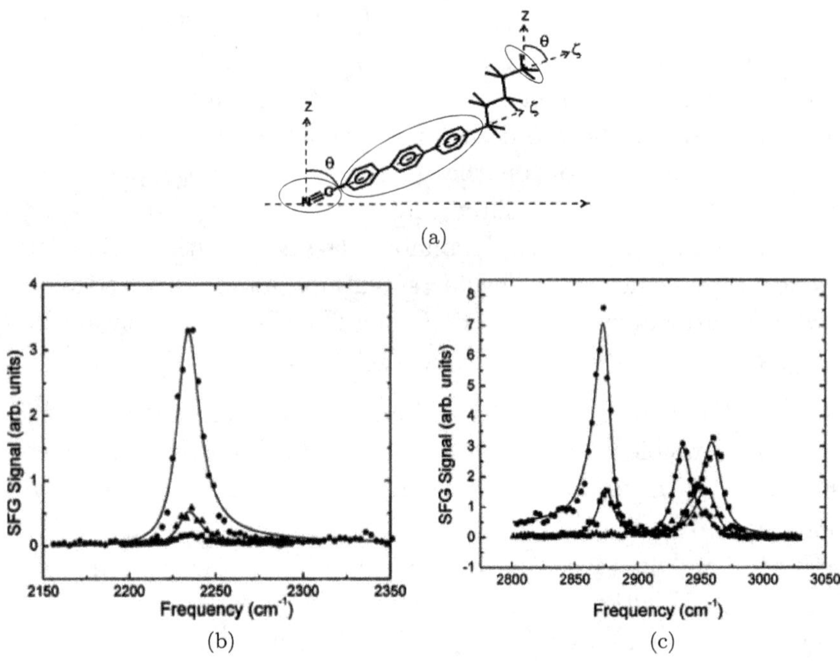

Fig. 6.5. (a) Molecular structure of 4"-n-pentyl-4-cyano-terphenyl (5CT). (b) and (c) SF spectra in the CN and CH stretch ranges, respectively. Circles, squares, and triangles are experimental data obtained with SSP, PPP, and SPS polarizations, respectively. Only CH_3 modes are prominent in (c) indicating the alkyl chains of 5CT are all-trans. Solid curves are fitting curves. Analysis of the spectra of different polarizations yield the average polar orientation angles of CN and CH_3 groups, both appearing to be around $50°$ (after Ref. [19]).

$CH_3(CH_2)_4$ alkyl chain. The orientations of the three sections can be separately determined following the above description. For the CN headgroup, the nonvanishing $\overset{\leftrightarrow}{\alpha}^{(2)}$ elements of the CN stretch are $\alpha^{(2)}_{z'z'z'}$ and $\alpha^{(2)}_{x'x'z'} = \alpha^{(2)}_{y'y'z'} = r\alpha^{(2)}_{z'z'z'}$ with z' along the CN bond. The measured spectra of the CN stretch from the three polarization combinations, SSP, SPS, and PPP, are shown in Fig. 6.5(b), from which three independent nonvanishing $\chi^{(2)}_{ad,ijk}$ (or $A^{(2)}_{ijk}$) elements and their two ratios at the stretch resonance can be extracted from spectral analysis, and then r, and θ_0 for the C $\equiv$N bond can be determined. It was found that $r = 0.25$ and $\theta_0 \approx 53°$. The spectra of three polarization combinations for the alkyl chain of

5CT in the CH stretch range are displayed in Fig. 6.5(c). Only the CH_3 stretches are prominent, indicating that the chains are in all-trans conformation. The three peaks at 2875, 2960, and 2940 cm^{-1} originate from symmetric stretch, antisymmetric stretch, and Fermi resonance between symmetric stretch and overtone bending modes of CH_3, respectively. The symmetric CH_3 stretch is polarized along the symmetry axis of the group and its spectra can be used to find the orientation of the symmetry axis. Similar to the CN case, the nonvanishing $\overset{\leftrightarrow}{\alpha}^{(2)}$ elements of CH_3 symmetric stretch are $\alpha^{(2)}_{z'z'z'}$ and $\alpha^{(2)}_{x'x'z'} = \alpha^{(2)}_{y'y'z'} = r\alpha^{(2)}_{z'z'z'}$ with z' along the symmetric axis of CH_3 and the spectral analysis led to $r = 2.5$ and $\theta_0 \approx 54°$. This angle is nearly the same as that of $C \equiv N$ as it should be from the known molecular conformation of 5CT. To find the orientation of the donor-acceptor para-terphenyl chromophore, SHG instead of SFG can be used because the three phenyl rings of the chromophore are successively twisted and the SH nonlinear optical polarizability of the chromophore can be approximated to be that of a long cylindrically symmetric unit with independent, nonvanishing elements $\alpha^{(2)}_{z'z'z'}$ and $\alpha^{(2)}_{x'x'z'} = \alpha^{(2)}_{y'y'z'} = r\alpha^{(2)}_{z'z'z'}$ with z' along the cylindrical axis. Three SHG measurements with polarization combinations PS (output P-polarized and input S-polarized), PP, and PM, can yield three independent $\chi^{(2)}$ elements or two ratios, $\chi^{(2)}_{PP}/\chi^{(2)}_{PS}$ and $\chi^{(2)}_{SM}/\chi^{(2)}_{PS}$, for the chromophore, where M denotes a linear polarization dividing S and P. The result allowed finding of $r = -0.05$ and $\theta_0 \approx 50°$ that are again consistent with the above-mentioned orientations of CN and alkyl chain knowing the conformation of 5CT. In this case, if we start the analysis by assuming $r = 0$, we would have two parameters for the orientation distribution to be determined. We can, for example, assume $f(\theta)$ to be a Gaussian function, $C\exp[-(\theta - \theta_0)^2/\sigma^2]$, and determine θ_0 and σ from the measurement.

For azimuthally anisotropic adsorbed layer, there are more independent $\chi^{(2)}$ elements that can be measured. For example, if the above 5CT layer has the chromophores align along $\hat{x}$, we would have 6 independent $\chi^{(2)}$ elements, or five ratios, that can be measured by SHG, allowing determination of r and 4 parameters describing the orientation distribution of the chromophores in θ and ϕ.[20]

We note that such orientation measurements cannot be taken too seriously because of the necessary approximations involved. First of all, an orientation distribution generally cannot be specified by only a few parameters. Assumption of its being a δ-function or a Gaussian function may not be able to sufficiently describe the orientation distribution even in the approximate sense. Even if the assumption is acceptable, deduction of $\chi^{(2)}$ elements from measured spectra has complication. As discussed in Sec. 2.1, it involves knowing the proper Fresnel coefficients to correctly describe the input and output fields in the adsorbed layer. This relies on knowledge of effective refractive indices for the adsorbed layer, which unfortunately could only be estimated. Finally, the measurement can determine the absolute orientation of up or down only by comparing with a known reference. Even with these limitations, information about approximate orientation of molecules at a surface or interface can be very useful in many applications.

6.4. Adsorption Isotherm

Impurities in a solution may prefer to appear at an interface to lower the free energy of the system. This could happen for impurities in solids. Carbon diffusion from copper bulk to surface to help nucleation of graphene growth by chemical vapor deposition is an example. More often encountered and relevant cases are adsorption of solute molecules from solutions to air/liquid and solid/liquid interfaces. Using SFVS as a probe, adsorbed molecules can be identified from their vibrational spectra and their adsorption geometry deduced from spectral dependence on beam geometry and polarizations. Another important piece of often aspired information is on how readily molecules would adsorb on a selective interface. This can be acquired from the adsorption isotherm that SFVS can also provide.

Measuring an adsorption isotherm is actually quite challenging, especially for adsorption at a buried interface; SFVS is probably the most viable means because it can monitor molecular adsorption *in situ* at buried interfaces. However, we still need to assume adsorption is homogeneous, and molecular orientation and conformation are independent of surface density of adsorbates.

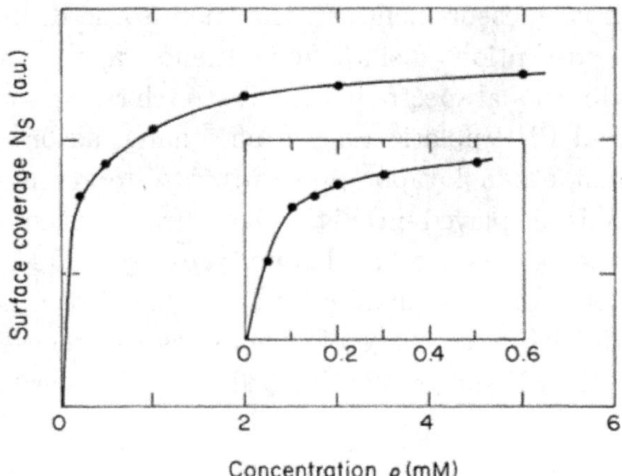

Fig. 6.6. Adsorption isotherm of p-nitrobenzoic acid adsorbed on fused silica from an ethanolic solution obtained from reflected SHG measurement. The inset is an expansion of the low concentration part (after Ref. [9]).

We can then deduce surface density of adsorbates from the spectral intensity of a vibrational mode associated with the adsorbates, usually with reference to the maximum surface coverage. If the molecular orientation changes with increase of adsorption, but can be measured by polarization-dependent SFVS, we can extract the surface density of adsorbates from the observed spectrum by taking into account the effect of molecular orientation on the spectrum in spectral analysis.

Here, we show a few examples of measurement of adsorption isotherms on adsorbates at various interfaces. The first isotherm was obtained by SHG on p-nitrobenzoic acid (p-NBA) adsorbed on fused silica from ethanol solution. The SHG intensity from p-NBA reflected from the silica side was measured as a function of p-NBA concentration (ρ) in ethanol. With the orientation of adsorbed p-NBA unchanged, the SFG intensity was proportional to the square of the surface density (N_S) of p-NBA on silica. The result of N_S versus ρ extracted from the experiment is presented in Fig. 6.6.[21] It displays the usual behavior of increase of N_S with increase of ρ, i.e., initial rapid increase followed by saturation. Adsorption energy of $\Delta G \sim -8$ Kcal/mol was deduced from the initial slope, assuming the adsorption process follows the simple Langmuir model.

To illustrate measurement of adsorption isotherm by SFVS, we take ethanol adsorption on silica from ethanol vapor as an example. Surface SF vibrational spectra in the CH stretch region were obtained SSP, SPS, and PPP polarizations from ethanol adsorbed on silica under different ethanol vapor pressures.[22] A representative set of SSP spectra is displayed in Fig. 6.7(a), from which the surface density of ethanol versus ethanol vapor pressure could be obtained. In this case, the orientation of adsorbed ethanol on silica actually changed with surface coverage of ethanol, as can be seen from the SSP, SPS, and PPP spectra of adsorbed ethanol under two different

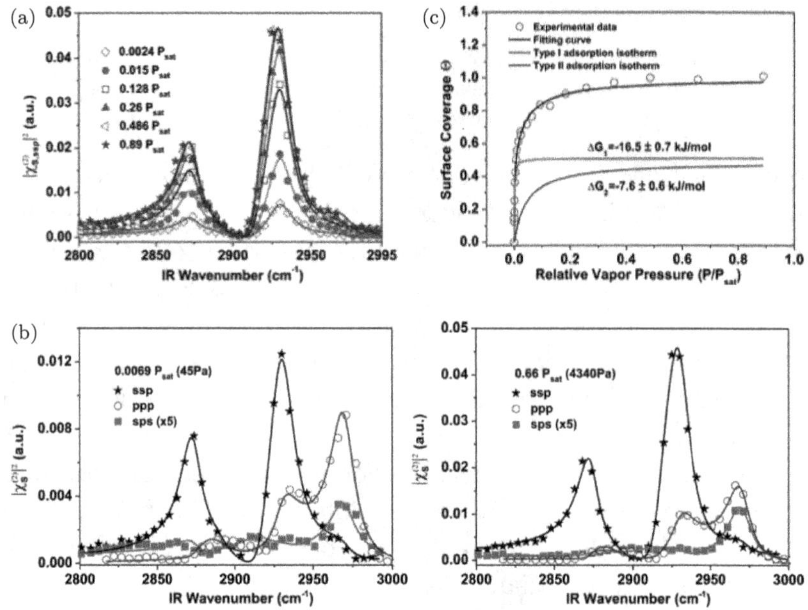

Fig. 6.7.   (a) SSP SF CH-stretch spectra of ethanol adsorbed at the vapor/silica interface from ethanol vapor with pressure ranging from $0.0024 P_{sat}$ to $0.89\ P_{sat}$. The saturation vapor pressure, P_{sat}, of ethanol at 22 °C is 6615 Pa. The solid curves are from fitting. (b) SSP, SPS, and PPP spectra of adsorbed ethanol under two different ethanol vapor pressures. Different peak ratios indicate that ethanol orientations in the two cases are different. (c) Adsorption isotherm of ethanol on silica deduced from the measured polarization-dependent SF spectra. The data can be fit by the simple Langmuir model with two adsorption sites, one with adsorption energy of $\Delta G = -15.5$ kJ/mol and a saturation coverage of $\theta_{sat} = 51\%$ and the other with $\Delta G = -7.6$ kJ/mol and $\theta_{sat} = 49\%$. The solid curves are the fitting curves (after Ref. [22]).

vapor pressures in Fig. 6.7(b); in addition to the spectral intensity change with vapor pressure, the spectral profile also varies because of change of ethanol orientation. Fortunately, the observed polarization dependence of the spectra allowed determination of the adsorbed ethanol orientation under different ethanol vapor pressures (Sec. 6.3). Knowing ethanol orientation, it was then possible to deduce the surface density of ethanol versus ethanol vapor pressure from the measured spectra, as plotted in Fig. 6.7(c). This adsorption isotherm could be fit by a simple Langmuir model with two adsorption sites specified by two different adsorption energies, $\Delta G \sim -7.6$ Kcal/mol and $\Delta G \sim -16.5$ Kcal/mol.

Molecular adsorption from a solution onto an interface not only depends on how the adsorbates interact with the interface, but to a larger extent depends on how they interact with solvent molecules. More highly soluble molecules obviously are less likely to adsorb at an interface. The interplay of molecular interactions with surrounding molecules and the interface determines whether one molecular species prefers to adsorb at an interface than the other. In fact, solutes adsorbed at an interface from solution can be considered as a competitive adsorption process with solute molecules replacing solvent molecules at the interface, and also as a co-adsorption process if solute molecules do not cover the whole interface. Competitive adsorption and co-adsorption are of interest to many disciplines because they directly influence interfacial properties and functions. Again, SFVS is an effective tool to probe these processes as we shall discuss in the next section.

6.5. Competitive Adsorption and Co-adsorption

Whenever there are two or more molecular species present around an interface, they would compete with one another to adsorb at the interface. Sometimes, one species wins and dominates the surface coverage; other times, two or more species coexist at the interface. Such competitive adsorption and co-adsorption processes occur ubiquitously in nature and in modern science and technology. Since SFVS can identify molecular species as well as their surface density and adsorption geometry at all interfaces accessible by light, it is uniquely suited to study these processes. We discuss here the cases of alcohols adsorbed from liquid mixtures on hydrophilic silica and hydrophobic silane-covered silica for illustration.[23]

Methanol, ethanol, propanol, and butanol (C1 to C4) can dissolve in water completely and form binary liquid mixtures from 0 to 100%. They also tend to coexist with water at an interface. At low concentration, alcohols can be considered as impurities in water that adsorb from water to an interface. During adsorption, they competitively replace water molecules at the interface, but their adsorption is not strong enough to completely replace water, resulting in co-adsorption with water at the interface. However, alcohols with longer chain lengths have more limited solubility in water, and can adsorb at an interface with full surface coverage even at low concentrations. We consider here only short-chain (C1 to C4) alcohols.[23] At the air/water interface, surface SFVS spectra show that alcohols preferentially emerge at the interface with their hydrophobic CH chains protruding out of water, and the adsorption strength increases from C1 to C4. Since alcohol molecules are amphiphilic with OH on one end and CH_3 on the other, it is interesting to know whether alcohols like to adsorb on hydrophobic or hydrophilic interfaces, or both.

Figures 6.8(a) and 6.8(b) show representative sets of SF CH stretch spectra for adsorbed C2-C4 alcohols adsorbed on hydrophilic fused silica from alcohol–water mixed solutions of different compositions.[23] (Deuterated water was used to avoid distortion of CH stretch spectra caused by overlapping OH stretch modes of water.) Spectra of adsorbed methanol from methanol–water mixtures could not be detected. It is believed that methanol molecules are adsorbed on silica in oppositely polar-oriented pairs with OH of one molecule hydrogen-bonded to silica. This appears to be also the case for ethanol adsorption from solution when ethanol concentration is sufficiently high. At low concentrations, the spectrum was found similar to those of adsorbed ethanol from vapor in Fig. 6.7(a), dominated by the symmetric CH_3 stretch (r^+) and Fermi resonance (FR-CH_3) at 2875 and 2930 cm^{-1}, respectively. At high concentrations, the r^+ mode decreased and the FR and r^- mode at 2975 cm^{-1} increased with increase of concentration as shown in Fig. 6.8(a). This observation can be understood knowing that ethanol adsorbs on silica in monolayer form at low concentrations, but with increasing concentrations, more ethanol molecules should adsorb with opposite orientation on the already present ethanol monolayer. An oppositely oriented ethanol pair, head to head (sketched in the inset of Fig. 6.8(a)) or twisted about the axis, has no inversion symmetry; their r^+ modes are still opposite in phase and cancel each other in SFVS, but the r^- modes actually enforce each other.[24] Thus, as more ethanol pairs appear on silica, the r^+ peak in the spectrum becomes weaker and the r^- peak stronger. In reality, the second adsorbed ethanol layer is not as ordered as the first layer, but the conclusion is still qualitatively true. The same situation exists in the cases of propanol and butanol adsorption on silica; because the molecules are longer, the second layer is more disordered and the spectral change with concentration was found to be less drastic. Such information about adsorbate arrangement and geometry cannot be obtained by other techniques.

It is seen from the sets of SSP and PPP spectra of ethanol in Fig. 6.8(a), the relative spectral profile remains unchanged with mixture composition, indicating that the orientation of adsorbed ethanol does not vary with surface density of ethanol. The surface coverage of ethanol for given bulk ethanol concentration in water can be deduced directly from the corresponding spectrum. The resulting adsorption isotherm is plotted on the right side of Fig. 6.8(a). The initial slope of the isotherm yields the adsorption free energy, $\Delta G \approx -3.7$ kcal/mol, implying that ethanol adsorbs more readily on silica than water. Similar measurements were made for butanol, 1-propanol, and 2-propanol adsorbed on silica from their mixed solutions with water. Their spectra and adsorption isotherms are given in Fig. 6.8(b); in all cases, ΔG was found to be roughly the same. Finally, it is interesting to note that spectra of adsorbed methanol could be detected at the interface of hydrophobic OTS(octadecyltrichlorosilane)-covered silica with methanol–water mixtures as displayed in Fig. 6.8(c).[23] The difference from the interface of hydrophilic silica with alcohol–water mixtures is that alcohol molecules should now adsorb at the interface with methyl groups pointing toward the interface. The adsorption isotherm yields an adsorption energy of $\Delta G = -7.1$ kJ/mol for methanol adsorption on OTS-covered silica, which is

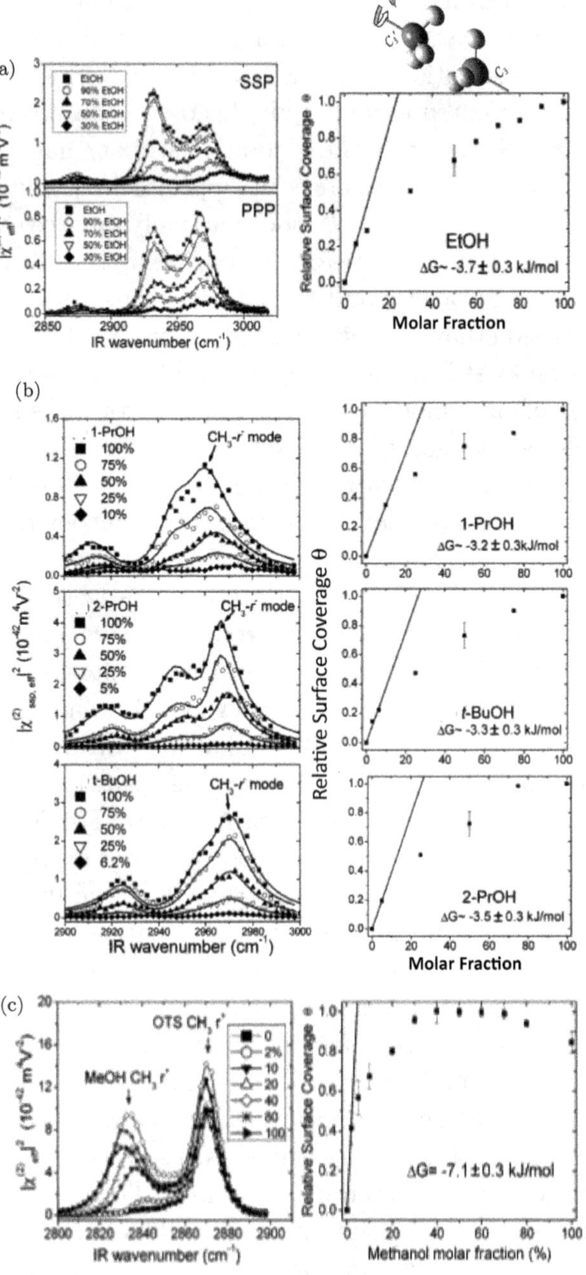

Fig. 6.8.   (*continued*)

Fig. 6.8. (*continued*) (a) A representative set of SF CH-stretch spectra, taken with SSP and PPP polarizations, of ethanol adsorbed at the silica/solution interface from ethanol-deuterated water mixtures of various compositions. The adsorption isotherm deduced from the spectra is displayed on the right. The solid curves are from fitting. The inset describes a head-to-head pair of methyl groups forming a dimer lack of inversion symmetry. (b) Similar sets of SSP SF spectra and associated isotherms for 1-propanol, 2-propanol, and butanol adsorbed at silica/solution interfaces. (c) A similar set of SSP SF spectra and associated isotherm for methanol adsorbed at the interfaces of OTS-covered silica immersed in methanol–water mixtures (after Ref. [23]).

almost 2 times larger than that for C2-C3 alcohol adsorption on bare silica. Study of adsorption of other alcohols on OTS-covered silica was not performed because of complication due to overlapping of OTS and alcohol spectra in the CH stretch range. Qualitative understanding of the above results was described in Ref. [23]. Briefly, three factors in consideration of free energy minimization govern the competitive adsorption process: interactions between molecules, interactions between molecules and substrate, and entropy. For short-chain alcohols and water, intermolecular interactions are dominated by hydrogen bonding; chain–chain interaction is relatively weak. Because there can be significantly more hydrogen bonds between water molecules than between water and alcohol, one finds that alcohol molecules tend to be pushed out from water assemblage, and preferentially adsorb at a water/substrate interface. Entropy change plays a secondary role, and should be nearly the same for different alcohols. Therefore, ΔG is expected to be roughly the same for different short-chain alcohols. Interaction strengths of water and alcohols with substrate contribute to the value of ΔG. On a silica surface, water interacts more strongly with silica via hydrogen-bonding than alcohols, but on OTS-covered silica, it is the reverse. Consequently, ΔG for alcohol adsorption from alcohol/water mixture is expected to be more negative for the latter case, as was indeed observed.

In the case of alcohol adsorption on OTS-covered silica, OTS is first adsorbed on silica and then alcohol molecules adsorbed on top of OTS. One may consider it as co-adsorption of OTS and alcohol on silica, but perhaps more correctly, sequential adsorption of OTS and alcohol. The OTS monolayer is highly packed; molecules adsorbed on it are not likely to induce conformational changes on OTS molecules. This is not true for more loosely packed monolayer, and SFVS is uniquely capable of probing such changes.[25] We take double-chain surfactant DOAC [dioctadecyldimethyl ammonium chloride, $(CH_3(CH_2)_{17})_2N^+(CH_3)_2Cl^-)$] monolayers on silica as an example. The chain density of a DOAC monolayer is half of that of the OTS monolayer. As a result, the chains have room for gauche defects as manifested by the appearance of the symmetric CH_2 stretch mode at ~ 2850 cm^{-1} in the top spectrum displayed in Fig. 6.9(a) for a DOAC monolayer on fused silica in air. When the monolayer was immersed in deuterated alkane liquid of increasing chain lengths from C10, C12, C14, to C18, the CH_2 mode became progressively weaker, and at C18, the chain conformation became all-trans (see Fig. 6.9(a)). This indicates that alkane molecules must have adsorbed on DOAC with their chains wedging in so that the increasing surface chain density or chain–chain interaction between alkane and DOAC effectively removed the gauche defects on the DOAC chains. Among the four alkanes, C18 has the longest chain length and provides strongest chain–chain interaction. When the DOAC monolayer was immersed in water, the CH stretch modes essentially all disappeared. They reappeared when the sample was taken out of water and dried. This indicates that the chains of DOAC being hydrophobic must have curled up in contact with water. It was however most interesting to note that when the DOAC monolayer was immersed in hexadecanol, the CH stretch modes also almost all disappeared (Fig. 6.9(b)), but if the monolayer was immersed in a solution of 0.1M deuterated hexadecanol in CCl$_4$, the alkyl chain spectrum of DOAC appeared with little gauche defects (Fig. 6.9(c)). The results suggest that hexadecanol molecules in pure hexadecanol must have formed an orientation-ordered, hydrogen-bonded array at the interface that prevented individual hexadecanol molecules from wedging into the chain array of the DOAC monolayer, while hexadecanol molelcules in diluted hexadecanol solution behaved just the opposite.

As described above, molecules adsorbed from a solution (or co-adsorbed with solvent molecules) at an interface may have their orientation and conformation changed with surface density. It is also possible that the change is drastic and phase-transition-like at some

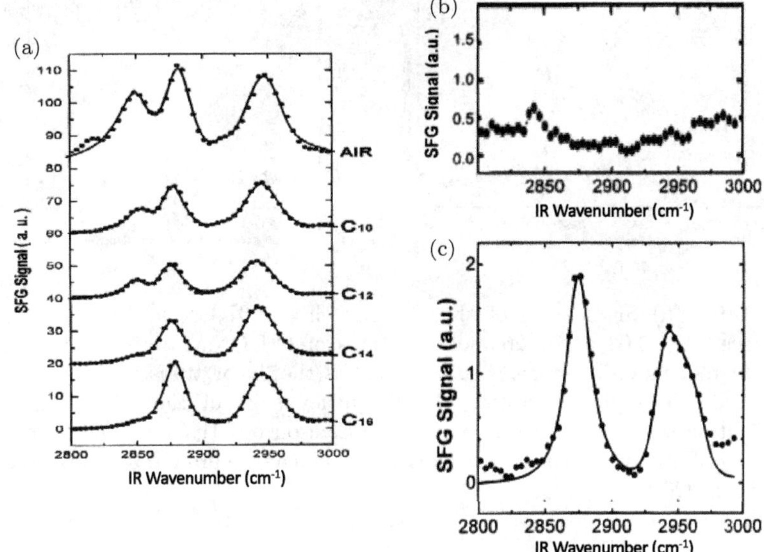

Fig. 6.9. SSP SF vibrational spectra of a DOAC monolayer: (a) at the silica/deuterated alkane interface with the alkane chain length indicated on each spectrum; (The spectrum for the DOAC monolayer at the silica/air interface is shown at the top for comparison.) (b) at the silica/deuterated hexadecanol (HD) interface; (c) at the interface of silica/solution of 0.5M HD in CCl_4. The alkyl chains of DOAC become all-trans in (a) if the alkane chain length is C16 or more due to strong chain–chain interaction. The same is true in (c) due to HD chains wedging into the DOAC chain array, but the DOAC chains curl up in (b) when in contact with pure HD liquid because HD molecules forming a hydrogen-bonded sheet can no longer have their chain wedge into DOAC (after Ref. [25]).

point. Such a case was reported by Eisenthal and coworkers on adsorption of acetonitrile (CH_3CN) from CCl_4 at the air/solution interface using SFVS.[26] They found that the CN stretch mode of CH_3CN shifted rapidly in frequency around 0.07 mole fraction in CCl_4 solution and became close to the one appearing in the bulk IR absorption spectrum, as presented in Fig. 6.10(a). The spectrum remained unchanged for CH_3CN above 0.07 mole fraction as the surface coverage of CH_3CN approached saturation. Analysis of the spectra revealed that the frequency shift was correlated with the sudden change of CN orientation tilt from 40° to 70° away from the surface normal (deduced from the measured SF polarization

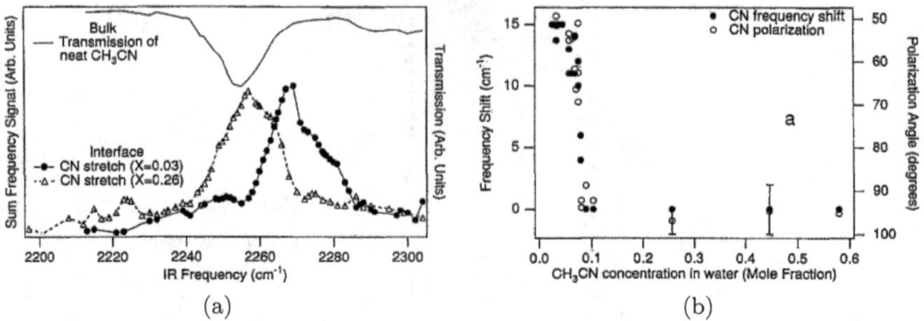

Fig. 6.10. (a) SF spectra of the CN stretch of CH_3CN at the air/solution interface with 0.03 and 0.26 mole fractions of CH_3CN in water, exhibiting a stretch frequency difference. The top trace is the linear transmission spectrum of CH_3CN in a bulk solution. (b) Frequency shift of the CN stretch and polarization of the SF output versus mole fraction of CH_3CN in water deduced from the SFVS measurement revealing a sudden change around 0.1 mole fraction (after Ref. [26]).

angle given in Fig. 6.10(b)). Such an orientation change is opposite to what one normally expects, i.e., adsorbed molecules ought to become more ordered and aligned as the surface density increases. The observation was interpreted qualitatively from the free energy minimization point of view: Neighboring adsorbed CH_3CN molecules oriented closer to the surface normal experience strong repulsive dipole–dipole interaction. If the surface density of CH_3CN is too high, the molecules may flip in orientation to reduce the dipole–dipole interaction, although it is not understood why the change is so sudden.

Molecular adsorption, including competitive adsorption and co-adsorption, at interfaces occurs ubiquitously and plays important roles in polymeric and biological systems, and SFVS can again serve as a unique tool for their investigation. We shall provide more details on the topics in Chapters 9 and 10. Electrochemistry is another area where molecular adsorption plays an essential role that we shall discuss more in Sec. 8.10.

6.6. Surfactant Monolayers

A molecular monolayer covering fully the surface of a material often completely changes the surface properties and functions of the material. This is particularly true for surfactant monolayers composed of amphiphilic long-chain molecules. Deposition of a long-chain surfactant layer on a hydrophilic surface can change the latter into hydrophobic, and vice versa. Such surfactant layers have found important applications in chemistry and material sciences. They resemble the structure of a half leaf of bio-membranes. IR-RAS and other optical techniques such as Brewster-angle microscopy and ellipsometry have been used to study surfactant monolayers, but SFVS has been proven to be able to provide more information.[27] To illustrate how SFVS can be employed to study surfactant monolayers, we use here the case of a pentadecanoic acid (PDA) monolayer on water as an example.[28]

Monolayers composed of insoluble organic molecules that typically constitute a hydrophilic head group and a long hydrophobic tail can be spread on water are generally labeled as Langmuir monolayers. For a Langmuir monolayer on water in a water trough, its surface pressure (or tension) can be measured and its surface density known from the number of molecules present in the measured surface area. With the number of molecules kept constant, the surface density of the monolayer can be varied by reducing the surface area or increasing the surface pressure. At a given temperature, the pressure-area $(\pi - A)$ plot is the 2D equivalent of the pressure-volume $(P - V)$ plot for a 3D system. An ideal $\pi - A$ curve is sketched in Fig. 6.11(a) and the measured one for PDA is given in Fig. 6.11(b). When A is sufficiently large (or π sufficiently low), the monolayer is in the 2D gas phase. As A reduces, the monolayer makes a phase transition to the liquid expanded (LE) phase by first going through a gas–LE coexistence region (not shown in Fig. 6.11(b)). Further reduction of A leads to another phase transition from the LE phase to a liquid condensed (LC) phase, again first going through an LE–LC coexistence region (characterized by constant π). Although the π-A diagram clearly displays the 2D phase transition behavior, no microscopic information on how the molecular structure of the monolayer varies along the π-A curve had been available before SFVS stepped in.

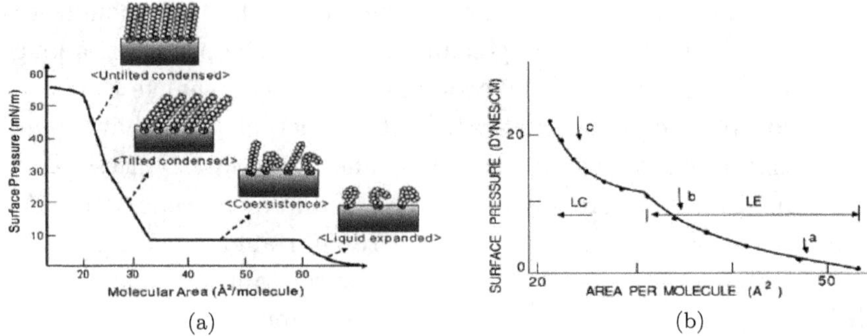

Fig. 6.11. (a) An idealized π-A isotherm for a Langmuir monolayer going through various 2D phases. The cartoons describe chain conformations in different phase regions (after Ref. [27]). (b) The measured π-A isotherm for a pentadecanoic acid Langmuir monolayer (after Ref. [28]).

As one would expect, the chain conformation of surfactant molecules could change with the surface packing density, or more specifically, with the chain density. Figure 6.12 shows the CH stretch spectra of a PDA Langmuir monolayer taken by SFVS with SSP and SPS polarization combinations.[28] The three sets of spectra are for surface areas of $A = 0.47$, 0.34 and 0.22 nm^2/molecule; the first two are in the LE phase and the last one in the LC phase. The spectra for $A = 0.22$ nm^2/molecule are typical of all-trans alkyl chains; the prominent peaks in the spectra are from the terminal methyl groups: the symmetric CH$_3$ stretch and its Fermi resonance with the bending mode at 2875 and 2940 cm^{-1} and the antisymmetric CH$_3$ stretch at 2960 cm^{-1}. For $A = 0.34$ and 0.47 nm^2/molecule, the symmetric and antisymmetric CH$_2$ stretch modes at 2850 and 2920 cm^{-1} stand out due to appearance of gauche defects on the chains. Thus, the LC and LE phases are distinguished not only by their surface chain densities but also by their resultant chain conformations. Presumably, at sufficiently large surface molecular density, chain–chain interactions between molecules would self-assemble surfactant molecules into LC domains with all-trans alkyl chains. The chain conformation is directly related to the chain density; as seen from the SF spectra for a double-chain Langmuir monolayer, DPPC (1,2-dipalmitoyl-sn-glydero-3-phosphocholine) in Fig. 6.12(c), the alkyl chains of DPPC also appear all-trans at a surface density of ≥ 0.20 nm^2/chain.[29]

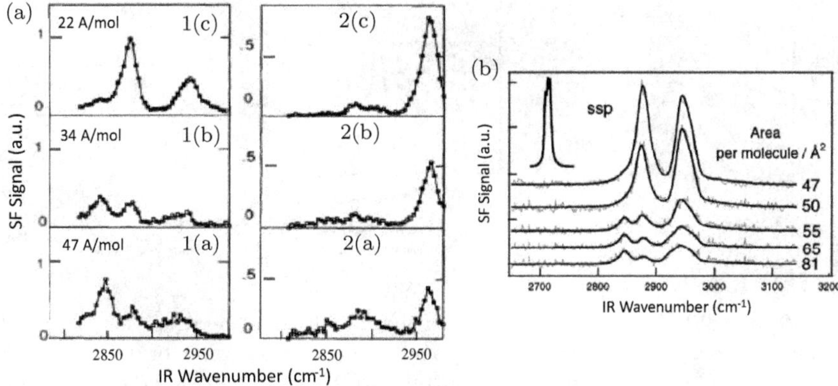

Fig. 6.12. (a) SF vibrational spectra of pentadecanoic acid (PDA) Langmuir monolayers of different surface coverages, marked by arrows a,b,c, on the π-A isotherm in Fig. 6.11(b), revealing conformational changes of the alkyl chains of PDA from all-trans in the LC phase to increasing Gauche defects with lower surface density in the LE phase. 1(a)–1(c) and 2(a)–2(c) were taken with SSP and SPS polarization combinations, respectively (after Ref. [28]). (b) SF vibrational spectra of DPPC at different surface coverages, showing that when the area per chain reaches $25/m^2$ (50 molecules/m^2), the chains become all-trans (after Ref. [29]).

The hydrophilic headgroups of Langmuir monolayers can also be probed by SFVS. Most relevant Langnuir monolayers have acidic headgroups. When immersed in water, the headgroups can be protonated or deprotonated by adjusting pH of water solution, and can be monitored by the vibrational spectrum of the headgroups. We take the SFVS study of a palmitic acid ($C_{15}H_{31}COOH$) monolayer on water as an example.[30] The COOH headgroups in water generally are partially protonated into COO^-. The stretch vibrations of C $-$OH and C $=$O of COOH at 1338 cm^{-1} and 1720 cm^{-1}, respectively, and that of COO^- at 1410 cm^{-1} at different pH of water can be observed, as displayed in Fig. 6.13. With Na$^+$ or K$^+$ present in water, another species, COO^--K^+, would appear at the interface with the stretch vibration of COO^- shifted to 1475 cm^{-1}, or in the case of $COO^-$$-Na^+$, shifted to 1434 cm^{-1}. Fitting of the observed spectra allowed deduction of surface densities of the different species and the charge state of the monolayer. The charge state is important because it governs the interfacial water structure underneath the monolayer and

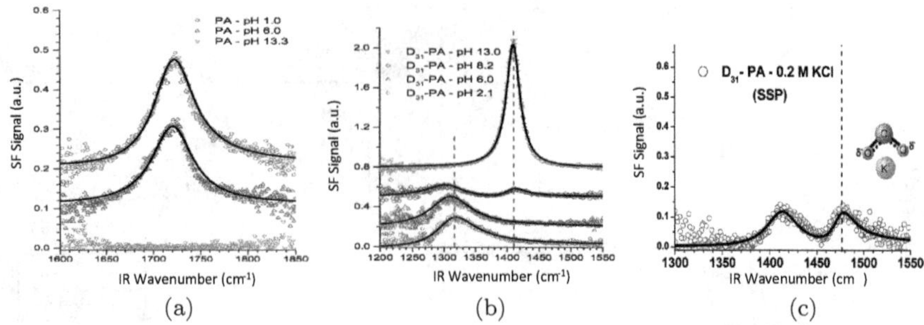

Fig. 6.13. SSP SF vibrational spectra in the CO stretch region for the head group of a palmitic (C15H31COOH) acid monolayer on water of different bulk pH: (a) C =O stretch mode of the neutral COOH head group at 1720 cm^{-1}, and (b) C-OH and COO- stretch modes at 1318 and 1410 cm^{-1}, respectively (after C.Y. Tang, Ph.D. Thesis, Department of Chemistry, The Ohio State University, 2010). (c) COO-K mode at 1475 cm^{-1} appearing together with the COO^{-} mode at 1410 cm^{-1} in the presence of KCl in water (after Ref. [30]).

dictates the interfacial properties and functions, as we shall see later in more details in Secs. 8.6, 8.7, and 8.9. It also affects the monolayer structure itself; for example, the π-A curve can be affected by charge repulsion of headgroups. The above discussion shows that SFVS can help paint a more complete picture of Langmuir monolayers on water.

We note in passing that SFVS can also be employed to study Langmuir–Blodgett monolayers, which are surfactant monolayers grafted onto solid substrates from Langmuir monolayers on water. The grafting process may modify (or deteriorate the quality of) the monolayer structure and SFVS can monitor the changes.

6.7. Surface Reactions

Generally speaking, there are two types of surface reactions: reactions through interaction of adsobates with surface and reactions through interaction of adsorbates with adsorbates with surface as an intermediate. Atomic and molecular adsorption on a substrate discussed in the previous section can be regarded as a surface reaction of the first type; research interest is on learning how atoms/molecules adsorb, desorb, and possibly dissociate/associate on a surface. The difficulty often encountered is that the surface structure is not well defined. This prompted the early development of surface science on crystalline surfaces in ultrahigh vacuum (UHV). In the second case, the main interest is on learning how particular surfaces can enhance or deter reactions between adsorbates. Again, for real understanding of reaction mechanism, we need to know microscopically how adsorbates situate on surfaces, and often studies of samples in UHV are imperative. With a plethora of techniques developed for surface studies in UHV,[1] there has been tremendous progress in understanding of surface reactions over the past half century. However, in many practically relevant cases, a material surface is exposed to ambient atmosphere, and its structure and properties are most likely not the same as in UHV considering that the pressure difference of the two is over 12 orders of magnitude. To bridge this so-called "pressure gap", it is important that we have viable tools capable of probing microscopic structures and properties of surfaces in real atmosphere. Optical techniques, particularly IRRAS and SF spectroscopy, partly satisfy the need, but based only on spectroscopic information, there is still difficulty to ascertain the structure of a surface on the atomic scale. Indeed, uncertainty in the microscopic structure of a surface in real environment is an unsettling problem. Techniques like X-ray scattering and spectroscopy that can provide such information *in situ* are not yet available. In IRRAS or SFS study of surface reactions in real environments, we normally have to trust the result as long as it is reproducible, and assume a reasonable surface structure to interpret the result. There has been effort to prepare a surface in a more trustworthy way: The surface is first cleaned and well characterized in UHV, and then the chamber is transformed into ambient conditions for adsorption and surface reaction measurements. It is assumed that the surface structure does not change in the transfer process. This assumption usually cannot be verified. The only justification is that after the measurements, when the sample is transferred back to UHV with all adsorbates desorbed, the surface structure is found to have returned to the original one.

In recent years, SF spectroscopy has been adopted to study molecular adsorption and surface reactions on model catalyst surfaces in UHV-compatible high-pressure chambers.[31] Most works have

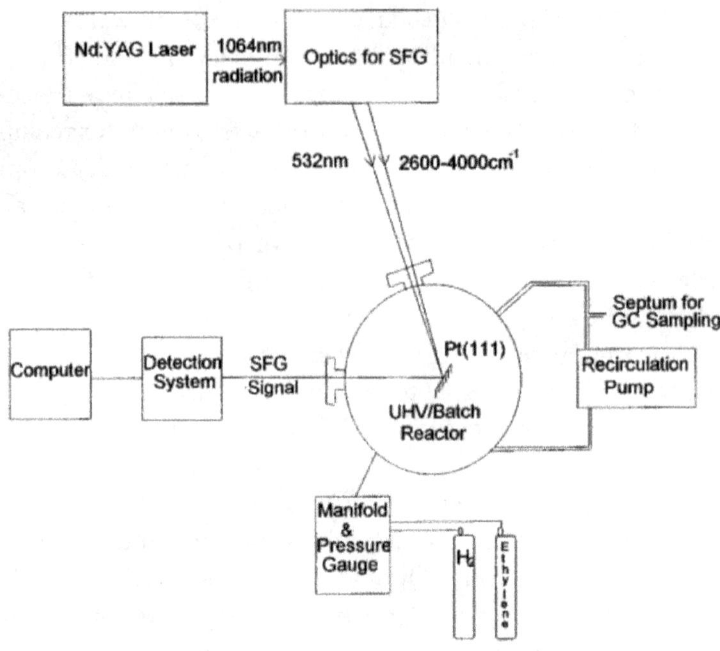

Fig. 6.14.   A UHV-compatible high-pressure chamber that allows simultaneous *in situ* SFVS and gas chromatographic measurements for surface characterization and probing of catalytic reactions over a wide range of gas pressure (after Ref. [31]).

been on small molecules adsorbed on, first, crystalline surfaces such as Pt(111), Rh(111), Pd(111), Pt(110), Pd(100) etc., and then, supported nano-metallic particles.[32] Focus is on adsorption sites, structures, geometries, surface reactions, and their dependences on temperature and gas pressure. Results are successful in providing information that cannot be expected from UHV studies and in narrowing the long-standing "pressure gap".

We take CO adsorption and oxidation on metal surfaces as an example. A representative experimental setup with a UHV-compatible high-pressure chamber is described in Fig. 6.14.[31] With the chamber evacuated to UHV, the sample can have its surface cleaned by argon bombardment and annealing, and characterized by Auger spectroscopy and LEED (low energy electron diffraction). The gas inlet then allows the chamber to be filled with the desired atmosphere, and the gas outlet is connected to a gas chromatograph for analysis of gas composition in the chamber. Optical windows on the chamber provide access for optical spectroscopy measurements.

Figure 6.15(a) presents the PPP SF spectra of a stretch mode of CO adsorbed on Pt(111) with different surface coverages prepared under different CO vapor pressures.[33] (Because the transmission Fresnel factor for S-polarized light is close to zero, SF output from SSP, SPS, and PSS polarization combination is very weak.) The sharp mode around 2090 cm^{-1} can be attributed to CO adsorbed on top of Pt. Its frequency is lower than that of the gaseous CO (2143 cm^{-1}) due to CO interaction with Pt, and increases with surface coverage from <2081 to 2097 cm^{-1} due to increasing CO–CO interaction. In comparison with Pt(111) in UHV, no new adsorbed CO species was detected. In an earlier study,[34] however, the observed spectral change with pressure was different. As seen in Fig. 6.15(b), a much weaker mode at 1845 cm^{-1} attributed to CO at bridge sites of Pt(111) appears at low pressures, and a broadband below the sharp peak grows increasingly stronger as pressure increases. It was believed that the surface had appreciable defects, and the broadband arose from CO adsorbs at defect sites. The spectrum was reversible when the gas pressure was reduced back to UHV. Similar results were obtained for CO adsorption on Rh(111) as seen in Fig. 6.15(c);[35] (a side band of the sharp peak also appears and grows stronger at higher pressure). In this case, if the sample was heated above 600K and CO pressure above 1 mbar, the spectrum became irreversible as adsorbed CO underwent the reaction $2CO \rightarrow C + CO_2$ and left C on the Rh surface. For CO adsorption on the (111) surface of other metals, for example, CO on Pd(111) in Fig. 6.15(d),[36] the global features of the spectral changes with gas pressure and sample temperature are similar, but details are different, including the intensity ratio of the CO stretch modes at top and bridge sites.

Nano-metallic particles or clusters are technologically more relevant because they are catalysts in real catalytic processes. SFVS is capable of probing CO adsorption on supported nanoparticles although their exposed facets are hard to define. Figure 6.16 is an example showing the stretch spectra of adsorbed CO on Pd

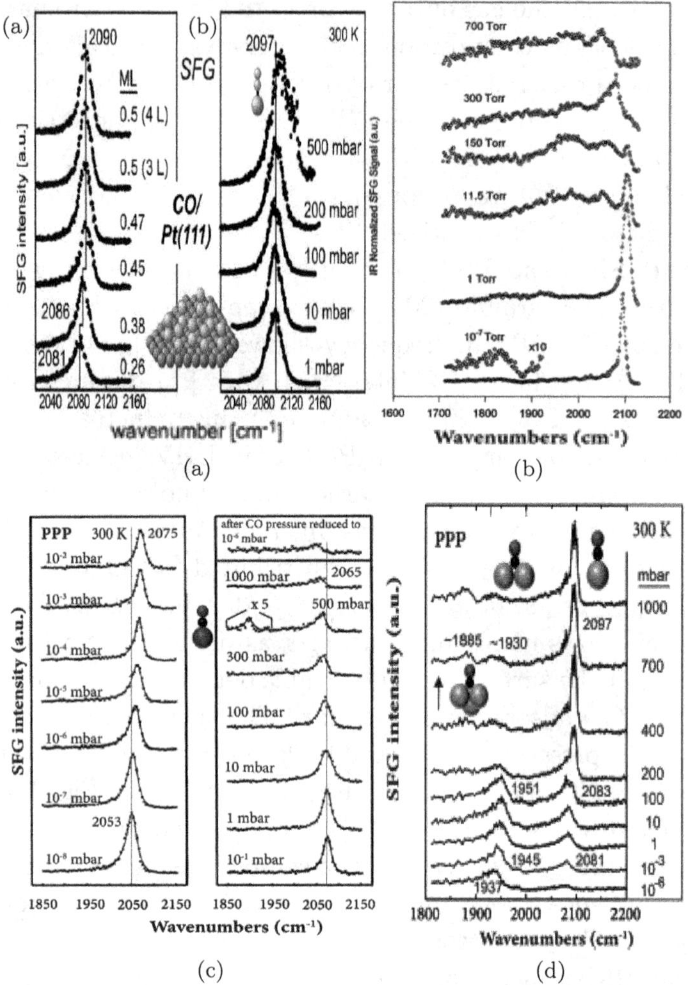

Fig. 6.15. (a) PPP SF vibrational spectra of CO on Pt(111) at 300K: left panels for submonolayer CO coverage in UHV and right panels for coverage under CO gas pressure range of 1 to 500 mbar. The single peak originates from CO sitting on top of Pt as sketched in the inset (after Ref. [33]). (b) A similar set of PPP SF spectra of CO on Pt(111) with surface defects (after Ref. [34]). (c) and (d) Similar sets of PPP SF spectra of CO on Rh(111) and Pd(111), respectively. The atop-CO mode dominates in all cases (after Refs. [35, 36]).

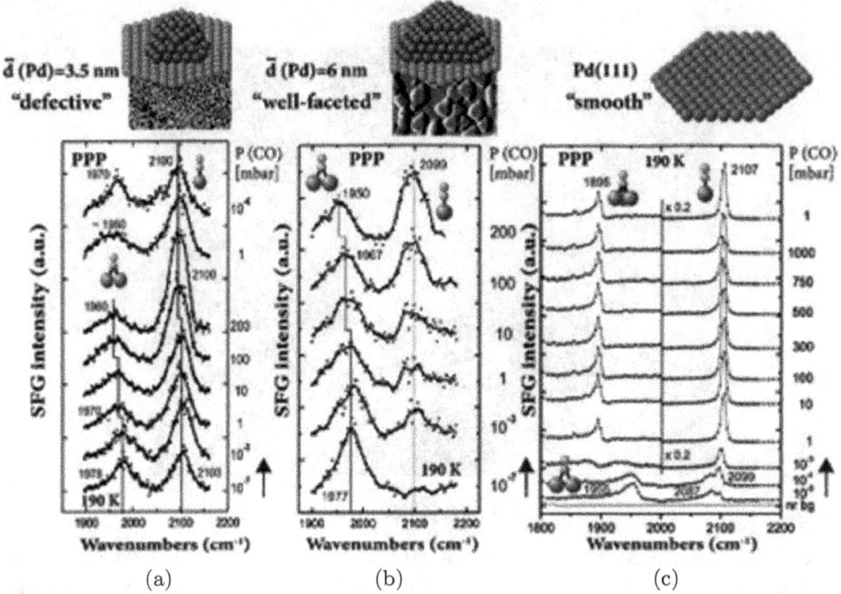

Fig. 6.16. PPP SF vibrational spectra at 190K of CO on (a) 3.5 nm and (b) 6 nm Pd nanoparticles supported by Al_2O_3/NiAl(110), and (c) crystalline Pd(111) under a wide range of gas pressures. The spectra of (a) and (b) are dominated by CO at top and bridge sites, but the spectra of (c) are dominated by CO at top and hollow sites for pressure >1 mbar (after Ref. [36]).

nanoparticles of two different sizes under different pressures (10^{-7}– 200 mbar) in comparison with those of CO on crystalline Pd(111), all taken at 190K.[36] The smaller nanoparticles were more defective and the larger nanoparticles were well faceted. In both cases, the stretch modes of CO at top and bridge sites stand out, but the two spectra and their variations with gas pressure are appreciably different. The spectra of CO on crystalline Pd(111) are even more different. In addition to the much sharper CO stretch at top sites, the bridge-site mode disappears at pressures >1 mbar and a new mode at hollow sites appears. This example illustrates that SFVS can be an effective tool to study supported nanoparticle catalysts.

Adsorbates on metals can form adsorbate/metal surface complexes with hybridized surface electronic states that dominate the surface electronic properties. The case of CO on Pt(111) is again

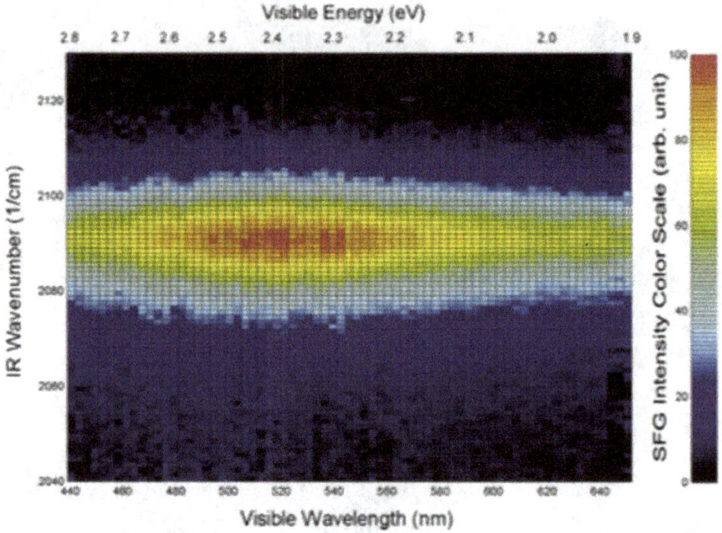

Fig. 6.17. Measured SF intensity for CO on Pt(111) as a function of IR and visible frequencies displaying a double resonance when the IR frequency is around the stretch resonance of atop CO at $\sim$2090 cm^{-1} and the visible frequency is around the electronic transition between the charge transfer states of atop CO/Pt(111) at $\sim$2.5 eV (after Ref. [37]).

taken here as an example.[37] Electronic transitions from occupied to unoccupied surface states can in principle be probed by optical spectroscopy, but the measurement is challenging because the signal is weak. With SF spectroscopy, however, it is possible to have the signal enhanced by electronic-vibrational double resonance. This is because the hybridized surface electronic states are originated from adsorbed CO and must be strongly coupled to the CO vibration, thus galvanizing the double resonance. Figure 6.17 shows the SF spectral distribution of such a resonance for CO on Pt(111); the SF signal from excitation of the CO stretch mode is seen to be resonantly enhanced when the visible input frequency approaches the transition from the highest occupied to the lowest unoccupied surface charge transfer state at $\sim$2.5 eV.

Surface catalysis speeds up chemical reactions with the surface acting as a promoter, and this is the backbone of many industrial chemical processes. In the past half century, it has been extensively investigated in UHV. The result unfortunately is not necessary the same as that in real environments because of the large pressure gap between the two. Study of catalysis in real environments at the microscopic level is desired. Optical techniques such as SFVS and infrared RAS can partly fill the need.

We use the case of CO oxidation on Pt(111) to illustrate how SFVS can be effective as a probe to monitor a catalytic process. In the experiment of Ref. [38], the Pt(111) surface was first well prepared in UHV (Fig. 6.14), and a gas mixture of CO, O_2, and He with selected partial pressures was then introduced into the chamber and continuously circulated in the system by a recirculation pump; the He gas was there to help the sample reach thermal equilibrium in the chamber. A gas chromatograph was used to sample the gas composition in the chamber through a septum. As seen in Fig. 6.18(a), the CO stretch spectrum from SFVS on CO/Pt(111) under an atmosphere of 100 torr of CO, 40 torr of O_2, and 600 torr of He is essentially the same as that without O_2, and He in the atmosphere if the sample temperature is not sufficiently high so that the CO oxidation reaction $(2CO + O_2 \rightarrow CO_2)$ rate is still very low. With increase of temperature beyond 650K, the CO oxidation rate picks up, as seen from the measured CO to CO_2 turnover rate (TOR); at the same time, the sharp peak of atop CO stretch in the spectrum reduces in strength and a broadband at its low frequency side emerges from CO adsorbed at defect sites on Pt(111). When the temperature reaches the ignition point, i.e., the reaction becomes self-sustained, the atop-CO peak disappears and the broadband becomes stronger. The surface CO coverage on Pt at a constant temperature could be adjusted by changing the CO/O_2 ratio in the gas mixture. Figures 6.18(b) and 6.18(c) show that the TOR varies inversely with the atop CO surface coverage, but increases linearly with the defective CO surface coverage. While the atop CO species on Pt(111) apparently dominate in UHV studies, the above results indicate that they are actually inhibitors for the catalytic CO oxidation reaction. Instead, it is CO adsorbed at defect sites mainly responsible for the reaction. The result of SF spectroscopy study on CO oxidation on Rh(111) is qualitatively similar, but quantitatively different, as described in Fig. 6.18(d).[35]

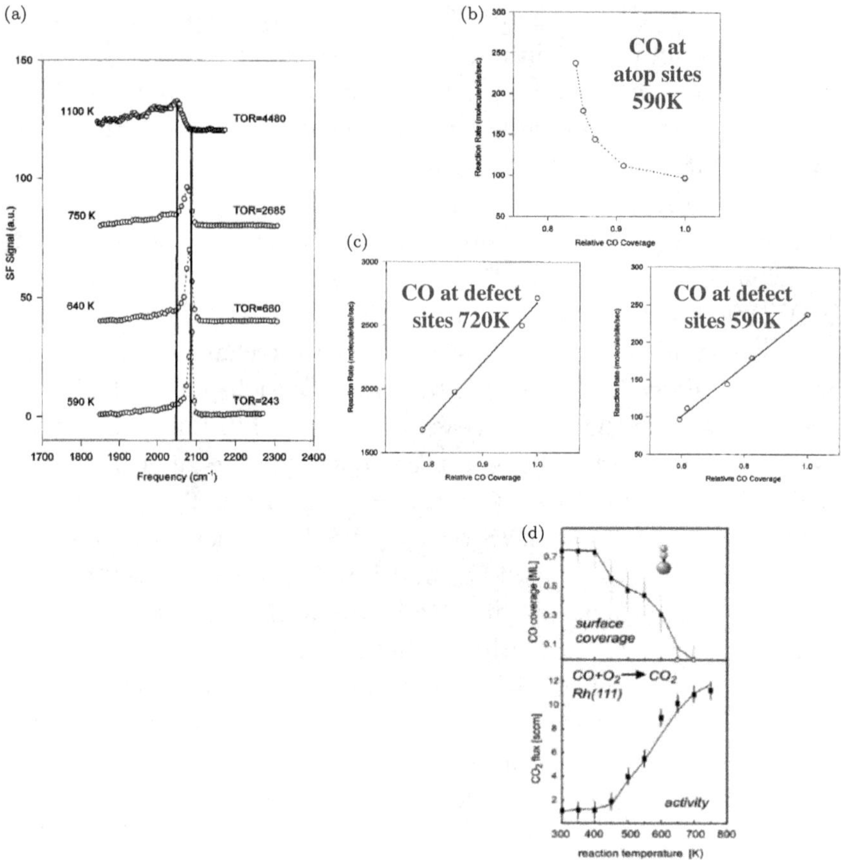

Fig. 6.18. (a) Temperature dependence of *in situ* PPP SF spectra of CO on Pt(111) during CO oxidation reaction under 40 torr of O_2/100 torr of CO/600 torr of He. TOR refers to turn-over rate of CO to CO_2. The sharp peak at 2090 cm^{-1} and the broadband peaked at 2050 cm^{-1} are attributed to CO at top sites and defect sites, respectively. The reaction became self-sustained above 760K, which is characterized by the disappearance of the sharp peak and prominence of the broadband. (b) TOR versus surface coverage of atop CO deduced from the spectra obtained under different relative CO/O_2 partial pressures, showing that as TOR increases, the atop CO coverage decreases. (c) TOR versus surface coverage of CO on defective sites at two different temperatures, showing that TOR is linearly proportional to the coverage (after Ref. [38]). (d) Similar results were found in SFVS studies of CO oxidation on Rh(111): Surface coverage of atop CO decreases and TOR increases as temperature increases (after Ref. [35]).

Knowledge about reaction pathway and possible existence of intermediate species is important for understanding a catalytic process and can be acquired by SFVS. We take hydrogenation of ethylene to ethane ($C_2H_4 + H_2 \rightarrow C_2H_6$) via reaction on Pt(111) as an example.[39] In UHV, as sketched in the top inset of Fig. 6.19, ethylene molecules are π-bonded to Pt at temperatures below 52K, but change to di-σ-bonded above 52K and lose an H to become ethylidyne (C_2H_3) above 240K (but below 500K). The observed SF CH stretch spectrum of a monolayer of ethylidyne on Pt(111) is given in Fig. 6.19(a). The study of hydrogenation of ethylene or ethylidyne was carried out at 295K with an atmosphere of 35 torr of C_2H_4, 100 torr of H_2, and 615 torr of Ar in a chamber like the one in Fig. 6.14. The gas chromatograph analyzing the gas mixture in the chamber detected ethane (C_6H_6) as a reaction product with a TOR of 11 molecules per surface Pt atom. Accordingly, the SF spectrum changed to the one shown in Fig. 6.19(b) that reveals two additional modes belonging to π-bonded and di-σ-bonded ethylenes. It was of interest to learn if one of them or both served as the intermediate species for ethylene hydrogenation. In the above experiment, if an ethylidyne monolayer was pre-deposited on Pt(111) before the gas mixture was introduced, the TOR remained nearly the same and the observed spectrum (Fig. 6.19(c)) was similar to that of Fig. 6.19(b) except that the di-σ-bonded ethylene mode was significantly weaker. This indicates that di-σ-bonded ethylene on Pt has little to do with hydrogenation of ethylene, but π-bonded ethylene is the intermediate species in the hydrogenation process, although in an earlier study, the reverse was believed to be true.

To simulate practical catalytic processes, supported nano-metallic particles should be used. Somorjai and coworkers replaced the crystalline sample by supported nano-metallic particles in the UHV-compatible high-pressure chamber for operando studies. Using SFVS, they observed again, during ethylene hydrogenation, the presence of ethylidyne and di-σ-bonded ethylene on Pt nanoparticles, suggesting that the catalytic process was similar to that occurring on Pt(111).[40] They also interrogated the effect of oxide supports on reactivity and found that during furfuryl hydrogenation

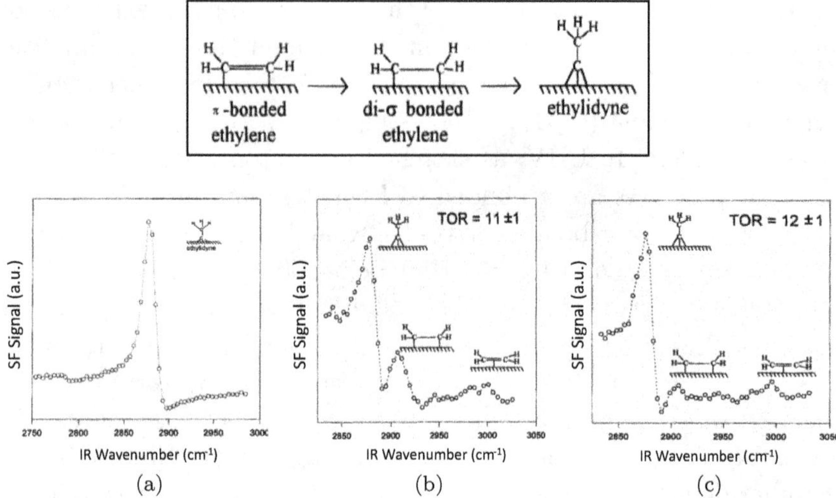

Fig. 6.19. Top frame: Schematic representation showing thermal evolution of ethylene on a metal surface in UHV. As temperature increases, ethylene dehydrogenation proceeds from π-bonded ethylene at low temperature, to di-σ-bonded ethylene at intermediate temperature, to ethylidyne at high temperature. Bottom row: PPP SF vibrational spectra of (a) a full monolayer of ethylidyne on Pt(111) at 300K in UHV, (b) species appearing on an initially clean Pt(111) at 295K during ethylene hydrogenation under an atmosphere of 35 Torr of C2H4, 100 Torr of H2, and 615 Torr of Ar, and (c) species appearing on Pt(111) during hydrogenation under the same condition as in (b), but the Pt(111) surface was originally dosed with a saturated coverage of ethylene. While the turn-over rates (TOR) in (b) and (c) are nearly the same, the di-σ-bonded ethylene mode is significantly weaker in (c) indicating that π-bonded ethylene is the intermediate species in the reaction (after Ref. [39]).

($C_5H_4O_2 + 2H_2 \rightarrow C_5H_6OH$), Pt nanoparticles on TiO_2 were more active as catalysts than on SiO_2 because the nanoparticles had activated oxygen vacancy centers on the surface of TiO_2.[41] The nanoparticles, however, are easily contaminated or structurally distorted, and cannot be well characterized during a catalytic process.

Catalytic reactions on oxides in real atmosphere are also of great interest for practical applications. Photo-catalytic reactions of water splitting and methanol dissociation on TiO_2 are good examples. Again, SFVS is an effective tool to identify and monitor surface species on oxides during reactions. Surfaces of TiO_2 are known to be reactive and able to dissociate methanol and split water molecules under UV irradiation; surface oxygen vacancies were believed to be the activity centers. In a SFVS study on anatase $TiO_2(101)$, the surface phonon spectrum (Sec. 7.2B) of anatase (mainly the stretch between surface Ti and an O underneath) was detected and used to track evolution of surface oxygen vacancy under the influence of UV irradiation and different adsorbed species.[42] In a separate study on different facets of rutile and anatase TiO_2, it was found that methanol and water adsorbed on TiO_2 in the molecular form with little dissociation, but in the presence of defects, methanol dissociation was observed with methoxy left on surface.[43]

SFVS study of catalytic processes at liquid–solid interfaces have been attempted with a flow cell that allows reaction products to be simultaneously measured during spectroscopy measurements.[44] As in the vapor/solid case, the deficiency of such measurements is also in their inability to well characterize a solid surface exposed to a real environment. Electrochemical reactions at water interfaces have been studied by SFVS and will be discussed in Sec. 8.10.

6.8. Summary and Prospects

This chapter describes how SF spectroscopy, mainly SFVS, can be used to probe molecular adsorption at interfaces and characterize adsorbates, with representative examples given for illustration. Topics covered are summarized as follows:

- Both IR reflection absorption spectroscopy (IRRAS) and SFVS can be used to probe molecular adsorption at interfaces with sub-monolayer sensitivity, but SFVS is more flexible and informative.
- Both IRRAS and SFVS can identify adsorbates and their conformation through their vibrational spectra, but the two techniques follow different selection rules.
- Polarization-dependent SFVS measurement of independent elements of SF nonlinear susceptibility of adsorbates provides information on average orientation of adsorbates.

- Surface coverage of adsorbates at an interface can be deduced from their SF vibrational spectra and, accordingly, the corresponding adsorption isotherm can be measured.
- Competitive adsorption and co-adsorption at interfaces can be studied by SFVS, yielding information on how intermolecular interactions affect adsorption kinetics as well as orientation and conformation of adsorbates.
- SFVS is most effective to be used to interrogate molecular structures of molecular monolayers on water (Langmuir films) and their changes with environment.
- With samples in a UHV-compatible high-pressure chamber, SFVS can be adopted to probe molecular adsorption on well prepared metal crystalline surfaces over a wide range of gas pressures, revealing possible differences between adsorptions in UHV and under high gas pressure. Similar studies on supported nano-metallic particles are possible.
- Catalytic reactions on crystalline metal surfaces and supported nano-metallic particles over a wide range of gas pressures and compositions can be monitored by SFVS, bridging the so-called pressure gap between catalytic studies under UHV and real atmosphere. SFVS study of catalytic reactions at liquid water interfaces is also possible.

SFVS is now a well-accepted tool to probe molecular adsorption and surface reactions at an interface. What could still be further developed is its capability for time-resolved measurement and possible microscopic imaging. So far, studies have been carried out on essentially all model systems. It would be most exciting to see if the technique could be applied to practical problems.

References

1. Somorjai, G. A.; Li, Y.: *Introduction to Surface Chemistry and Catalysis.* 2nd ed. J. Wiley: New York, 2010.
2. McIntyre, J. D. E.; Aspnes, D. E.: Differential Reflection Spectroscopy of Very Thin Surface Films. *Surf. Sci.* 1971, 24, 417–434.
3. Hoffman, F. M.: Infrared Reflection-Absorption Spectroscopy of Adsorbed Molecule. *Surf. Sci. Rep.* 1983, 3, 109–192.

4. Chabal, Y. J.: Surface Infrared Spectroscopy. *Surf. Sci. Rep.* 1988, 8, 211–357.

5. Jackson, J. D.: *Classical Electrodynamics*. 3rd ed. J. Wiley: New York, 1999.

6. Chen, C. K.; de Castro, A. R. B.; Shen, Y. R.: Surface-Enchanced Second Harmonic Generation. *Phys. Rev. Lett.* 1981, 46, 145–148.

7. Chen, C. K.; Heinz, T. F.; Ricard, D.; Shen, Y. R.: Detection of Molecular Monolayers by Optical Second Harmonic Generation. *Phys. Rev. Lett.* 1981, 46, 1010–1012.

8. Tom, H. W. K.; Heinz, T. F.; Shen, Y. R.: Studies of Molecular Adsorbates at Interfaces by Optical Second Harmonic Generation. *Laser Chem.* 1983, 3, 279–292.

9. Heinz, T. F.; Tom, H. W. K.; Shen, Y. R.: Determination of Molecular-Orientation of Monolayer Adsorbates by Optical Second Harmonic Generation. *Phys. Rev. A* 1983, 28, 1883–1885.

10. Tom, H. W. K.; Mate, C. M.; Zhu, X. D.; Crowell, J. E.; Heinz, T. F.; Somorjai, G. A.; Shen, Y. R.: Surface Studies by Optical Second Harmonic Generation — the Adsorption of O_2, CO, and Sodium on the Rh(111) Surface. *Phys. Rev. Lett.* 1984, 52, 348–351.

11. Tom, H. W. K.; Zhu, X. D.; Shen, Y. R.; Somorjai, G. A.: Investigation of the Si(111)-(7x7) Surface by Second Harmonic Generation — Oxidation and the Effects of Surface Phosphorus. *Surf. Sci.* 1986, 167, 167–176.

12. Heinz, T. F.; Chen, C. K.; Ricard, D.; Shen, Y. R.: Spectroscopy of Molecular Monolayers by Resonant Second Harmonic Generation. *Phys. Rev. Lett.* 1982, 48, 478–481.

13. Rupprechter, G.: Sum Frequency Generation and Polarization-Modulation Infrared Reflection Absorption Spectroscopy of Functioning Model Catalysts from Ultrahigh Vacuum to Ambient Pressure. *Adv. Catal.* 2007, 51, 133–263.

14. Shrestha, M.; Luo, M.; Li, Y.; Xiang, B.; Xiong, W.; Grassian, V. H.: Let There Be Light: Stability of Palmitic Acid Monolayers at the Air/Salt Water Interface in the Presence and Absence of Simulated Solar Light and a Photosensitizer. *Chem. Sci.* 2018, 9, 5716–5723.

15. Tyrode, E.; Liljeblad, F. D.: Water Structure Neat to Ordered and Disordered Hydrophobic Silane Monolayers: A vibrational Sum Frequency Spectroscopy Study. *J. Phys. Chem. C* 2013, 117, 1780–1790.

16. Tom, H. W. K.: Studies of Surfaces Using Optical Second Harmonic Generation. Ph.D. Thesis, University of California at Berkeley, 1984.

17. Balajka, J.; Hines, M. A.; DeBendetti, W. J. I.; Komora, M.; Pavelee, J.; Schmid, M.; Deibold, U.: High-Affinity Adsorption Leads to

Molecularly Ordered Interfaces on TiO_2 in Air and Solution. *Science* 2018, 361, 786–789.

18. Liu, W. T.: (Private coommunications).

19. Zhuang, X.; Miranda, P. B.; Kim, D.; Shen, Y. R.: Mapping Molecular Orientation and Conformation at Interfaces by Surface Nonlinear Optics. *Phys. Rev. B* 1999, 59, 12632–12640.

20. Chen, W.; Feller, M. B.; Shen, Y. R.: Investigation of Anisotropic Molecular Orientational Distributions of Liquid-Crystal Monolayers by Optical Second Harmonic Generation. *Phys. Rev. Lett.* 1989, 63, 2665–2668.

21. Heinz, T. F.; Tom, H. W. K.; Shen, Y. R.: Determination of Molecular-Orientation of Monolayer Adsorbates by Optical Second Harmonic Generation. *Phys. Rev. A* 1983, 28, 1883–1885.

22. Xu, H; Zhang, D.; Tian, C.; Shen, Y. R.: Structure of the Submonolayer of Ethanol Adsorption on a Vapor/Fused Silica Interface Studied by Sum Frequency Vibrational Spectroscopy. *J. Phys. Chem. A* 2015, 119, 4573–4580.

23. Zhang, L. N.; Liu, W. T.; Shen, Y. R.; Cahill, D. G.: Competitive Molecular Adsorption at Liquid/Solid Interfaces: A Study by Sum-Frequency Vibrational Spectroscopy. *J. Phys. Chem. C* 2007, 111, 2069–2076.

24. Liu, W.; Zhanag, L.; Shen, Y. R.: Interfacial Layer Structure at Alcohol/Silica Interfaces Probed by Sum-Frequency Vibrational Spectroscopy. *Chem. Phys. Lett.* 2005, 412, 206–209.

25. Miranda, P. B.; Pflumio, V.; Sajio, H.; Shen, Y. R.: Interaction between Surfactant Monolayers and Alkanes or Alcohols at Solid/Liquid interfaces. *J. Am. Chem. Soc.* 1998, 120, 12092–12099.

26. Zhang, D.; Gutow, J. H.; Eisenthal, K. B.; Heinz, T. F.: Sudden Structural-Change at an Air Binary-Liquid Interface — Sum Frequency Study of the Air Acetonitrile-Water Interface. *J. Chem. Phys.* 1993, 98, 5099–5101.

27. Sung, W.; Kim, D.; Shen, Y. R.: Sum-Frequency Vibrational Spectroscopic Studies of Langmuir Monolayers. *Curr. Appl. Phys.* 2013, 13, 619–632.

28. Guyot-Sionnest, P.; Hunt, J. H.; Shen, Y. R.: Sum-Frequency Vibrational Spectroscopy of a Langmuir Film — Study of Molecular-Orientation of a Two-Dimensional System. *Phys. Rev. Lett.* 1987, 59, 1597–1600.

29. Roke, S.; Schins, J.; Muller, M.; Bonn, M.: Vibrational Spectroscopic Investigation of the Phase Diagram of a Biomimetic Lipid mMonolayer. *Phys. Rev. Lett.* 2003, 90, 128101.

30. Tang, C. Y.; Allen, H. C.: Ionic Binding of Na+ versus K+ to the Carboxylic Acid Headgroup of Palmitic Acid Monolayers Studied by Vibrational Sum Frequency Generation Spectroscopy. *J. Phys. Chem. A* 2009, 113, 7383–7393.

31. Su, X.; Cremer, P. S.; Shen, Y. R.; Somorjai, G. A.: High-Pressure CO Oxidation on Pt(111) Monitored with Infrared-Visible Sum Frequency Generation. *J. Am. Chem. Soc.* 119, 3994–4000.

32. Li, X.; Rupprechter, G.: Sum Frequency Generation Spectroscopy in Heterogeneous Model Catalysis: A Minireview of CO-related Processes. *Catal. Sci. & Tech.* 2020. DOI: 10.1039/d0cy01736a.

33. Rupprechter, G.; Dellwig, T.; Unterhalt, H.; Freund, H. J.: High-Pressure Carbon Monoxide Adsorption on Pt(111) Revisited: A sum Frequency Generation Study. *J. Phys. Chem. B* 2001, 105, 3797–3802.

34. Su, X.; Cremer, P. S.; Shen, Y. R.; Somorjai, G. A.: Pressure Dependence (10^{-10}-700 Torr) of the Vibrational Spectra of Adsorbed CO on Pt(111) Studied by Sum Frequency Generation. *Phys. Rev. Lett.* 1996, 77, 3858–3860.

35. Pery, T.; Schweitzer, M. G.; Volpp, H. R.; Wolfrum, J.; Ciossu, L.; Deutschmann, O.; Warnatz, J.: Sum-Frequency Generation *In Situ* Study of CO Adsorption and Catalytic CO Oxidation on Rhodium at Elevated Pressures. *P Combust. Inst.* 2002, 29, 973–980.

36. Unterhalt, H.; Rupprechter, G.; Freund H.-J.: Vibrational Sum Frequency Spectroscopy on Pd(111) and Supported Pd Nanoparticles: CO Adsorption from Ultrahigh Vacuum to Atmospheric Pressure. *J. Phys. Chem. B* 2002, 106, 356–367.

37. Chou, K. C.; Westerberg, S.; Shen, Y. R.; Ross, P. N.; Somorjai, G. A.: Probing the Charge-Transfer State of CO on Pt(111) by Two-Dimensional Infrared-Visible Sum-Frequency Generation Spectroscopy. *Phys. Rev. B* 2004, 69, 153413.

38. Su, X.; Cremer, P. S.; Sen, Y. R.; Somorjai, G. A.: High-Pressure CO Oxidation on Pt(111) Monitored with Infrared-Visible Sum Frequency Generation. *J. Am. Chem. Soc.* 1997, 119, 3994–4000.

39. Cremer, P. S.; Su, X. C.; Shen, Y. R.; Somorjai, G. A.: Ethylene Hydrogenation on Pt(111) Monitored *In Situ* at High Pressures Using Sum Frequency Generation. *J. Am. Chem. Soc.* 1996, 118, 2942–2949.

40. Aliaga, C.; Park, J. Y.; Yamada, Y.; Lee, H. S.; Tsung, C. K.; Yang, P.; Somorjai, G. A.: Sum Frequency Generation and Catalytic Reaction Studies of the Removal of Organic Capping Agents from Pt nanoparticles by UV-Ozone Treatment. *J. Phys. Chem. C* 2009, 113, 5150–6155.

41. Baker, L. R.; Kennedy, G. J.; Van Spronsen, M.; Hervier, A.; Cai, X.; Chen, S.; Wang, L. W.; Somorjai, G. A.: Furfuraldehyde Hydrogenation

on Titanium Oxide-Supported Platinum Nanoparticles Studied by Sum Frequency Generation Vibrational Spectroscopy: Acid–Base Catalysis Explains the Molecular Origin of Strong Metal–Support Interactions. *J. Am. Chem. Soc.* 2012, 134, 14208–14216.

42. Cao, Y.; Chen, S.; Li, Y.; Gao, Y.; Yang, D.; Shen Y. R.; Liu, W. T.: Evolution of Anatase Surface Active Sites Probed by In situ Sum-Frequency Phonon Spectroscopy. *Sci. Ad.* 2016, 2, e1601162.

43. Yang, D.; Li, Y.; Liu, X.; Cao, Y.; Gao, Y.; Shen, Y. R.; Liu, W. T.: Facet-Specific Interaction between Methanol and TiO_2 Probed by Sum Frequency Vibrational Spectroscopy. *Proc. Nat. Acad. Sci.* 2018, 115, E3888–E3894.

44. Thompson, C. M.; Carl, L. M.; Somorjai, G. A.: Sum Frequency Generation Study of the Interfacial Layer in Liquid-Phase Heterogeneously Catalyzed Oxidation of 2-Propanol on Platinum: Effect of the Concentrations of Water and 2-Propanol at the Interface. *J. Phys. Chem. C* 2013, 117, 26077–26083.

Review Articles

- Rupprechter, G.: Sum Frequency Generation and Polarization-Modulation Infrared Reflection Absorption Spectroscopy of Functioning Model Catalysts from Ultrahigh Vacuum to Ambient Pressure. *Adv. Catal.* 2007, 51, 133–263.

- Sung, W.; Kim, D.; Shen, Y. R.: Sum-Frequency Vibrational Spectroscopic Studies of Langmuir Monolayers. *Curr. Appl. Phys.* 2013, 13, 619–632.

- Savara, A.; Weitz, E.: Elucidation of Intermediates and Mechanisms in Heterogeneous Catalysis Using Infrared Spectroscopy. *Ann. Rev. Phys. Chem.* 2014, 65, 249–273.

- Liu, F.; Wang, H.; Sapi, A.; Tatsumi, H.; Zherebetskyy, D.; Han, H. L.; Carl, L. M.; Somorjai, G. A.: Molecular Orientations Change Reaction Kinetics and Mechanism: A Review on Catalytic Alcohol Oxidation in Gas Phase and Liquid Phae on Size-Controlled Pt Nanoparticles. *Catalysis* 2018, 8, 226.

- Li, X.; Rupprechter, G.: Sum Frequency Generation Spectroscopy in Heterogeneours Model Catalysis: A Mini Review of CO-Related Catalysis. *Catal. Sci. Technol.* 2021, 11, 12–26.

• Yu, C. C.; Imoto, S.; Seki, T.; Chiang, K. Y.; Sun, S. M.; Bonn, M.; Nagata, Y.: Accurate Molecular Orientation at Interfaces Determined by multimode Polarization-Dependent Heterodyne-Detected Sum-Frequency Generation Spectroscopy Via Multidimensional Orientation Distribution Function. *J. Chem. Phys.* 2022, 156, 094703.

Chapter 7

Interfacial Structures of Bulk Materials

Functions of a material are often governed by the surface structure of the material and its interaction with the environment. Structures of surface and bulk are generally different even for the same material. The reason is that atoms or molecules at the surface and in the bulk experience different surroundings; surface structure changes when the environment changes. More appropriately, we ought to talk about, not surface structure, but interfacial structure of a material with its adjoining medium. It is the interfacial structure that determines the properties and functions of the material in a given environment.

Finding microscopic interfacial structures and properties is experimentally challenging. The surface signal is often difficult to detect in the presence of an overwhelming bulk signal. Surface–specific tools that can suppress the bulk signal are needed. Sum frequency spectroscopy is one such tool and is arguably the most viable one. We discuss in this chapter how SF spectroscopy can be used to probe interfacial structures of various types of materials: solids, liquids, liquid crystals, and ionic liquids.

7.1. General Considerations

We consider an interface as a microscopically thin layer between two media, the structure of which is different from the bulk structures of both media. (The interface of a material in vacuum is the bare surface of the material.) The interfacial structure is characterized by symmetry and resonances different from the bulks. Resonances from electronic transitions describe electronic structure and resonances from vibrational transitions describe atomic or molecular arrangement. Different symmetries and resonance features of surface and bulk allow separate measurements of the two. Surface–specific tools are generally needed to probe an interface unless surface and bulk resonances are distinguishable as in the case of photoemission spectroscopy.

As mentioned in Secs. 2.1 and 2.2, SF spectroscopy is surface–specific if contribution from bulk electric–dipole nonlinearity is forbidden and contribution from bulk electric–quadrupole nonlinearity is negligible. The latter is often the case and is what we shall assume in the following discussion unless specifically noted. In SF spectroscopy, the surface nonlinear susceptibility $\overset{\leftrightarrow}{\chi}_S^{(2)}$ is the response coefficient that reflects the symmetry and resonances of an interface. The technique aims at measuring the independent, nonvanishing elements of $\overset{\leftrightarrow}{\chi}_S^{(2)}$ for the interface. Both electronic and vibrational resonances can be probed. The resonances are likely to appear in different spectral ranges for an interface of two significantly different materials, yielding specific information about the interface in relation to one material or the other or both.

In comparison with other surface–specific probes, SF spectroscopy is more versatile, and can provide structural information over a broad spectral range. As we shall see in the following sections, it is applicable to a wide variety of interfaces. Currently, however, the spectral range at low frequencies is limited to $\sim$20 THz because of lack of sufficiently strong IR puplses. This limits SF vibrational spectroscopy to materials composed of lighter elements. As is usually the case for spectroscopy, extracting structural information about an interface from $\overset{\leftrightarrow}{\chi}_S^{(2)}$ spectra is not straightforward. Fortunately, the rapid progress of theoretical calculations, namely, molecular dynamics simulations, in recent years has made it less problematic.

7.2. Interfaces of Crystalline Solids

7.2.1. *Surfaces in Ultrahigh vacuum*

Surfaces in UHV can be cleaned and well-characterized. Surface science experiments carried out in UHV are certainly most trustworthy. In early days, to establish the credibility of SHG/SFG spectroscopy, it was necessary to demonstrate that the technique could also detect familiar surface effects and phenomena in UHV. Surface reconstruction of metals and semiconductors in response to external conditions is well known. Si(111) with surface structures of (2×1) and (7×7) is a representative case. Heinz *et al.* were able to use SHG to observe the reconstructed surface structures, the transition from Si(111)(2×1) to Si(111)(7×7) upon thermal annealing, and the time dependence of the transition.[1]

The metastable (2×1) surface structure of Si(111) has a single mirror plane passing through the $[2\bar{1}\bar{1}]$ direction and that of the equilibrium (7×7) surface has the full 3m azimuthal symmetry of the bulk crystal. Accordingly, the nonzero elements of $\overset{\leftrightarrow}{\chi}_S^{(2)}$ for normally incident input and reflected SHG output are $\chi_{S,xxx}^{(2)}, \chi_{S,xyy}^{(2)}, \chi_{S,yyx}^{(2)} = \chi_{S,yxy}^{(2)}$ for the (2×1) surface and $\chi_{S,xxx}^{(2)} = -\chi_{S,xyy}^{(2)} = -\chi_{S,yxy}^{(2)} = -\chi_{S,yyx}^{(2)}$ for the (7×7) surface, where x and y are along $[2\bar{1}\bar{1}]$ and $[01\bar{1}]$ directions, respectively. From the description in Sec. 2.1, the SH output has the expression $S \propto |\hat{e}(2\omega) \cdot \overset{\leftrightarrow}{\chi}_S^{(2)} : \hat{e}(\omega)\hat{e}(\omega)|^2 |E(\omega)|^4$, with $\hat{e}$ denoting beam polarization. It shows for given $\overset{\leftrightarrow}{\chi}_S^{(2)}$ how S depends on input/output polarizations. Figure 7.1 presents the polar plots of the experimental data and theoretical fits of SHG versus input polarization from Si(111)(2×1) and Si(111)(7×7) with no selection of output polarization (top frames) and output polarization along $[2\bar{1}\bar{1}]$ and $[01\bar{1}]$(middle and bottom frames). The patterns for the two surface structures are clearly different. In the experiment, the initially cleaved Si surface revealed the (2×1) structure; upon thermal annealing, the (7×7) surface began to emerge at $\sim$275K in a few seconds, and a well-defined (7×7) structure was observed

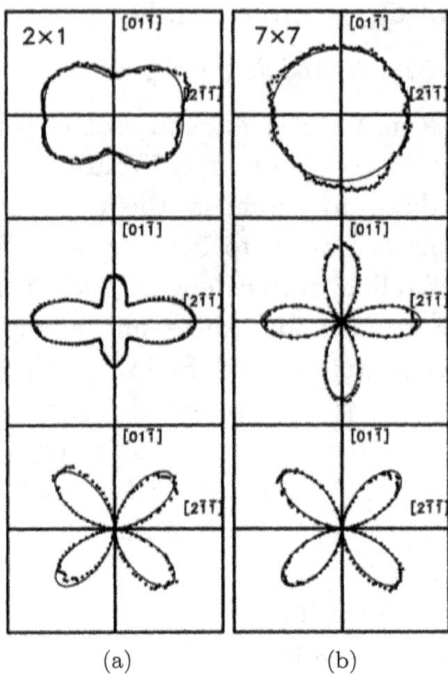

Fig. 7.1. Reflected SH intensity from (a) Si(111)-(2 × 1) and (b) Si(111)-(7 × 7) surface as a function of the direction of the linear polarization of the normally incident input beam. The top panels display the total SH signal with no output polarization selection; the middle and bottom panels show, respectively, the SH signal polarized along the $[2\bar{1}\bar{1}]$ and $[01\bar{1}]$directions. The dotted curves are experimental data, while the solid curves are theoretical fits (after Ref. [1]).

at 600K. When the surface was oxidized or covered by an amorphous Si monolayer, the SHG signal was almost completely suppressed.[2] This indicates that bulk contribution was negligible.

Microscopic surface studies of metals and semiconductors in UHV mainly focus on atomic/molecular adsorption and associated effects that were discussed in Sec. 6.8. Probing surface properties of such materials themselves by SF spectroscopy is generally difficult because strong bulk contribution as well as contribution from mobile carriers to SFG could complicate the matter. For example, a gradient of carrier population can create a built-in field in a surface layer that dominates the SF output. On metals, IR-visible SF spectroscopic measurements by many groups have found that reflected SF output in the visible around 500 nm from Pt is weak, but very strong for many other metals; the frequency dependence in the visible could also be significant. Presumably this is the result of different electronic structures for different metals and different resonant contributions from their interband electronic transitions, but detailed understanding is not yet available. Whether the signal comes from the surface or the bulk or both is also not known. Actually, SHG from metals was studied extensively both theoretically and experimentally in early years.[3-9] It was found that electric–quadrupole (including magnetic–dipole) contributions from both intraband and interband transitions contribute to SHG. One would expect that both surface and bulk contribute to SHG in reflection.

7.2.2. *Surface phonons*

Surface phonons directly reflect the structure of a surface lattice as the bulk phonons do with a bulk lattice. In early years, He atomic scattering (HAS) and electron energy loss spectroscopy (EELS) were the main techniques to detect surface phonons. The former has a spectral range limited to 30 meV ($\sim$250 cm^{-1}), and the latter is usually dominated by the macroscopic Fuchs–Kliewer surface phonon–polariton resonances that appear when the real part of the bulk dielectric constant approaches -1.[10] More recently, SF spectroscopy has been developed into an effective means to probe surface phonons, not only for surfaces in vacuum or gas, but also for buried interfaces.

The first demonstration of SF spectroscopy of surface phonons was on diamond C(111) in UHV.[11] No C-C stretch mode was observed when the surface had the (1×1) structure, but two bands at $\sim$1350 and 1475 cm^{-1} appeared when thermal annealing transformed the surface structure to (2×1), as shown in Fig. 7.2(a) for SSP, PPP, and SPS polarizations. Apparently, the (1×1) surface hardly breaks the inversion symmetry of the C-C bonding network, but the (2×1) surface does strongly. The result supports Pandry's buckled π-bonded chain model for the (2×1) surface sketched in Fig. 7.2(b); the 1350 and 1475 cm^{-1} modes can be assigned to surface phonons associated with C-C bonds that connect first and second

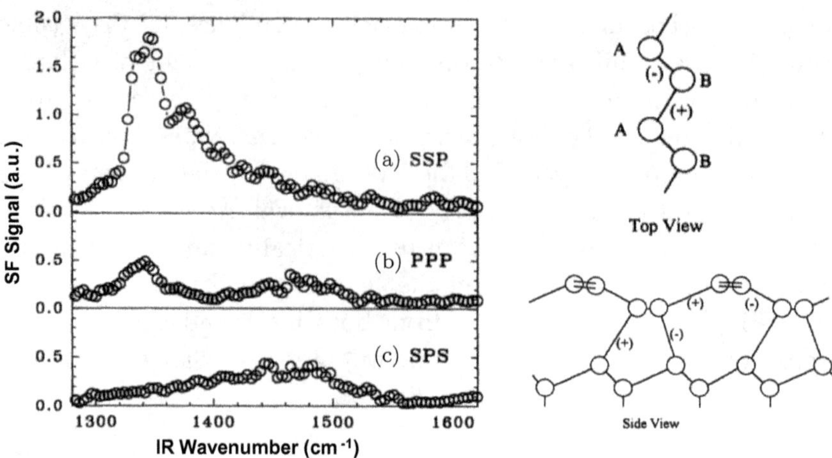

Fig. 7.2. SF spectra of diamond C(111)-(2 × 1) surface in air in the C–C stretching region taken with SSP, PPP, and SPS polarization combinations. They can be explained by the buckled π-bonded chain model structure described on the right (after Ref. [11]).

surface bilayers and that connect the sublayers of the top bilayer, respectively.

Even if a crystal lacks inversion symmetry, it is still possible for SF spectroscopy to detect surface phonons as long as the bulk contribution to SFG can be suppressed or discriminated with proper beam polarizations and geometry. We take α-quartz(0001) as an example.[12] As discussed in Sec. 4.4, it has the $D_3(32)$ point symmetry that has no inversion center. With SSP input/output polarizations, SFG detects the nonvanishing bulk nonlinear susceptibility element $\chi^{(2)}_{B,yyx} = -\chi^{(2)}_{aaa}\cos 3\phi$, where $\hat{x} - \hat{y}$ is the surface plane, the three-fold symmetry axis of the crystal, $\hat{c}$, is along the surface normal, and $\hat{a}$ is along a two-fold symmetric axis of the crystal separated from $\hat{x}$ by angle ϕ. It is seen that for $\phi = \pm\pi/6$ and $\pi/2$, SFG from the bulk should vanish, but the surface SSP SFG originating from the surface nonlinear susceptibility element $\chi^{(2)}_{S,yyz}$ should not because of broken inversion symmetry along z. The experimental result is described in Fig. 7.3(a). Three modes are observed at $\phi = 31.8°$ in the spectral range from 850 to 1150 cm⁻¹; only the one at 1064 cm⁻¹ can be identified as an allowed bulk phonon mode from the selection rule[12] (Sec. 4.4). At $\phi = 30°$, this bulk mode is almost completely suppressed, while the two other modes at 880 and 980 cm^{-1} hardly change indicating that they must come from surface phonons or vibrations. Supported by ab initio calculation on the reconstructed surface structure of an unhydrated α-quartz(0001), the 880 cm^{-1} mode could be assigned to the symmetric stretch of the strained Si-O-Si surface groups. The 980 cm^{-1} mode was attributed to the Si-OH stretch, knowing that the surface exposed to air was partially hydrogenated. The polarization dependence of the mode (Fig. 7.3(b)) yielded a bond orientation angle less than 30° away from the surface normal, again indicating that the reconstructed surface structure was different from the bulk-terminated one, which would have bond angles at 47° and 67° (see schematic representation of the surface structure in Fig. 7.3(c)). As further support of the assignment, the 880 cm^{-1} mode was seen to have enhanced and the 980 cm^{-1} mode diminished (Fig. 7.3(b)) when H is desorbed upon thermal annealing. It was also seen that the Si-OH mode was greatly reduced when an OTS layer covered the surface (Fig. 7.3(d)).

SF spectroscopic detection of surface phonons of other crystals have also been reported, namely, Al_2O_3,[13] TiO_2,[14] $SrTiO_3$,[15] and others. Because the currently available IR pulses do not have sufficient energy below $\sim$650 cm^{-1}, studies of surface phonons have so far been limited to those involving light atoms, mostly oxides and possibly fluorides. We would expect that surface reconstruction depends on the environment the surface is exposed to. It should be interesting to learn how a surface structure changes with molecular adsorption[2,13,14] or/and in contact with another medium. In the latter case, detection of spectral variation of surface vibrations of fused silica, and hence the silica surface reconstruction, in aqueous solution in response to pH and salt concentration has recently been explored.[16] This will be described in more detail in Sec. 8.9.

One example that shows response of surface phonons to molecular adsorption was reported on anatase $TiO_2(101)$,[14] which has been extensively studied because of its practical relevance. The surface is known to have oxygen vacancies at the surface and the surface structure is sketched in Fig. 7.4(a). There are two active surface

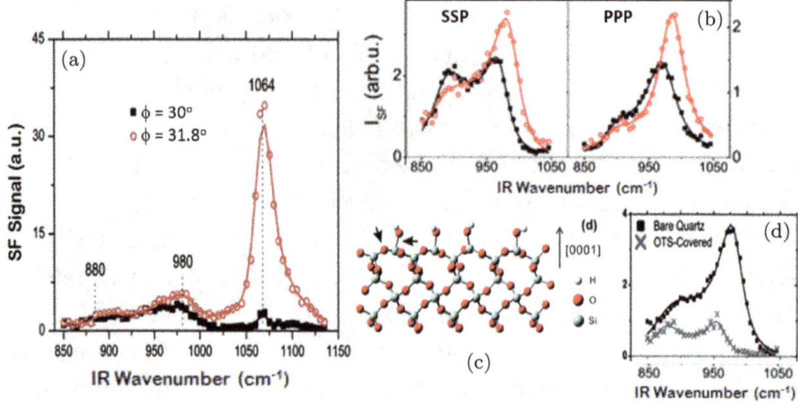

Fig. 7.3.   (a) SSP SF spectra of α-quartz(0001) in air taken at azimuthal angles $\phi = 30°$ and 31.8°, revealing the presence of the surface Si–OH stretch mode at 980 cm^{-1} and the surface Si–O–Si mode at 880 cm^{-1}, as well as suppression of the 1064 cm^{-1} bulk phonon mode at $\phi = 30°$. The curves are guide for eyes. (b) SSP and PPP SF spectra of bare α-quartz(0001) in air taken at $\phi = 30°$ after being baked at 100°C (black squares) and then rehydroxylated (red open circles), showing that the surface structure change is reversible. (c) Schematic representation of a possible surface structure of α-quartz(0001) composed of Si–OH and Si–O–Si as indicated by arrows. (d) Comparison of SSP SF spectra of bare and silane (OTS)-covered α-quartz(0001) at $\phi = 30°$ showing that the Si–OH mode is greatly suppressed by adsorption of silane (after Ref. [12]).

sites, one associated with the five-coordinated Ti(5c) atoms and the other with the oxygen vacancy sites, V_O^{Surf}, originally occupied by two-coordinated oxygen atoms O(2c). The population of V_O^{Surf} is expected to be in equilibrium with that of the subsurface vacancies V_O^{Sub} via diffusion. SF spectroscopy measurement identified a surface phonon mode at 860 cm^{-1} (Fig. 7.4(b)) that came mainly from the stretch vibration of Ti(5c) with respect to the three-coordinated oxygen, O(3c), below. Upon UV irradiation, the surface phonon mode dropped in intensity (Fig. 7.4(c)); the reduction depended on the ambient gas the surface was exposed to, more in methanol and water, but least in oxygen. The decrease was due to increase of V_O^{Surf} that transformed more Ti(5c) to Ti(4c) and hence reduced the number of Ti(5c)-O(3c) bonds. Molecular adsorption at the oxygen vacancies altered the equilibrium between V_O^{Surf} and V_O^{Sub} favoring less V_O^{Surf}. It was obvious that the change in oxygen atmosphere was

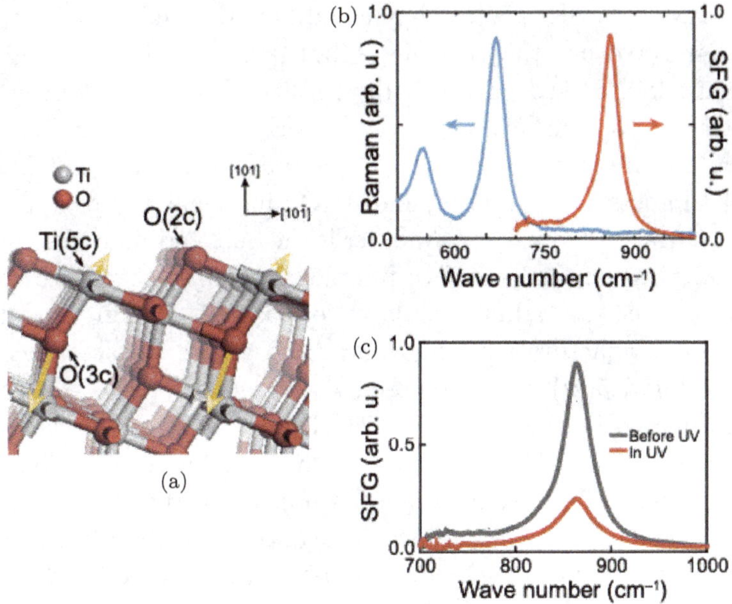

Fig. 7.4. (a) Structure of anatase(101) near the surface. Ti(5c) and O(2c) denote the 5-coordinated titanium and 2-coordinated oxygen at the surface and O(3c) the 3-coordinated oxygen in subsurface. Yellow arrows describe the displacements of Ti(5c) and O(3c) associated with the surface phonon. (b) Raman (blue) and SF (red) vibrational spectra from anatase(101) taken with all beams P-polarized. The two Raman modes are the highest-frequency transverse optical phonon modes of bulk anatase. The 850 cm^{-1} mode appearing only in the SF spectrum is the surface phonon mode. (c) Surface phonon spectrum before and after UV irradiation, showing that the surface phonon mode is greatly reduced by UV-induced oxygen vacancies on the surface (after Ref. [14]).

less because oxygen could replenish the oxygen vacancies created by UV irradiation. The measurement also showed that molecular adsorption occurred at the oxygen vacancy sites rather than on Ti(5c) as suggested by some researchers.

7.2.3. *Complement to X-ray determination of surface structures*

X-ray diffraction, reflection, and absorption are usually employed to find crystalline surface structures. However, the result is often not unique, but depends on the model chosen for interpretation of the X-ray data. SF spectroscopy could help determine which model is correct. We use the R-plane ($1\bar{1}02$) of α-alumina (α-Al$_2$O$_3$) as an example.[17] Based on the X-ray results, three models have been proposed for the surface structure of (α-Al$_2$O$_3$)($1\bar{1}02$) (Fig. 7.5). One is a bulk-terminated structure with the topmost Al layer missing and equal numbers of AlO, Al$_2$O, and Al$_3$O functional groups appearing on the surface (Fig. 7.5(a)). Another is the bulk-terminated structure with Al$_3$O as the only surface functional groups (Fig. 7.5(b)). The third one has an additional layer of oxygen covering the bulk-terminated structure with AlO and Al$_3$O as the surface functional groups (Fig. 7.5(c)). When exposed to air, AlO and Al$_2$O are likely to be protonated into AlOH$_2$ and Al$_2$OH, but Al$_3$O is not because the valence electrons of O are already largely shared with three Al atoms. Knowing which hydroxylated species appear on the surface can tell us which model is correct. Since X-ray is not sensitive to hydrogen, it is difficult for X-ray techniques to detect the hydroxylated species. SF vibrational spectroscopy however can readily do so.

The Im $\chi^{(2)}_{S,yyz}$ spectrum of the protonated (α-Al$_2$O$_3$)($1\bar{1}02$) surface, extracted from SF vibrational spectroscopy measurements with SSP polarizations and the incidence plane set at an azimuthal angle of $\gamma = 180°$ from the [$1\bar{1}01$] direction in the surface plane, is shown in Fig. 7.5(d).[17] The spectrum is composed of three OH stretch modes; the 3680 and 3520 cm^{-1} modes can be assigned to the dangling OH and bonded OH (with H bonded to a neighboring O) of AlOH$_2$, respectively, and the 3365 cm^{-1} mode to the bonded OH of Al$_2$OH. We can then immediately recognize that the bulk-terminated structure with the missing Al layer (Fig. 7.5(a)) is the correct structural model because the other models do not have AlOH$_2$ and Al$_2$OH as surface functional groups. Further measurements with different input/output polarization combinations and azimuthal angles (γ) were carried out for quantitative determination of the orientations of the three OH bonds, from which the atomic spacing of the surface lattice could be deduced.[17] As expected, it was different from that of the bulk lattice since surface structure should relax to adjust to the ambient condition.

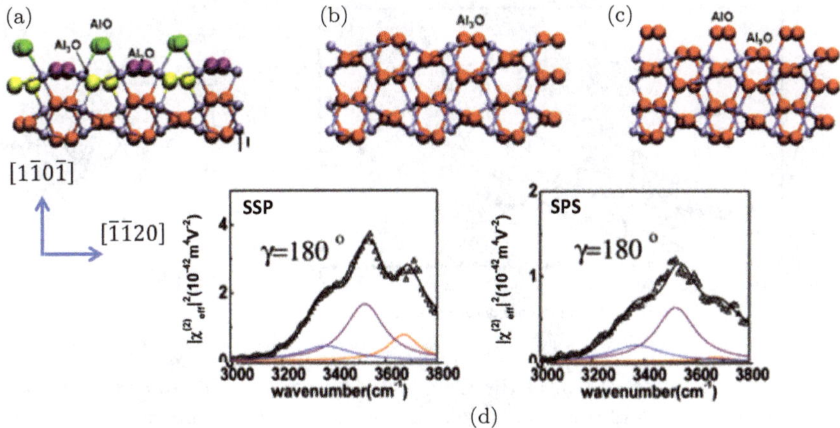

Fig. 7.5. Three different structural models for α-alumina $(\alpha$-$Al_2O_3)(1\bar{1}02)$ surface from X-ray studies: (a) bulk-terminated structure with the topmost Al layer missing and equal numbers of AlO, Al_2O, and Al_3O functional groups on the surface; (b) bulk-terminated structure with Al_3O being the only surface functional group; (c) bulk-terminated structure with an additional oxygen layer at the top and the surface functional groups are AlO and Al_3O. (d) SSP and SPS SF spectra of $(\alpha$-$Al_2O_3)(1\bar{1}02)$ in the OH stretch range, showing the existence of three modes from spectral decomposition (solid curves) in support of the proposed model in (a); γ denotes the angle between the incidence plane and the $[1\bar{1}0\bar{1}]$ direction in the glide plane $(\bar{1}\bar{1}20)$ (after Ref. [17]).

7.2.4. *Interfaces of heterostructures*

Solid/solid interfaces are notoriously difficult to interrogate because the bulk material often forbids a probe beam to access such an interface. The problem is however less troublesome for SF spectroscopy. Usually the optical absorption length of a solid is ≥ 10 nm and an epitaxially grown thin film of a few nm is expected to have the same structure as a bulk. If a solid/solid interface can be prepared with an epitaxially grown thin film on one side, we should be able to use SFG spectroscopy to probe the interface.

The first demonstration was carried out by SHG spectroscopy on an interface formed by 30 μm of CaF_2 on Si(111).[18] The tunable input was incident from the CaF_2 side, and the reflected SH output was detected. SHG from bulk Si and CaF_2 was found to be negligible. The recorded SHG spectrum is displayed in Fig. 7.6(a). Two prominent

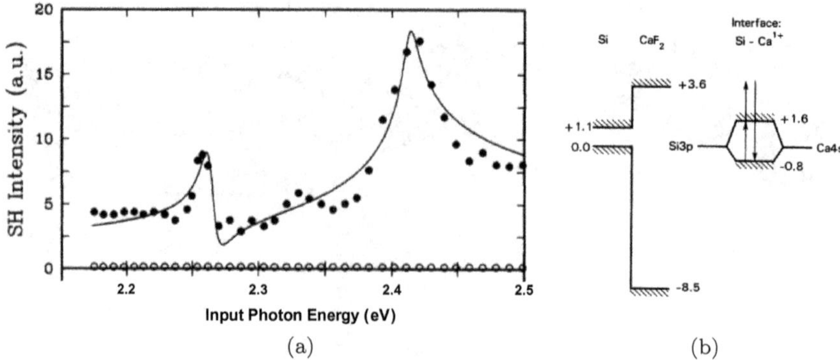

Fig. 7.6. (a) SH spectrum of the CaF_2/Si(111) interface. The filled circles are experimental data and the solid curve, a theoretical fit. (b) Band gap structure of the interface. The interfacial states arising from hybridized Si(3p) dangling-bond and Ca^+(4s) orbitals are shown on the right (after Ref. [18]).

peaks were observed in the 2.2–2.5 eV range. The stronger one at 2.4 eV was assigned to the transition between interfacial states formed by hybridization of the dangling Si(3p) and Ca^+(4s) orbitals, and the weaker one to its n = 1 exciton.

Interfaces of heterostructures have recently received increasing attention because of the exciting new development in several areas. For example, the interface of two different complex oxides can have structure and functional properties very different from the bulks, leading to emergence of new phases, enhancement of ferroelectricity, modification of ferromagnetization, formation of 2D electron gas, appearance of novel magnetic states, and so on with potential applications in magnetoelectric devices.[19] Interfacial excitons and phonons have been observed in heterointerfaces of 2D materials, but they can also exist at interfaces of bulk materials. Surface superconductivity induced by interfacial phonons has been proposed.[20] In all cases, to understand interfacial properties and phenomena, knowledge about interfacial structure is needed. Currently, no effective surface–specific tool is available for effective probing of buried solid/solid interfaces except perhaps SFG spectroscopy, but even the latter is yet to be further developed.

SF spectroscopy to probe electronic transitions of heterointerfaces may be more challenging because the spectrum often appears as broad bands without clear resonant features and can only be interpreted with help of theoretical calculations. Probing interfacial phonons is easier since the phonon modes often show up as narrow bands with frequencies that could be estimated. Indeed, interfacial phonons of a heterojunction between two complex oxides have been detected and their change with change of the heterostructure observed.[21]

7.2.5. *Miscellaneous*

There has been a boom in research on new materials, especially quantum materials, in recent years. Current research effort is more on their bulk properties, but increasing attention is being paid to their interfacial properties. Multiferroic materials discussed above serve as a good example. Interfaces of topological insulators are particularly interesting because they have robust metallic surface states protected by topology and time-reversal symmetry in contrast to the band structure of an ordinary insulator. Photoemission spectroscopy has been employed to probe surface topological band structures. One would expect surface–specific SHG/SFG could also be used to study structure and properties of such interfaces. Indeed, SHG measurements with a fixed input frequency on $Bi_2Se_3(111)$, which is a topological insulator with inversion symmetry, have been reported.[22,23] It was found that a space-charge region was formed when the interface was doped by Se vacancies, resulting in not only band bending and carrier redistribution in the region, but also a significant change of the surface nonlinear susceptibility, $\chi_S^{(2)}$. The latter is not understood although it probably came from change of the surface electronic structure. Obviously, we can expect to learn more about topological interfaces using SFG spectroscopy, especially if interfacial phonons can be detected and electron–phonon coupling can be measured through doubly resonant SF spectroscopy.

SHG/SFG can also probe surface magnetization of magnetic crystals, which are relevant for magnetic devices but lack effective means to study. We discussed in Sec. 4.2 that the nonlinear susceptibility of a magnetic crystal consists of two parts, $\overset{\leftrightarrow}{\chi}_B^{(2)} = \overset{\leftrightarrow}{\chi}_{Bi}^{(2)} + \overset{\leftrightarrow}{\chi}_{Bc}^{(2)}$, where $\overset{\leftrightarrow}{\chi}_{Bi}^{(2)}$ reflects the crystal structure symmetry and $\overset{\leftrightarrow}{\chi}_{Bc}^{(2)}$ reflects symmetry of the spin (magnetic moment) arrangement in the crystal. Different symmetries of $\overset{\leftrightarrow}{\chi}_{Bi}^{(2)}$ and $\overset{\leftrightarrow}{\chi}_{Bc}^{(2)}$ lead to different sets of nonvanishing elements for $\overset{\leftrightarrow}{\chi}_{Bi}^{(2)}$ and $\overset{\leftrightarrow}{\chi}_{Bc}^{(2)}$, allowing separate determination of $\overset{\leftrightarrow}{\chi}_{Bi}^{(2)}$ and $\overset{\leftrightarrow}{\chi}_{Bc}^{(2)}$. Since bulk and surface have different symmetries, we can expect to have surface nonlinear susceptibility given by $\overset{\leftrightarrow}{\chi}_{S}^{(2)} = \overset{\leftrightarrow}{\chi}_{Si}^{(2)} + \overset{\leftrightarrow}{\chi}_{Sc}^{(2)}$ with different sets of nonvanishing elements between $\overset{\leftrightarrow}{\chi}_{Si}^{(2)}$ and $\overset{\leftrightarrow}{\chi}_{Bi}^{(2)}$, $\overset{\leftrightarrow}{\chi}_{Sc}^{(2)}$ and $\overset{\leftrightarrow}{\chi}_{Bc}^{(2)}$, and $\overset{\leftrightarrow}{\chi}_{Si}^{(2)}$ and $\overset{\leftrightarrow}{\chi}_{Sc}^{(2)}$. It is then possible to probe surface magnetization by measuring $\overset{\leftrightarrow}{\chi}_{Sc}^{(2)}$ separately. This was proposed earlier and illustrated by derivation of nonvanishing elements of $\overset{\leftrightarrow}{\chi}_{S}^{(2)}$ for SHG from face center cubic (fcc) crystals with and without surface magnetization along symmetric crystal axes.[24] Subsequently, the idea was verified experimentally.[25]

Surface of a substrate plays an important role in layer-by-layer growth of crystals. Epitaxial growth usually takes a few monolayers to reach the bulk crystal structure, depending on the lattice mismatch between the crystal and substrate. In some cases, a thin crystalline film grown on a substrate can have a structure that does not exist in nature. For example, a thin Co film grown on GaAs (110) appears to be body center cubic (bcc) while bulk Co normally has a hexagonal close packed (hcp) structure.[26] Rutherford backscattering is commonly used to monitor layer-by-layer growth, but we expect that SHG/SFG can be used to monitor the growth *in situ* through nonlinear optical response as well. This possibility however has not yet been seriously explored. A special case of growth of polar-oriented pyroelectric ice films on Pt(111) monitored by SF vibrational spectroscopy will be discussed in Sec. 8.12. Surface-induced bulk ordering generally occurs in liquid crystal films and will be described in Sec. 7.5.

7.3. Interfacial Structures of Liquids

Liquids are certainly as important as solids in our modern life, but they are generally less understood at the atomic/molecular level than solids. Our knowledge about liquid interfaces is particularly thin because of limited capability of currently available techniques to probe microscopic structure of liquid interfaces despite recent advances of X-ray spectroscopy, photoemission spectroscopy, and others. In this respect, SF vibrational spectroscopy has been proven to be most versatile and capable of providing crucial information.

Unlike in solids, molecules in liquids are relatively free to rotate and translate and are in a state of dynamic equilibrium. Structural information comes from average molecular orientation and arrangement that are random in bulk, but can be ordered at a surface or interface. Bulk and surface molecular conformations may also be different. As described in Chapter 2, SF vibrational spectroscopy (SFVS) allows *in situ* probing of an interfacial structure. It has become a unique and most viable tool for studies of liquid interfaces.

7.3.1. *Polar liquids*

Liquids composed of polar molecules are likely to have molecules polar-oriented at a surface or interface. The polar orientation depends on how the molecules interact with the opposing medium. The first experiment demonstrating effectiveness of SFVS to probe liquid surfaces was on methanol liquid exposed to air.[27] It recorded the first vibrational spectra of any liquid interface ever studied. While it was probably expected that the methyl groups of methanol should point toward air at the surface because of their hydrophobicity and possible strong hydrogen bonding of the OH end groups, there was no experimental verification before the SFVS work. The CH stretch spectra of the methyl groups taken by reflected SFVS with SSP, PPP, and SPS polarizations are shown in Fig. 7.7(a).[28] The prominent peak at 2832 cm^{-1} can be assigned to the symmetric stretch of CH_3 and the peak at 2951 cm^{-1} and the shoulder at 2925 cm^{-1} to the Fermi resonances of the symmetric stretch mode with the overtones of the bending modes. The antisymmetric stretch of CH_3 is too weak to be seen although it was observed in the Raman spectrum of bulk methanol at 2974 cm^{-1}. The strong SSP spectrum and the nearly vanishing SPS spectrum suggest right away that the methyl groups have their symmetric axes aligned along the surface normal. Analysis to extract the orientation of the methyl groups (Sec. 6.3) indicated that the axis are tilted by $\sim30°$ (or an angular

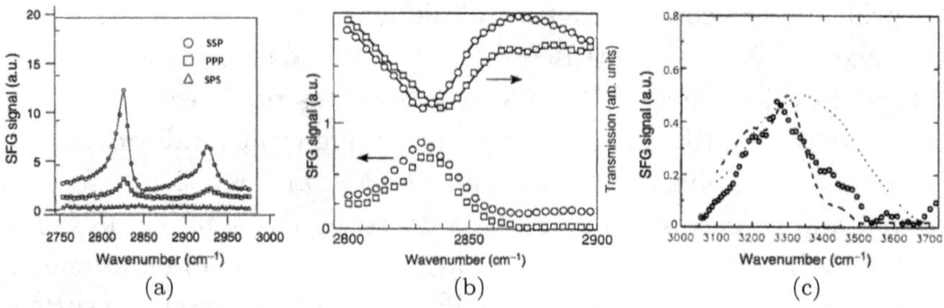

Fig. 7.7. (a) Reflected SSP, PPP, and SPS SF spectra of the vapor/methanol interface in the CH stretch range (after Ref. [28]). (b) Bulk IR absorption spectra (top) and surface SSP SF spectra (bottom) of the CH_3 symmetric stretch mode of methanol for pure methanol (circles) and 1:1 methanol–water mixture (squares). The IR spectra show a frequency shift between the two, but the SF spectra do not, indicating that the latter is from the surface (after Ref. [27]). (c) SSP SF spectrum from the vapor/methanol interface in the OH stretch region (open circles) in comparison with the IR absorption spectra of liquid (dotted line) and crystalline methanol (dashed line) (after Ref. [29]).

spread of $\sim30°$) from the surface normal. Bulk contribution was found negligible on comparison with the spectrum obtained from transmitted SFVS (Sec. 2.2). It was further confirmed by comparing the SF spectra with the IR absorption spectra of pure methanol and 1:1 methanol–water mixture presented in Fig. 7.7(b). The IR spectra from the bulks display a band shift between the two, but the surface-dominated SF spectra of the two are nearly the same because methanol molecules fully cover the surface of the mixed solution with a structure resembling the surface of pure methanol.[27] This also illustrates the fact that liquid mixtures usually have very different surface and bulk compositions. The SF surface spectrum of OH stretches of methanol liquid was also observed. As seen in Fig. 7.7(c),[29] the broad OH stretch band is significantly narrower than the IR absorption spectrum of bulk methanol, but close to that of solid methanol. It is an indication that the OH terminal groups of methanol are hydrogen-bonded to form a more ordered hydrogen bonding network at the surface than in the bulk, compatible with the high orientation order of the protruding methyl groups at the surface. The SFVS results here give a fairly comprehensive, hitherto unavailable, picture about the surface structure of methanol liquid.

We should expect other alcohols to have similar orientations at their air/liquid interfaces and SFVS could provide verification. Indeed, as seen in Fig. 7.8, the CH stretch spectra of alcohols $C_nH_{2n+1}OH$ with $n = 1$ to 8 resemble that of methanol except for shifts of mode frequencies because the methanol C is now bonded with C instead of O and the appearance of the antisymmetric stretch of CH_3 at 2960 to 2970 cm^{-1} in the PPP and SPS spectra. (The antisymmetric CH_3 stretches are still weak in the SSP spectrum because their vibrational plane is close to the surface and SSP SFVS is insensitive to vibration in the surface plane.) As the alkyl chain length increases, the CH_2 stretch modes and their Fermi resonances at $\sim$2850 cm^{-1} and 2890–2920 cm^{-1} become increasingly visible due to appearance of gauche defects in the chain that break the near-inversion symmetry of the $(CH_2)_n$ chain. (For $n > 14$, the defects would be eliminated again by chain–chain interactions as mentioned in Sec. 6.6).

Through molecular interaction, the orientation order of surface molecules is expected to extend to two to three monolayers deep. The bonded OH spectrum of methanol, for example, suggests that the OH end groups of polar-oriented methanol molecules at the air/liquid interface must have hydrogen-bonded not only to OH of neighbors in the surface monolayer, but also to OH of molecules below the surface monolayer, setting up a somewhat disordered second monolayer with methyl groups pointing away from the surface. The same should be true for other alcohols although the second monolayer is expected to be more disordered for longer chain length.

The polar orientation of surface molecules depends on the medium they face. At a hydrophilic alcohol/silica interface, for example, alcohol molecules in the surface monolayer have their OH end groups bonded to silica and methyl groups polar-oriented away from the interface. This was verified by SFVS. Figures 7.9(a) and 7.9(b) show, for comparison, the SSP, SPS, and PPP CH stretch spectra of C1-C4 alcohol/air and alcohol/silica interfaces, respectively.[30] The two sets of spectra are very different. We focus on the SPS spectra that can be more easily analyzed. The pronounced CH_3 symmetric stretch and its Fermi resonance modes in the alcohol/air spectra

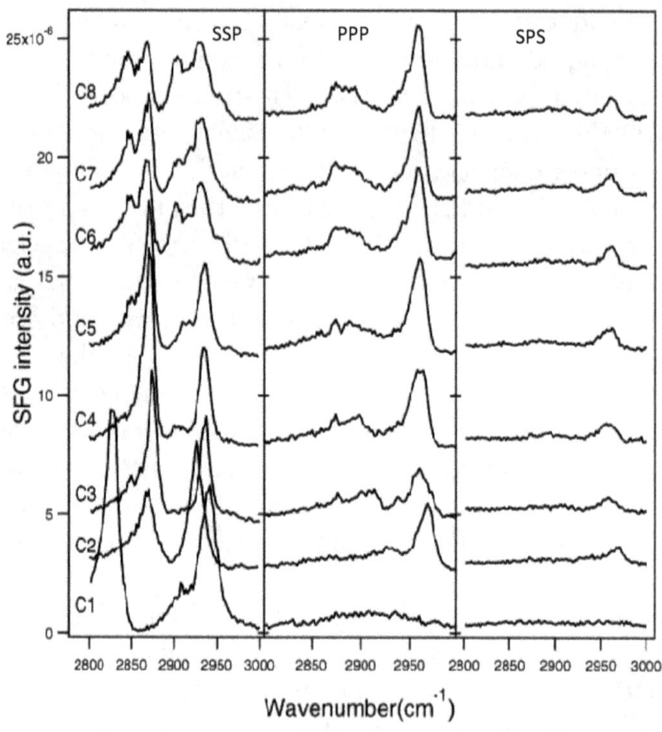

Fig. 7.8. SF vibrational spectra taken with SSP, PPP, and SPS polarization combinations in the CH stretch range for alcohol($C_n H_{2n+1}OH$)/vapor interfaces ($n = 1$–8). The solid lines are guides for eyes (after Ref. [28]).

become nearly invisible in the alcohol/silica spectra for methanol, and are reduced to less than half for C2–C4. The CH_3 antisymmetric stretch modes and their Fermi resonances, however, are strongly enhanced. The result can be readily understood knowing that there can be a second ordered alcohol monolayer with opposite orientation under the first surface alcohol monolayer. Opposing CH_3 symmetric stretches of the two monolayers have their contributions to SFVS cancel each other, but the antisymmetric stretches generally enforce each other (See Sec. 6.5). In real cases, because the second monolayer is more disordered, the result is not so drastic.

In special cases, orientation-ordered molecular interaction can be so strong that molecular orientation order at an interface of a liquid

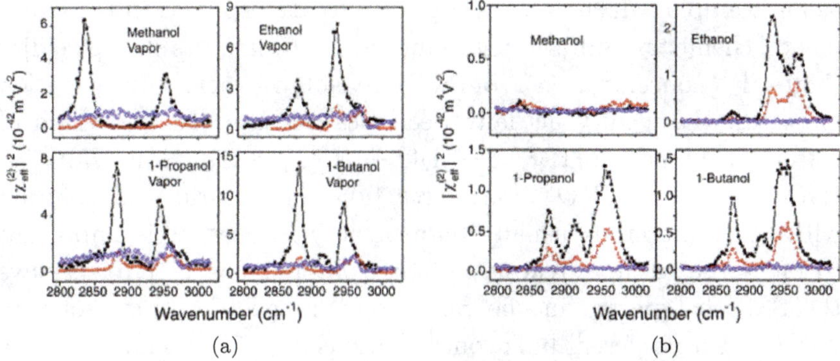

Fig. 7.9. SF spectra of interfaces of C1–C4 liquid alcohols with (a)vapor and (b) silica in the CH stretching range taken with SSP (black squares), PPP (red crosses), and SPS (blue open circles) polarization combinations (after Ref. [30]).

can be extended over many monolayers away from the interface. This can happen to liquid crystals as we shall discuss later.

Surface structure of a liquid can be affected by the opposing medium, but the reverse is also true. The interface has a composite structure resulting from molecular interaction and rearrangement of both sides, including possible inter-penetration of molecules to the opposite side. At a gas/liquid interface, the liquid surface structure may depend on molecular adsorption at the interface and vice versa. At a liquid/liquid interface, the interfacial structure may consist of varying molecular arrangement and composition over a finite interfacial layer thickness. At a liquid/solid interface, while molecular orientation and arrangement of the liquid surface tends to adjust more readily to the solid surface structure, the solid surface may also be altered by the liquid. We shall give a more detailed discussion on this aspect of liquid interfaces later using water as an example in Chapter 8. In all such cases, SFVS can be an effective tool to probe the composite interfacial structures as long as it can access the interfacial vibrational resonances.

Surfaces of mixed liquid solutions are of common interest because their surface functional properties can be varied by adjusting bulk composition, but we need surface characterization to monitor changes, and SFVS is a viable probe. We take sulfuric acid solutions as an example, which is a relevant case to atmospheric and environmental chemistry. Sulfuric acid solution is known for its strong acidity (high H^+ concentration), but as the sulfuric acid concentration increases, the acidity actually decreases because of suppression of dissociative reactions $H_2SO_4 + H_2O \rightarrow HSO_4^- + H_3O^+$ and $HSO_4^- + H_2O \rightarrow SO_4^{2-} + H_3O^+$. Since reactions of sulfuric acid solution with other materials usually happen at interfaces, it is important to know the surface structure of the solution. Figure 7.10 displays the SSP SF spectra in the SO stretch range of the air/solution interface with different bulk concentrations of sulfuric acid.[31] At mole fraction x < 0.4 ($\sim$14M), only one prominent peak at $\sim$1055 cm^{-1} stands out; it grows and slightly red-shifts as x increases. This mode can be assigned to the symmetric stretch of SO_3 of HSO_4^- at the interface; the antisymmetric stretch at 1220 cm^{-1} was not seen, but

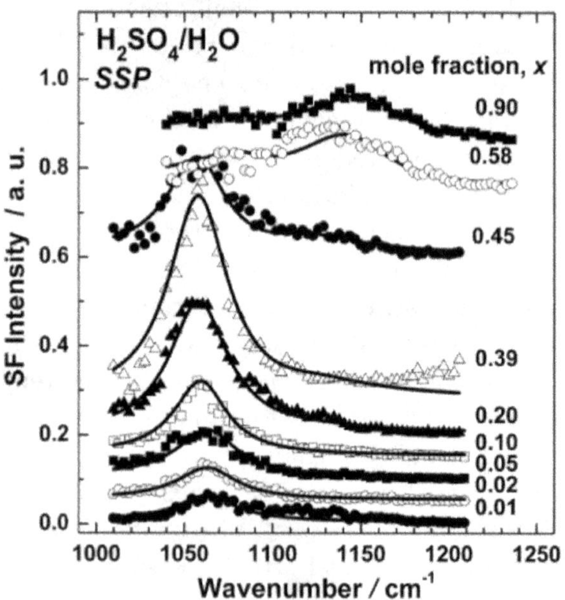

Fig. 7.10.   SSP SF spectra of interfaces of air/sulfuric acid solutions with different sulfuric acid mole fractions x. The SO_3 symmetric stretch of $H_2SO_3^-$ at 1055 cm^{-1} increases with x up to 0.4. It then decreases rapidly to zero with further increase of x, while the SO_2 symmetric stretch of H_2SO_4 at 1150 cm^{-1} emerges. At $x > 0.6$, only the1150 cm^{-1} survives showing that H_2SO_4 no longer dissociates. The solid curves are guides for eyes (after Ref. [31]).

was detected in the SPS spectra, indicating that the symmetric axis of the surface SO_3 was close to the surface normal. The hydrogen-bonded OH spectral intensity was found to decrease with increasing x, reflecting the decrease of H_2O concentration and possible surface structural change of H_2O. Figure 7.10 also shows that increase of x beyond 0.4 sees reduction of the 1055 cm^{-1} mode and emergence of a new mode at 1150 cm^{-1} that can be attributed to the symmetric stretch of SO_2 of H_2SO_4. Above x~0.6, the 1055 cm^{-1} mode practically disappears, showing that H_2SO_4 is no longer dissociated at the surface. This correlates well with dissociation of H_2SO_4 in the bulk that also stops above x = 0.6 as noticed by Raman spectroscopy. The above results and interpretation have been substantiated by molecular dynamics (MD) simulations on cases with x $\leq$ 0.3.[32] In highly diluted solutions (x < 0.05), SO_4^{2-} from dissociation of HSO_4^- in the bulk was detected by Raman observation of the symmetric and antisymmetric modes of SO_4 at 981 and 1104 cm^{-1}, respectively, but seemed too weak to be detected at the surface by SFVS.

Many other interfaces of liquids have been interrogated by SFVS, which is now well established as the only effective and versatile probe for liquid interfaces. In some cases such as alcohols, the key surface structural information can be read off directly from the spectra if the modes are identifiable, but generally, spectra could be complex and theoretical help is needed to extract structural information from them. Fortunately, recent advances in MD simulations have facilitated such tasks.

7.3.2. *Nonpolar liquids*

Nonpolar liquids consist of nonpolar molecules with inversion symmetry. Therefore, in bulk nonpolar liquids, both the second order nonlinear polarizability, $\overset{\leftrightarrow}{\alpha}^{(2)}$, of molecules and the second order nonlinear susceptibilities, $\overset{\leftrightarrow}{\chi}^{(2)}$, of bulk liquids vanish under electric–dipole (ED) approximation. One would expect reflected SFG from a nonpolar liquid interface to be very weak and probably dominated by the electric quadrupole (EQ) contribution from the bulk. This is certainly true if the surface molecules are as randomly oriented as in the bulk, but may not be if they are orientation-ordered.

Nonpolar molecules orientation-ordered at an interface experience an asymmetric environment in the surface normal direction; their $\overset{\leftrightarrow(2)}{\alpha}$ affected by local field and molecular interactions with neighbors is no longer ED forbidden, and so is $\overset{\leftrightarrow(2)}{\chi_S}$, although they may still be weak and comparable to EQ bulk contribution in reflected SFG. It turns out that because the detectable vibrational modes of a nonpolar liquid can be divided by symmetry into IR-active and Raman-active groups, it is possible to obtain $\overset{\leftrightarrow(2)}{\chi_S}$ spectra of nonpolar liquids from phase-sensitive reflective and transmitted SFVS measurement.[33] As we shall see, this comes from the fact that only bulk IR-active modes contribute to the reflected SFVS and their contribution can be extracted from the transmitted SFVS. In the following discussion, we shall focus on vibrational resonances characterized by the imaginary part of nonlinear susceptibilities measured by phase-sensitive SFVS (Sec. 3.5).

As described in Secs. 2.1 and 2.2 (or more rigorously in Ref. [2.2]), SFVS measures

$$\overset{\leftrightarrow(2)}{\chi_{S,eff}} \equiv \overset{\leftrightarrow(2)}{\chi_{SD}} - \hat{z} \cdot \overset{\leftrightarrow(2)}{\chi_{q3}} - \frac{\overset{\leftrightarrow(2)}{\chi_{BBQ}}}{i\Delta k_z^{II}} \tag{7.1}$$

with $\overset{\leftrightarrow(2)}{\chi_{BBQ}} \equiv [\overset{\leftrightarrow(2)}{\chi_{q1}} \cdot i\vec{k}_1^{II} + \overset{\leftrightarrow(2)}{\chi_{q2}} \cdot i\vec{k}_2^{II} - i(\vec{k}_1^{II} + i\vec{k}_2^{II}) \cdot \overset{\leftrightarrow(2)}{\chi_{q3}}]$, where $\Delta k_z^{II} = \Delta k_{tz}^{II}$ for transmitted SFG and $\Delta k_z^{II} = \Delta k_{rz}^{II}$ for reflected SFG. (EQ contribution from the interface layer is neglected here or can be lumped into $\overset{\leftrightarrow(2)}{\chi_{SD}}$.) In explicit tensorial notation with the Cartesian coordinates, we have

$$(\chi_{BBQ}^{(2)})_{\alpha\beta\gamma} = \sum_\kappa \chi_{q1,\alpha(\bar{\kappa}\beta)\gamma}^{(2)}(ik_{1,\kappa}^{II}) + \chi_{q2,\alpha\beta(\bar{\kappa}\gamma)}^{(2)}(ik_{2,\kappa}^{II}) -$$
$$i(k_{1,\kappa}^{II} + k_{2,\kappa}^{II})\chi_{q3,(\bar{\kappa}\alpha)\beta\gamma}^{(2)}] \tag{7.2}$$
$$(\hat{z} \cdot \overset{\leftrightarrow(2)}{\chi_{q3}})_{\alpha\beta\gamma} = \chi_{q3,(\bar{z}\alpha)\beta\gamma}^{(2)}$$

where the bracketed sub-indices indicate the EQ field component with the one under a bar referring to the direction of the wave vector component. For example, $\chi^{(2)}_{q3,(\bar{z}\alpha)\beta\gamma}$ refers to the SFVS process in which $E_{2\gamma}(\omega_2)$ excites an ED vibrational transition and $E_{1\beta}(\omega_1)$ excites a virtual ED transition from the vibrational excited state to electron excited states, followed by a virtual EQ transition back to the ground state with emission of $E_{3\alpha}(\omega_3)$. It is seen that $(\hat{z} \cdot \overleftrightarrow{\chi}^{(2)}_{q3})_{\alpha\beta\gamma} = \chi^{(2)}_{q3,(\bar{z}\alpha)\beta\gamma}$ vanishes if the vibrational transition is IR-inactive. With proper beam geometry and polarizations, it is possible to have $\frac{\overleftrightarrow{\chi}^{(2)}_{BBQ}}{i\Delta k_z^{II}}$ vanish in reflected SFG, and $\chi^{(2)}_{q3,(\bar{z}\alpha)\beta\gamma}$ extracted from transmitted SFG measurement, thus allowing deduction of $\chi^{(2)}_{SD,\alpha\beta\gamma}$.[33]

Consider the case of SFVS with SSP, SPS, and PSP polarizations. It can be shown from Eqs. (7.1) and (7.2) that SFVS measures (Ref. [2.2])

$$(\chi^{(2)}_{S,eff})_{SSP} = \hat{S} \cdot \overleftrightarrow{\chi}^{(2)}_{S,eff} : \hat{P}\hat{P} = (\chi^{(2)}_{SD})_{yyz} - \chi^{(2)}_{q3,(\bar{z}y)yz}$$

$$+ \frac{\cos\theta_2^{II}}{k_{2,z}^{II}\Delta k_z^{II}}(\chi^{(2)}_{q1,y(\bar{z}y)z} - \chi^{(2)}_{q3,(\bar{z}y)yz})(\vec{k}_1^{II} \times \vec{k}_2^{II}) \cdot \hat{y}$$

$$(\chi^{(2)}_{S,eff})_{SPS} = (\chi^{(2)}_{SD})_{yzy} - \chi^{(2)}_{q3,(\bar{z}y)zy} + \frac{\cos\theta_1^{II}}{k_{1,z}^{II}\Delta k_z^{II}}(\chi^{(2)}_{q2,zy(\bar{z}y)}$$

$$- \chi^{(2)}_{q3,(\bar{z}y)yz})(\vec{k}_2^{II} \times \vec{k}_1^{II}) \cdot \hat{y}$$

where θ_1^{II} and θ_2^{II} are the incident angles of $E_1(\omega_1)$ and $E_2(\omega_2)$, respectively, in liquid. On reflected SSP SFVS, if we set $\vec{k}_1^{II} \parallel \vec{k}_2^{II}$, we measure $(\chi^{(2)}_{S,eff})_{R,SSP} = (\chi^{(2)}_{SD})_{yyz} - \chi^{(2)}_{q3,(\bar{z}y)yz}$. On transmitted SPS–SFVS, if we set $\theta_1^{II} = 0(k_{1x}^{II} = 0)$ so that $E_{1y}(\omega_1) = 0$ we measure $(\chi^{(2)}_{S,eff})_{T,SPS} \propto \chi^{(2)}_{q2,zy(\bar{z}y)} - \chi^{(2)}_{q3,(\bar{z}y)yz}$. Since $\chi^{(2)}_{q2,zy(\bar{z}y)} = 0$ for IR-active and Raman-inactive modes and $\chi^{(2)}_{q3,(\bar{z}y)yz} = 0$ for IR-inactive and Raman-active modes, $\chi^{(2)}_{q3,(\bar{z}y)yz}$ can be extracted directly from the transmitted SPS $(\chi^{(2)}_{S,eff})_{T,SPS}$ spectrum by retaining only the IR-active and Raman-inactive modes in the spectrum. Knowing $\chi^{(2)}_{q3,(\bar{z}y)yz}$, we can then obtain $\chi^{(2)}_{SD,yyz}$ from the reflected SSP $[(\chi^{(2)}_{SD})_{yyz} - \chi^{(2)}_{q3,(\bar{z}y)yz}]$ spectrum. We can also find $\chi^{(2)}_{SD,yzy}$ from the measured $(\chi^{(2)}_{SD})_{yzy} - \chi^{(2)}_{q3,(\bar{z}y)zy}$ with reflected SPS–SFVS and the symmetric relation $\chi^{(2)}_{q3,(\bar{z}y)zy} = \chi^{(2)}_{q3,(\bar{y}z)zy} = \chi^{(2)}_{q3,(\bar{z}y)yz}$.

We take the vapor/benzene interface as an example.[33] The measured spectra of $\mathrm{Im}[(\chi^{(2)}_{SD})_{yyz} - \chi^{(2)}_{q3,(\bar{z}y)yz}]$, $\mathrm{Im}[\chi^{(2)}_{SD})_{yzy} - \chi^{(2)}_{q3,(\bar{z}y)yz}]$, and $-\mathrm{Im}\,\chi^{(2)}_{q3,(\bar{z}y)yz}$ are shown in Fig. 7.11(a). There are three IR-active modes at 3036, 3071, and 3091 cm^{-1} and two Raman-active modes at 3049 and 3062 cm^{-1} known for bulk benzene. These modes are associated with the CH stretch vibrations of the benzene ring. The deduced $\mathrm{Im}\,\chi^{(2)}_{SD,yyz}$ spectrum is plotted in Fig. 7.11(b), but the $\mathrm{Im}\,\chi^{(2)}_{SD,yzy}$ spectrum is very weak (not plotted). Although

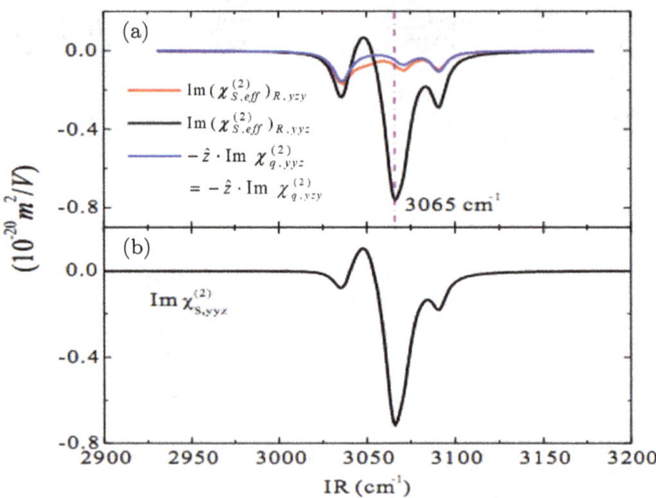

Fig. 7.11.　(a) Spectra of $-\hat{z}\cdot\mathrm{Im}\vec{\chi}^{(2)}_{q3,yyz}$ (blue line), $\mathrm{Im}(\chi^{(2)}_{S,eff})_{R,yyz}$ (black line), and $\mathrm{Im}(\chi^{(2)}_{S,eff})_{R,yzy}$ (red line) deduced from analysis of the SF vibrational spectra of the benzene/vapor interface. (b) A true surface spectrum of the benzene/vapor interface: spectrum of $\mathrm{Im}\,\chi^{(2)}_{S,yyz}$ (denoted as $\mathrm{Im}\,\chi^{(2)}_{SD,yyz}$ in the text) obtained by subtracting the spectrum of $-\hat{z}\cdot\mathrm{Im}\vec{\chi}^{(2),yyz}_{q}$ from the spectrum of $\mathrm{Im}(\chi^{(2)}_{S,eff})_{R,yyz}$ (after Ref. [33]).

benzene is nonpolar, its ED surface spectrum of Im $\chi^{(2)}_{SD,yyz}$ is quite strong and was found to dominate over the EQ bulk contribution of $-$Im $\chi^{(2)}_{q3,(\bar{z}y)yz}$ in the reflected SPP–SFVS, but Im $\chi^{(2)}_{SD,yzy}$ was much weaker. Both IR- and Raman-active modes are present in the spectra, but the Raman-active mode at 3065 cm^{-1} (shifted by 3 cm^{-1} from the bulk frequency) is particularly prominent in the Im $\chi^{(2)}_{SD,yyz}$ surface spectrum and nearly absent in Im $\chi^{(2)}_{SD,yzy}$ (Fig. 7.11(a)). Because it is a CH stretch vibration in the ring plane and is excited by the z component of $E_2(\omega_2)$, the spectrum indicates that the average orientation of the benzene ring plane must be close to the surface normal. Vibrational spectra of other nonpolar liquids obtained by SFVS have also been reported, but the spectral analyses were less extensive.[34,35]

7.3.3. *Surface freezing*

Interfacial molecules of a liquid can be ordered through their interaction with an opposing medium. The interfacial molecular layer being more ordered than the bulk liquid may undergo an order–disorder transition at a higher temperature than the bulk liquid. Such a surface freezing phenomenon is known to occur in many liquids. X-ray diffraction/reflection and linear optics such as ellipsometry have been used to study surface freezing, but they are limited in their capability to probe detailed molecular structure of the interfacial layer. Again, SFVS can be an effective complementary tool for the study.

We discuss here the case of n-eicosane $(C_{20}H_{42})$ liquid that has a bulk freezing temperature at $T_B = 35.6°C$ and a surface freezing temperature at $T_S = 38.6°C$, and was first studied by X-ray diffraction and ellipsometry.[34] Surface SF vibrational spectra of eicosane could be readily observed although eicosane is nonpolar.[35] The SSP and SPS spectra in the CH stretch range of the air/eicosane interface at 40°C (above T_S) and 37°C (above T_B, but below T_S) are presented in Fig. 7.12. Characteristic of long alkyl chains, the spectra show altogether 5 CH stretch modes: CH_3 symmetric stretch at $\sim$2875 cm^{-1}, degenerate CH_3 antisymmetric stretches at $\sim$2960 cm^{-1}, CH_2 symmetric and antisymmetric stretches at $\sim$2850

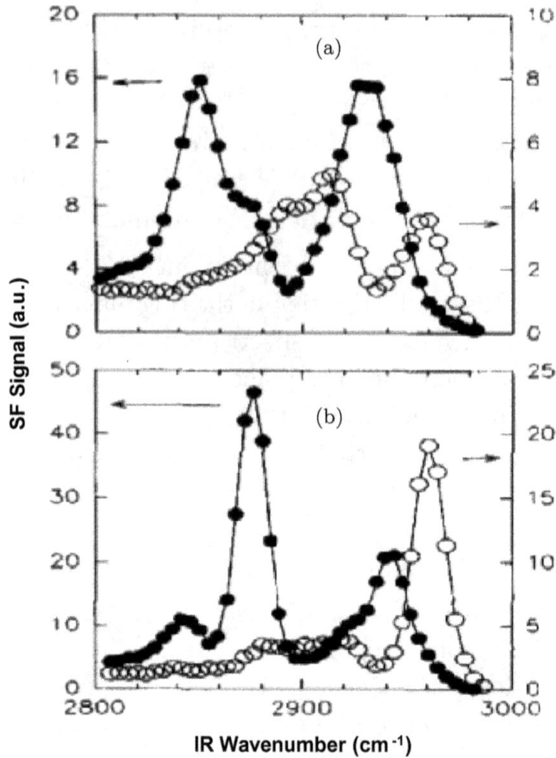

Fig. 7.12. Reflected SF spectra in the CH stretching range for the n-eicosane/vapor interface at (a) 40°C and (b) 37°C. The bulk melting temperature of n-eicosane is at 35.6°C and the surface melting temperature is at 37.5°C. Solid circles are for SSP polarization and open circles are for SPS polarization (after Ref. [35]).

and $\sim$2920 cm^{-1}, and Fermi-resonance between CH$_3$ symmetric stretch and bending overtones at $\sim$2940 cm^{-1}. It is seen that the CH$_2$ modes are prominent above T_S, but greatly suppressed below T_S, indicating that the chains of eicosane are gauche-defective above T_S, but become nearly all-trans below T_S. While the spectral analysis could be carried out more rigorously following the description in Sec.7.2(B), we can see qualitatively from comparison of SSP and SPS spectra that the chains are oriented close to the surface normal both above and below T_S (Sec. 6.3). Surface freezing of eicosane is therefore believed to be a surface phase transition process that as T drops to T_S, the surface layer through chain-chain interaction suddenly becomes more densely packed, and the defective CH_2 chains become nearly all-trans.

7.4. Liquid Crystals

The liquid crystalline phase is a state of matter between liquid and solid, in which a material can flow like liquid, and has no 3D positioning order, but exhibits long-range orientation order and optical birefringence. Liquid crystals (LC) are often composed of anisotropic organic molecules that are strongly coupled and correlated on the macroscopic scale. The direction of the orientation order (known as the director) of a single LC domain depends on its boundary condition, but can be easily affected by external perturbation. Unlike other materials, the bulk structure of LC in terms of its director orientation is also easily affected by its interfacial structure. An LC film sandwiched between two substrates can be prepared to have its director pointing in a selected direction with the desired optical birefringence. An applied electric or magnetic field can change the director and the optical birefringence, and accordingly modify light transmitted through the LC film. This is the basic principle underlying LC display devices that have been the corner stone of the LC industry in the past 40 years.

Without the boundary or field effect, the director of an LC bulk can point in any direction, but in the presence of an interface, the director can be guided to point along a selected direction. For instance, if a nematic LC film of 8CB (4'-cyano-4-octylbiphenyl, $CN(C_6H_4)_2(CH_2)_7CH_3$, Fig. 7.13(a)) is sandwiched between substrates coated by a surfactant layer of DMOAP, [n, n-dimethyl-n-octadecyl-3-amino-propytrimethoxysily-chloride, $CH_3(CH_2)_{17}(Me)_2 N^+(CH_2)_3Si(OMe)_3Cl^-$], the director of 8CB is perpendicular to the surface plane, known as homeotropic alignment. If the substrates are coated by polyimide and rubbed, the director is inclined to the surface plane along the rubbing direction, known as homogeneous alignment. Obviously, LC molecular interactions with the coated surfaces are responsible for the LC bulk alignment, and it should be interesting to have a detailed microscopic understanding of the aligning mechanism. We show in this section how SHG/SFG studies can provide a clear picture on surface-induced bulk alignment of LC at the molecular level. We start with the question how an LC monolayer is oriented and arranged on various substrates and follow with the question how the monolayer induces alignment of an LC film on it.

We take 8CB as a representative LC material. An 8CB molecule consists of a CN head group, a somewhat twisted biphenyl chromophore core, and an alkyl chain (Fig. 7.13(a)). It has a conformation of CN parallel to the core axis and the all-trans alkyl chain tilted away from the core plane by $\sim$35°. Similar to the case of a 5CT monolayer on water described in Sec. 6.3, nonresonant SHG can be used to probe orientation and arrangement of 8CB cores and SF vibrational spectroscopy to probe CN and alkyl chains for 8CB monolayers. We discuss here only the SHG measurements. They were performed on 8CB monolayers on water, clean glass, and DMOAP- and MAP [n-methyl-aminopropyltrimethoxy silane, $MeNH(CH_2)_3Si(OMe)_3$]-coated glasses.[36] In all cases, the LC monolayer was found to be azimuthally isotropic with the core axis of molecules tilted at $\sim$67° from the surface normal. It is however known that LC films sandwiched between clean glass plates and MAP-coated glass plates are homogeneously aligned and the one between DMOAP-coated plates are homeotropically aligned. The different bulk alignments obviously must come from the different surface coatings and the homeotropic alignment seems to result from the surface being decorated by the long alkyl chains of DMOAP.

We need to know how the "orientation-anchored" surface LC monolayer (ML) affects orientation of additional LC molecules adsorbed on it and vice versa. Figure 7.13(b) shows the SH response to evaporation of 8CB molecules onto different substrates; it displays the SH signal versus 8CB evaporation time.[37] The signal was calibrated to yield corresponding surface density of 8CB directly adsorbed on substrates. In all cases, it is seen that the signal saturates toward 1 ML. Afterwards, the signal decreases very gradually and the change is appreciable only after $\sim$10 ML. The result indicates that, first, beyond 1ML, additional adsorbed 8CB molecules must have appeared in quadrupole pairs (Fig. 7.13(c)), which contributes

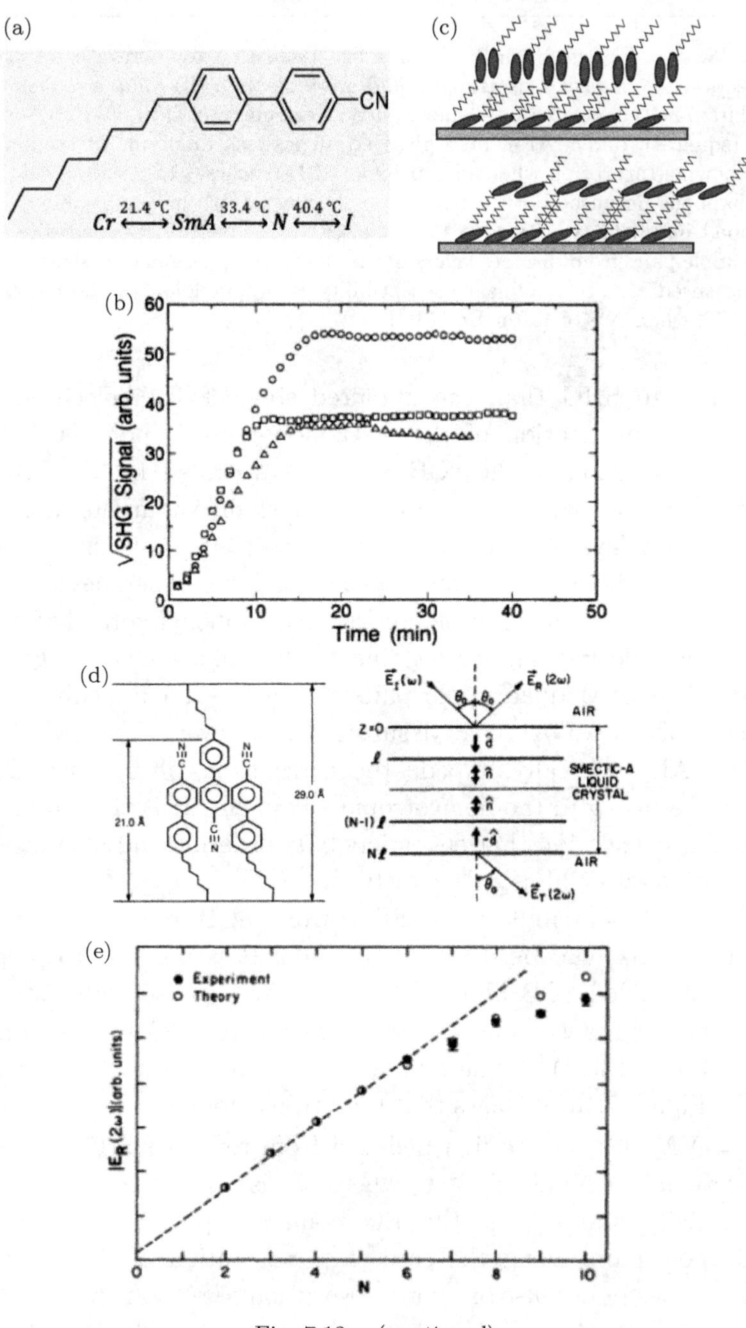

Fig. 7.13. (*continued*)

Fig. 7.13. (a) Sketch of an 8CB molecule. Transition temperatures of crystal ↔smectic-A ↔nematic ↔isotropic liquid are indicated. (b) Square root of SHG signal from 8CB deposited by evaporation on clean glass (circles), DMOAP-coated glass (squares), and MAP-coated glass (triangles), as functions of evaporation time showing saturation when adsorption of 8CB reaches ∼1ML (after Ref. [37]). This happens because above 1ML, 8CB molecules adsorb in quadrupole pairs as described in (c) in two possible buildups. (d) In a freely suspended film, 8CB quadrupole pairs from discrete layers in the smectic-A phase and as shown in (e), the measured surface nonlinear susceptibility is proportional to the number of layers, N, when $N \leq 6$ (after Ref. [38]).

much less to SHG than the unpaired single 8CB molecules, and second, the orientation of the 8CB monolayer is not affected by further adsorption of the 8CB layer. As sketched in Fig. 7.13(c), 8CB quadrupole pairs can adsorb through chain–chain interaction on an 8CB-monolayer-coverd substrate in two configurations, one with cores closely along the surface normal and the other having cores incline toward the surface plane. Usually, biphenyl cores like to be close to a dielectric surface, leading to the homogeneous alignment of adsorbed 8CB quadrupole pairs if there is no impediment. In the case of a DMOAP-coated substrate, however, the alkyl chains of DMOAP physically impede the cores from tilting toward the surface, resulting in the homeotropic alignment. It is the long-range molecular correlation that engenders bulk ordering and alignment of LC multilayers or films on a substrate.

The EQ contribution to SHG from 8CB quadrupole pairs, although weak, can be detected by SHG. It is possible to prepare freely suspended 8CB films in the smectic-A phase, with oriented quadrupole pairs forming discrete bilayers parallel to the surface plane (Fig. 7.13(d)).[38] The number of bilayers, N, in a film can be varied. Figure 7.13(e) shows that the surface nonlinear susceptibility, $\chi_S^{(2)} = N N_S \alpha_Q^{(2)}$, of the film deduced from reflected SHG is closely proportional to N until $N \geq 6$, where N_S is the surface quadrupole pair density and $\alpha_Q^{(2)}$ is the EQ nonlinear polarizability of the quadrupole pairs along the core axis. Saturation of SHG beyond $N = 6$ comes in because of strong absorption of 8CB at the measured SH frequency and multiple reflections at the film surfaces. From the measurement, the value of $\alpha_Q^{(2)}$ could be deduced, and was found to be $\sim 7\%$ of the ED polarizability $\alpha_D^{(2)}$ for single 8CB molecules.

Practical LC display devices consist of an LC film sandwiched between rubbed polymer-coated substrates in homogeneous alignment along the rubbing direction. How the rubbed polymer surface could induce the homogeneous bulk alignment was a mystery that was solved by SHG/SFG study. As will be discussed in Sec. 9.5(D), rubbing can lead to oriented surface polymer structure for LC molecules to adsorb on with orientation along the rubbing direction. Here, we describe the use of SHG to probe the geometry of an adsorbed LC monolayer on a rubbed polymer surface, from which we can successfully predict the bulk LC alignment.[39] We consider the case of an 8CB film sandwiched between two polyimide-coated glass plates rubbed in the same direction.

As mentioned in Sec. 6.3, SHG measurement on an azimuthally anisotropic monolayer allows determination of five parameters to characterize the orientation distribution of molecules in the monolayer. Application of the technique on an 8CB monolayer adsorbed on a polyimide-coated substrate rubbed with different strengths yielded results presented in Fig. 7.14. The polar angle distribution was found to center at $\sim 80°$ of a Gaussian profile having a $1/e$ width of $\sim 7°$. It did not change much with the rubbing strength (Fig. 7.14(a)). The azimuthal distribution, however, did depend on the rubbing strength; stronger rubbing created larger anisotropy (Fig. 7.14(b)). The deduced orientation distribution of the adsorbed 8CB monolayer now acted as the boundary condition in the standard Landau-de Gennes formalism for LC to predict bulk alignment of a nematic 8CB film with known bulk orientation order parameter and elastic constants. The calculated result on dependence of the tilt angle of the 8CB director from surface (known as the bulk pretilt angle) on the rubbing strength is plotted in Fig. 7.14(c) in comparison with tilt angles measured by ellipsometry;[38] the rubbing strength is described by the leading antisymmetric anisotropic parameter d_1 of the azimuthal orientation distribution in Fig. 7.14(b). The agreement between theory and experiment is good. This work proves that through SHG probing of adsorbed LC monolayers, we can have

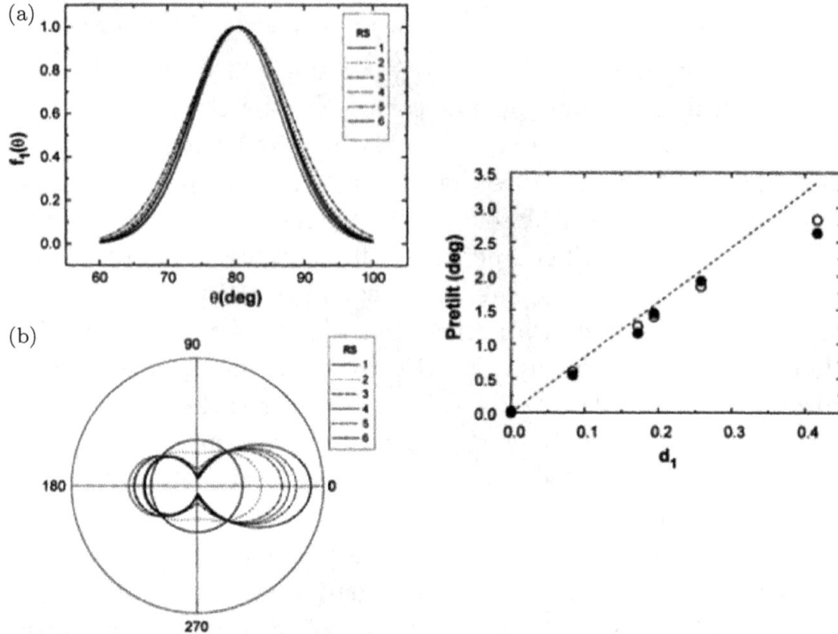

Fig. 7.14. (a) Polar and (b) azimuthal orientation distributions of an 8CB monolayer adsorbed on a rubbed polyimide-coated glass with different rubbing strengths, extracted from SHG measurements. While the polar orientation distribution is hardly affected by rubbing, the azimuthal orientation distribution becomes more anisotropic with stronger rubbing. (c) Pretilt angles versus rubbing strengths (represented by parameter d_1 deduced from the SHG measurement in (b)) (open circles) in comparison with those directly measured by ellipsometry (filled circles) (after Ref. [39]).

a quantitative understanding of surface-induced bulk alignment of LC films in practical LC display devices.

7.5. Ionic Liquids

An ionic liquid (IL) is composed of cations and anions in liquid form. Molten salts are IL, but require high temperature to get melted. It has been found that many organic ILs exist below 100°C. They have remarkable properties such as low vapor pressure, high viscosity, high thermal stability and heat capacity, moderate thermal and electric conductivity, selective high solubility for molecules, variable acidity, and others. They have been considered in many potential

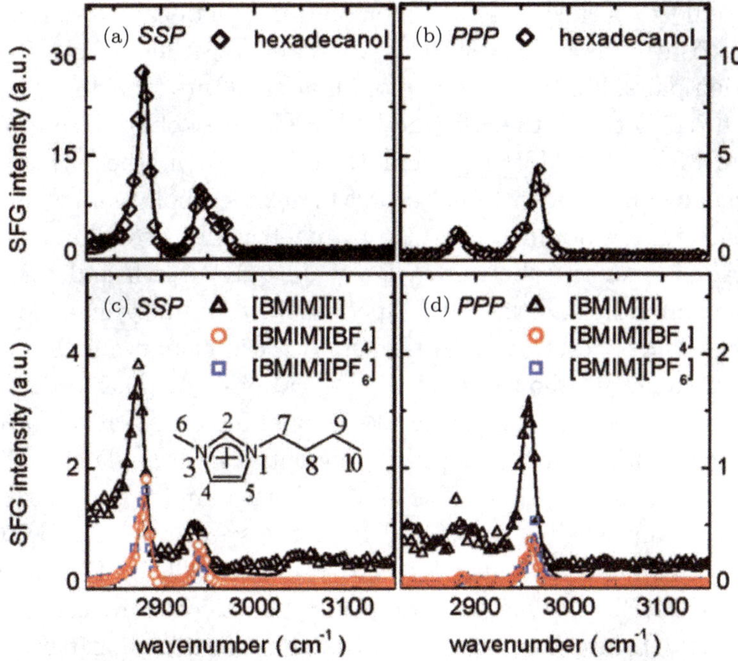

Fig. 7.15. (a) SSP and (b) PPP SF Spectra of a hexadecanol monolayer on water in the CH stretch range in comparison with (c) SSP and (d) PPP spectra of air/ionic liquid interfaces of BMIM(I), BMIM(BF$_4$), and BMIM(PF$_6$). Spectra of all four liquids are similar and are characteristics of the protruding all-trans alkyl chains of molecules polar oriented at the interfaces (after Ref. [40]).

applications, for example, as media for heat transfer and storage, as electrolytes in batteries, as materials for separation of polymers in plastic wastes, as catalysts, and so on. To understand surface functional properties of ILs, we need to know their surface structures and how they are affected by surface charges.

Because of the low vapor pressure of IL, it is possible to characterize their free surface in UHV, but for buried IL interfaces, SF spectroscopy is arguably the only effective probe. Surface studies of IL however have not yet been extensive. We describe here how SFVS can be used to characterize interfaces of IL taking representative ILs composed of 1-butyl-3-methylimidazolium (BMIM, see the inset in Fig. 7.15) cations and different anions as examples.[40]

We can consider an IL as a liquid with two components that co-adsorb or competitively adsorb at an interface (Sec. 6.5). However, one would expect that strong Coulomb interaction between ions is likely to attract cations and ions as close together as possible to form ion pairs, limited only by molecular structure blocking the way. Figure 7.15 presents the SSP and PPP SF spectra of air/IL interfaces of BMIM(I), BMIM (BF_4), and BMIM (PF_6) in the CH stretch range in comparison with the spectra of hexadecanol. In all cases, the SSP spectra are dominated by the symmetric CH_3 stretch mode and the Fermi resonance between CH_3 symmetric stretch and overtone bending modes, and the PPP spectra by the antisymmetric CH_3 stretch. The spectra are also similar to those of butynol (Fig. 7.9). Spectral analysis showed that BMIM cations are polar-ordered at the air/IL interface, very much like alcohol molecules at air/alcohol interfaces, with their chains protruding out at a tilt angle of $\sim$42°. In comparison, the tilt angle of hexadecanol was found to be $\sim$ 23°. The surface chain densities estimated from the spectra for BMIM were 1/2 of that of hexadecanol for BMIM(I) and 1/3 for BMIM(BF_4) and BMIM(PF_6). This indicates that the anions are coadsorbed at the air/IL interface with BMIM cations, forming ion pairs with BMIM; the larger area per molecule occupied by BF_4 and PF_6 than iodine(I) lowers the BMIM surface chain density. Unfortunately, surface spectra of these anions were not measured (or not observed), limiting our knowledge on their adsorption at the surface.

The situation of IL/solid interfaces is more complicated as chemical interaction between ionic molecules and solid plays an essential role but charge interaction between ions may also be important. We can perhaps still consider cations as the major component of IL with anions as the minor component more or less freely tagging along. We discuss here the cases of interfaces of BMIM(PF_6) with hydrophobic (silane-coated) and hydrophilic (bare) silica substrates.[41] In Figs.7.16(a)–7.16(c), we display the reflected SSP SF spectra in the CH stretch range of BMIM(PF_6)/air, BMIM(PF_6)/ODS-coated silane, and BMIM(PF_6)/silica, respectively, (where ODS denotes deuterated octadecylsilane). The spectra of BMIM(PF_6)/air and BMIM(PF_6)/ODS-coated silane in Figs. 7.16(a) and 7.16(b) are fairly similar except that the antisymmetric CH_3 stretch at $\sim$2970 cm^{-1} of the latter is more pronounced because

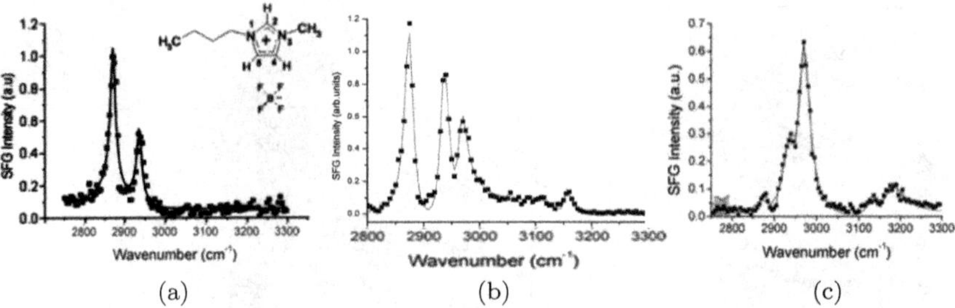

Fig. 7.16. SSP SF spectra in the CH stretch range of BMIM(BF$_4$) interfaces with (a) vacuum, (b) deuterated octadecylsilane (ODS)-coated silica, and (c) bare silica. Similarity and difference in the molecular orientations and arrangements of BMIM at the three interfaces are reflected in their spectra. See the text for more detailed description (after Ref. [41]).

the chains are tilted and more ordered. This is what one would expect knowing that BMIM molecules should be polar-oriented at the interface with their butyl chains pointing toward the interface through chain–chain interaction with ODS. Anions should be close to the BMIM cores to form ion pairs. The spectrum of BMIM (PF$_6$)/silica (Fig. 7.16(c)) is very different. It resembles more the spectrum of the butanol/silica interface,[30] indicating that the BMMI cores are chemically adsorbed on silica with their butyl chains pointing toward the IL bulk. Compared to the spectrum of BMIM (PF$_6$)/air, the CH$_3$ symmetric stretch and Fermi resonance modes are greatly reduced, but the CH$_3$ antisymmetric stretch mode is significantly enhanced just like the case of butanol/silica versus butanol/air. As explained in Sec. 7.4(A), this is due to the existence of a partially polar-ordered monolayer of BMIM next to the surface monolayer with chains facing each other; SFGs from the two layers interfere destructively for the symmetric stretch, but constructively for the antisymmetric mode at $\sim$2970 cm^{-1}.[30] (This $\sim$2970 cm^{-1} mode was interpreted in Ref. [41] as coming from the methyl group attached to the imidazolium ring). With the imidazolium rings adsorbed at the interface, their ring stretch modes around 3180 cm^{-1} could also be observed in the BMIM/silica spectrum. Anions are again expected to co-adsorb and form ion pairs with imidazolium cores at the interface.

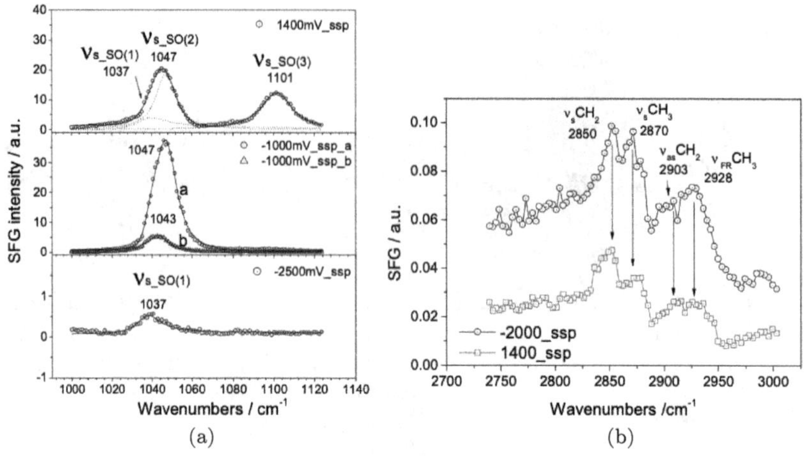

Fig. 7.17. SSP SF vibrational spectra of an electrochemical interface of BMIM(Otf)/Pt with positive and negative biased potentials. (a) CH stretch modes of BMIM. (b) SO stretch modes of SO_3 of Otf. The positive potential on Pt enhances the SO stretches and the negative potential enhances CH stretches (after Ref. [42]).

Charges on a solid substrate interfacing with IL certainly can affect preferential adsorption of cations and anions and their adsorption geometry at the interface. We consider the case of BMMI(Otf, trifluromethane sulfonate, SO_3-CF_3) in contact with a Pt electrode.[42] In this case, both CH stretches of BMMI cations and SO stretches of Otf anions can be detected. As seen in Fig. 7.17(a), the SSP spectra show that the SO stretches reduce significantly in intensity when the potential of Pt switches from +1.4v to –2.5v, and the CH stretches (Fig. 7.17(b)) increase correspondingly. This indicates that anions adsorb more on positive Pt and cations adsorb more on negative Pt. Spectral changes with the applied potential are however difficult to understand. Ions desorbed from Pt could still be around the electrode with net polar orientation and contribute to SFG. Obviously, careful and systematic studies of IL interfaces are still needed to acquire better understanding of their structure.

7.6. Summary and Prospects

Sum frequency spectroscopy through measurement of surface non-linear susceptibility $\overset{\leftrightarrow}{\chi}_S^{(2)}$ can probe interfacial structure of a bulk material. The technique has been used to explore different materials with unique advantages. In particular, for liquid interfaces and buried interfaces, it is currently the only viable probing technique.

- Surface reconstruction of crystalline solids can be monitored.
- Surface phonons of crystalline solids and their changes with surface reconstruction and environment can be observed.
- Surface structural information from $\overleftrightarrow{\chi}_S^{(2)}$ can complement X-ray studies on identifying correct microscopic surface structures.
- Electronic and phonon structures of buried solid/solid interfaces can be probed, and so can the interfaces of multiferroics, magnetic, and topological crystals.
- SF vibrational spectroscopy (SFVS) is a unique probe for liquid interfaces, and has provided many opportunities to interrogate liquid interfacial structures that can easily change with the environment or opposing medium.
- SFVS can be used to find the microscopic structural change on surface freezing transition of liquids. It could also be used to probe surface melting of solids.
- SFVS can be used to study orientation anchoring of liquid crystals (LC) at an interface and was able to provide molecular-level understanding of surface-induced bulk alignment of LC films in practical liquid crystal display devices.
- SFVS also provide unique opportunities to probe interfacial structures of ionic liquids.

More SF spectroscopy works on interfaces of exotic solid materials can be anticipated in the future. Being the only effective probe for interfaces of fluid media, SFVS is expected to continue leading the way in exploration of liquid and liquid-like interfaces. There are also areas yet to be looked into, for example, interfacial structures of micro-fluidity, advancing surfaces of growing crystals, and dynamics of interfacial structures. Currently, applications of SF spectroscopy are limited to the IR frequency range above $\sim$16 THz because of the lack of sufficiently energetic IR pulses below 16 THz. The situation may change with new developments of THz sources in the future.

References

1. Heinz, T. F.; Loy, M. M. T.; Thompson, W. A.: Study of Si(111) Surfaces by Optical Second Harmonic Generation — Reconstruction and Surface Phase-Transformation. *Phys. Rev. Lett.* 1985, 54, 63–66.
2. Heinz, T. F.; Loy, M. M. T.; Thompson, W. A.: Study of Symmetry and Disordering of Si(111)-7 × 7 Surfaces by Optical Second Harmonic Generation. *J. Vac. Sci. Technol. B* 1985, 3, 1467–1470.
3. Rudnick, J.; Stern, E. A.: Second Harmonic Radiation from Metal Surfaces. *Phys. Rev. B* 1971, 4, 4274–4290.
4. Jha, S.; Warke, C. S.: Interband Contributions to Optical Second Harmonic Generation at a Metal Surface. *Phys. Rev.* 1967, 153, 751–759.
5. Sipe, J. E.; So, V. C. Y.; Fukui, M.; Stegeman, G. I.: Analysis of Second Harmonic Generation at Metal Surfaces. *Phys. Rev. B* 1980, 21, 4389–4402.
6. Brown, F.; Parks, R. E.; Sleeper, A. M.: Nonlinear Optical Reflection from a Metallic Boundary. *Phys. Rev. Lett.* 1965, 14, 1029–1031.
7. Tom, H. W. K.; Aumiller, G. D.: Observation of Rotational Anisotropy in the Second Harmonic Generation from a Metal Surface. *Phys. Rev. B* 1986, 53, 8818–8821.
8. van Driel, H. M.: Second Harmonic Generation from Metal Surfaces: Beyond Jellium. *Appl. Phys. A* 1994, 59, 545–552.
9. Wang, F. X.; Rodriguez, F. J.; Albers, W. M.; Ahorinta, R.; Sipe, J. E.; Kauranen, M.: Surface and Bulk Contributions to the Second Order Nonlinear Optical Response of a Gold Film. *Phys. Rev. B* 2009, 80, 233402.
10. Fuchs, R.; Kliewer, K. L.: Optical Modes of Vibration in an Ionic Crystal Slab. *Phys. Rev.* 1965, 140, A22076–A2088.
11. Chin, R. P.; Huang, J. Y.; Shen, Y. R.; Chuang, T. J.; Seki, H.: Interaction of Atomic Hydrogen with the Diamond C(111) Surface Studied by Infrared-Visible Sum-Frequency-Generation Spectroscopy. *Phys. Rev. B* 1995, 52, 5985–5995.
12. Liu, W. T.; Shen, Y. R.: Surface Vibrational Modes of Alpha-Quartz(0001) Probed by Sum-Frequency Spectroscopy. *Phys. Rev. Lett.* 2008, 101, 016101.
13. Tong, Y.; Wirth, J.; Kirsch, H.; Wolf, M.; Saalfrank, P.; Campen, R. K.: Optically Probing Al-O and O-H Vibrations to Characterize Water Adsorption and Surface Reconstruction on Alpha-Alumina: An Experimental and Theoretical study. *J. Chem. Phys.* 2015, 142, 054704.

14. Cao, Y.; Chen, S.; Li, Y.; Gao, Y.; Yang, D.; Shen, Y. R.; Liu, W. T.: Evolution of Anatase Surface Active Sites Probed by *In Situ* Sum-Frequency Phonon Spectroscopy. *Sci. Adv.* 2016, 2, e1601162.

15. Liu, W. T.: (unpublished).

16. Li, X.; Brigiano, F. S.; Pezzotti, S.; Liu, X.; Chen, W.; Chen, H.; Li, Y.; Li, H.; Shen, Y. R.; Gaigeot, M-P.; Liu, W. T.:(to be published).

17. Sung, J. H.; Zhang, L. N.; Tian, C. S.; Waychunas, G. A.; Shen, Y. R.: Surface Structure of Protonated R-Sapphire ($1\bar{1}02$)Studied by Sum-Frequency Vibrational Spectroscopy. *J. Am. Chem. Soc.* 2011, 133, 3846–3853.

18. Heinz, T. F.; Himpsel, F. J.; Palange, E.; Burstein, E.: Electronic-Transitions at the CaF_2/Si(111) Interface Probed by Resonant 3-Wave-Mixing Spectroscopy. *Phys. Rev. Lett.* 1989, 63, 644–647.

19. Yu, P.; Chu, Y. H.; Ramesh, R.: Emergent Phenomena at Multiferroic Heterointerfaces. *Phil. Trans. R. Soc. A* 2012, 370, 4856–4871.

20. Li, Z. X.; Wang, F.; Yao. H.; Lee, D. H.: What Makes the T_c of Monolayer FeSe on SrTiO3 so High: A Sign-Problem-Free Quantum Monte Carlo Study. *Sci. Bull.* 2016, 61, 925–930.

21. Liu, W. T.: (private communications, unpublished).

22. McIver, J. W.; Hsieh, D.; Drapcho, S. G.; Torchinsky, D. H.; Gardner, D. R.; Lee, Y. S.; Gedik, N.: Theoretical and Experimental Study of Second Harmonic Generation from the Surface of the Topological Insulator Bi_2Se_3. *Phys. Rev. B* 2012, 86, 035327.

23. Shi, S.; Zhang, Y.; Yao, M.; Ji, F.; Qian, D.; Qiao, S.; Shen, Y. R.; Liu, W. T.: Surface and Bulk Contributions to the Second Harmonic Generation in Bi_2Se_3. *Phys. Rev. B* 2016, 94, 205307.

24. Pan, R. P.; Wei, H. D.; Shen, Y. R.: Optical Second Harmonic Generation from Magnetized Surfaces. *Phys. Rev. B* 1989, 39, 1229–1234.

25. Reif, J.; Zink, J. C.; Schneider, C. M.; Kirschner, J.: Effects of Surface Magnetism on Optical Second Harmonic Generation. *Phys. Rev. Lett.* 1991, 67, 2878–2881.

26. Prinz, G. A.: Stabilization of Bcc Co Via Epitaxial-Growth on GaAs. *Phys. Rev. Lett.* 1985, 54, 1051–1054.

27. Superfine, R.; Huang, J. Y.; Shen, Y. R.: Nonlinear Optical Studies of the Pure Liquid Vapor Interface — Vibrational-Spectra and Polar Ordering. *Phys. Rev. Lett.* 1991, 66, 1066–1069.

28. Wang, H. F.; Gan, W.; Lu, R.; Rao, Y.; Wu, B. H.: Quantitative Spectral and Orientational Analysis in Surface Sum Frequency Vibrational Spectroscopy. *Int. Rev. Phys. Chem.* 2005, 24, 191–256.

29. Stanners, C. D.; Du, Q.; Chin, R. P.; Cremer, P.; Somorjai, G. A.; Shen, Y. R.: Polar Ordering at the Liquid-Vapor Interface of N-Alcohols (C-1-C-8). *Chem. Phys. Lett.* 1995, 232, 407–413.

30. Liu, W; Zhang, L.; Shen, Y. R.: Interfacial Layer Structure at Alcohol/Silica Interfaces probed by Sum-Frequency Vibrational Spectroscopy. *Chem. Phys. Lett.* 2005, 412, 206–209.

31. Miyamae, T.; Morita, A.; Ouchi, Y.: First Acid Dissociation at an Aqueous H_2SO_4 Interface with Sum Frequency Generation Spectroscopy. *Phys. Chem. Chem. Phys.* 2008, 19, 2010–2013.

32. Ishioyama, T.; Morita, A.; Miyamae, T.: Surface Structure of Sulfuric Acid Solution Relevant to Sulfate Aerosol: Molecular Dynamics Simulation Combined with Sum Frequency Generation Measurement. *Phys. Chem. Chem. Phys.* 2011, 13, 20965–20973.

33. Sun, S. M.; Tian, C. S.; Shen, Y. R.: Surface Sum-Frequency Vibrational Spectroscopy of Nonpolar Media. *Proc. Nat. Acad. Sci.* 2015, 112, 5883–5887.

34. Matsuzaki, K.; Nihonyanaki, S.; Yamaguchi, S.; Nagata, T.; Tahara, T.: Quadrupolar Mechanism for Vibrational Sum Frequency Generation at Air/Liquid Interfaces: Theory and Experiment. *J. Chem. Phys.* 2019, 151, 064701.

35. Sefler, G. A.; Du, Q.; Miranda, P. B.; Shen, Y. R.: Surface Crystallization of Liquid N-Alkanes and Alcohol Monolayers Studied by Surface Vibrational Spectroscopy. *Chem. Phys. Lett.* 1995, 235, 347–354.

36. Guyot-Sionnest, P.; Hsiung, H.; Shen, Y. R.: Surface Polar Ordering in a Liquid Crystal Observed by Optical Second Harmonic Generation. *Phys. Rev. Lett.* 1986, 57, 2963–2966.

37. Mullen, C. S.; Guyot-Sionnest, P.; Shen, Y. R.: Properties of Liquid Crystal Monolayers on Silane Surfaces. *Phys. Rev. A* 1989, 39, 3745–3747.

38. Hsiung, H.; Shen, Y. R.: Probing the Struture of Freely Suspended Smectic-A Films by Optical Second Harmonic Generation. *Phys. Rev. A* 1986, 34, 4303–4309.

39. Zhuang, X.; Marrucci, L.; Shen, Y. R.: Surface-Monolayer-Induced Bulk Alignment of Liquid Crystals. *Phys. Rev. Lett.* 1994, 73, 1513–1516.

40. Jeon, Y.; Sung, J.; Bu, Y.; Vaknin, D.; Ouchi, Y.; Kim D.: Interfacial Restructuring of Ionic Liquid Crystals Determined by Sum-Frequency Generation Spectroscopy and X-ray Reflectivity. *J. Phys. Chem. C* 2008, 112, 19649–19654.

41. Badelli, S.: Interfacial Structure of Room-Temperature Ionic Liquids at the Solid-Liquid Interface as Probed by Sum-Frequency Generation Spectroscopy. *J. Phys. Chem. Lett.* 2013, 4, 244–252.

42. Zhou, W.; Inoue, S.; Iwahashi, T.; Kanai, K.; Seki, K.; Miyamae, T.; Kim, D.; Katayama, Y.; Ouchi, Y.: Electrochemical Double-Layer Structure of Pt Electrode/Ionic Liquids Interface Studied by *In Situ*

IR-Visible Sum Frequency Generation Spectroscopy. *ECS Trans.* 2009, 16, 545–557.

Review Articles

- Shen, Y. R.; Waychunas, G. A.: Sfg Studies of Oxide-Water Interfaces: Protonation States, Water Polar Orientations, and Comparison with Structure Results from X-Ray Scattering. In Wieckowski, A.; Korzeniewski, C.; Braunschweig, B. (eds), *Vibrational Spectroscopy at Electrified Interfaces.* J Wiley & Sons, Hoboken, New Jersey 2013, Chapter 2.
- Baldelli, S.: Interfacial Structure of Room-Temperature Ionic Liquids at the Solid-Liquid Interface as Probed by Sum-Frequency Generation. *J. Phys. Chem. Lett.* 2013, 4, 244–252.

Chapter 8

Interfaces of Water and Ice

Water is the most abundant and important liquid on earth. It covers three-quarters of the Earth's surface and constitutes two-thirds of our body. It creates and sustains life on Earth and is involved in all aspects of our lives. Naturally, it is the most studied substance in our civilization, but sadly it still remains the most mysterious liquid on Earth. Ice is equally mysterious. It has at least 18 known phases; no other solids have nearly as many.

While we do not yet have good understanding of bulk water, our knowledge of water interfaces is far worse. Yet, water interfaces play a pivotal role in many areas of our modern lives. They are instrumental to soil formation and weathering in geoscience, pollution and nutrient circulation in environmental science, membrane formation and protein hydration in life science, corrosion, plating, and cleaning in industrial processing, and washing and sanitizing in home maintenance. Knowing how to manipulate and control water interfaces may also be the key to solving the two impending crises facing the world, shortage of energy and water resources: energy crisis could be eradicated by photo-catalytic water splitting and water crisis by purification and desalination of water. Much like the case of other liquid interfaces, the difficulty of studying water interfaces is in lack of surface-specific tools. X-ray spectroscopy, photoemission spectroscopy, scanning force microscopy, and others have been employed, but information acquired so far has been limited. Since 1993 when the first set of vibrational spectra of water interfaces by SF spectroscopy was recorded, theoretical and experimental studies of water interfaces have boomed. From the vibrational spectra in the OH stretch range, we have been able to learn fairly well the basic microscopic structures of various water interfaces. This is one of the areas in which surface SF spectroscopy has created a major impact. We provide in this chapter a brief account of SF vibrational spectroscopic (SFVS) studies of water and ice interfaces.

8.1. General Considerations

For all problems involving water interfaces, it is the reactivity or functionality of water at an interface one would like to know and be able to manipulate. This means that we need to be acquainted with physical and chemical properties of water interfaces, but they require knowledge about water interfacial structures at the molecular level. Even a small variation in the geometric arrangement of interfacial water molecules could drastically change the interfacial properties. For instance, one expects completely different interfacial properties and functionalities if water molecules at an interface are flipped from hydrogen to oxygen facing the interface.

Fondness to be hydrogen (H-) bonded with neighbors is the unique characteristic of water molecules and is believed to be the reason for all the mysterious properties of water. In bulk water, thermal randomization breaks long-range order, but local H-bonding coordination may persist in dynamic equilibrium; water molecules are mobile and relatively free to reorient even with H-bonding. At an interface, because of boundary restriction, water molecules are less mobile and not easily reoriented so that a longer-range and better orientation-ordered H-bonding network is expected. Depending on the surface structure of the opposing medium, the surface H-bonding network of water can vary drastically. Such structural information of water is difficult to extract with the existing surface tools except SFVS. Indeed, most of the current structural information on water interfaces has come from SFVS studies. (See the list of review articles in References.)

In the literature, interfacial water structure is often discussed with the assumption that the surface structure of the opposing medium is known. Interfacial water structure is expected to change when the water solution is modified by solutes. Actually, an interface of water with an opposing medium is an integrated system. Not only is the surface water structure affected by the opposing medium, but the surface structure of the opposing medium can also be affected by contact with water. It is the structure of the whole interface we need to probe in order to have real understanding of the interface. Moreover, *in situ* probing of interfacial structural changes is desired because of fluidity of the system. Obviously such measurements are extremely challenging, particularly for buried water interfaces. Among currently available techniques, SFVS is poised to be the only viable one for such studies.

We should however note that there is still limit in SFVS's capacity to probe water interfaces. Firstly, it is known to be surface-specific only for SSP polarization combination; the EQ bulk contribution may not be negligible for other polarizations (to be described in Sec. 8.2). Secondly, surface spectra of water and opposing media may tangle up and be difficult to resolve. Thirdly, interfaces may not be accessible because of bulk absorption. Some progress has been made in this area and will be discussed in Sec. 8.7. Fourth, the available IR sources allow only detection of vibrational modes above $\sim$20 THz, limiting studies to water interfaces with media comprising lighter elements. Finally, to gain more detailed information on interfacial water structure, measurement of complex $\overset{\leftrightarrow}{\chi}_S^{(2)}$ is desired, but phase sensitive SFVS for different types of buried water interfaces is yet to be developed. Sum frequency spectra of water interfaces are generally complicated; to extract detailed interfacial structure from the spectra requires theoretical help.

8.2. Neat Water/Air Interfaces

Water/air interfaces are of paramount importance to atmospheric and environmental sciences, and have been extensively investigated by SFVS. The first surface SF vibrational spectra in the OH stretch range for the neat water/air interface were taken in 1993.[1] They were later reproduced repeatedly by many research groups. A set of more accurately measured $|\overset{\leftrightarrow}{\chi}^{(2)}_{S,eff}|^2$ spectra with SSP, SPS, and PPP polarizations are displayed in Fig. 8.1(a),[2] where $\overset{\leftrightarrow}{\chi}^{(2)}_{S,eff}$ is the effective surface nonlinear susceptibility defined in Eq. (2.19). It is seen that the SSP-polarized spectrum measuring $|(\chi^{(2)}_{S,eff})_{yyz}|^2 = |(\chi^{(2)}_{S,eff})_{xxz}|^2$ is much stronger than the others, e.g., one order of magnitude stronger than the SPS spectrum that measures $|(\chi^{(2)}_{S,eff})_{yzy}|^2 = |(\chi^{(2)}_{S,eff})_{xzx}|^2 \approx |(\chi^{(2)}_{S,eff})_{zyy}|^2 = |(\chi^{(2)}_{S,eff})_{zxx}|^2$. This immediately suggests that the surface water molecules are well polar-oriented with respect to the surface normal along z. In fact, it has been experimentally tested and conclusively argued that in the SSP spectrum of $(\chi^{(2)}_{S,eff})_{yyz}$, the ED surface nonlinear susceptibility $(\chi^{(2)}_{SD})_{yyz}$ should dominate over the EQ bulk contribution by one order of magnitude, but the same is not true for SPS and PPP spectra. So far, all theoretical and experimental studies of water interfaces have focused on SSP spectra. Fortunately, the vibrational spectrum of $(\chi^{(2)}_{SD})_{yyz}$ alone can already provide a comprehensive understanding of the structure of the water/vapor interface.

As discussed in Sec. 3.4, to unambiguously characterize surface vibrational resonances, Im $\chi^{(2)}_{S,eff}$ spectra are needed, but in early years, only $|\chi^{(2)}_{S,eff}|^2$ spectra for water were measured. To find $\text{Im}(\chi^{(2)}_{S,eff})_{yyz}$ spectra for the water/air interface, the broad band in Fig. 8.1(aA) was assumed to comprise discrete resonant profiles (Sec. 3.4) to fit the measured spectra. Unfortunately, such fitting was not unique; different research groups came up with different $\text{Im}(\chi^{(2)}_{S,eff})_{yyz}$ spectra from essentially the same $|(\chi^{(2)}_{S,eff})_{yyz}|^2$ spectrum, creating a great deal of confusion. The problem ended with the arrival of phase-sensitive SFVS that allows direct measurement of $\text{Im}(\chi^{(2)}_{S,eff})_{yyz}$.[3] Two representative $\text{Im}(\chi^{(2)}_{S,eff})_{yyz}$ spectra for the neat water/vapor interface are described in Fig. 8.1(b).[4,5] We note, from Eq. (2.19), that $\overset{\leftrightarrow}{\chi}^{(2)}_{S,eff} \equiv \overset{\leftrightarrow}{F}^{I-II}(\omega_3) \cdot \overset{\leftrightarrow}{\chi}^{(2)}_S : \overset{\leftrightarrow}{F}^{I-II}(\omega_1)\overset{\leftrightarrow}{F}^{I-II}(\omega_2)$. To find the characteristic $\chi^{(2)}_{SD,yyz} (\approx \chi^{(2)}_{S,yyz})$ for the interface, we need

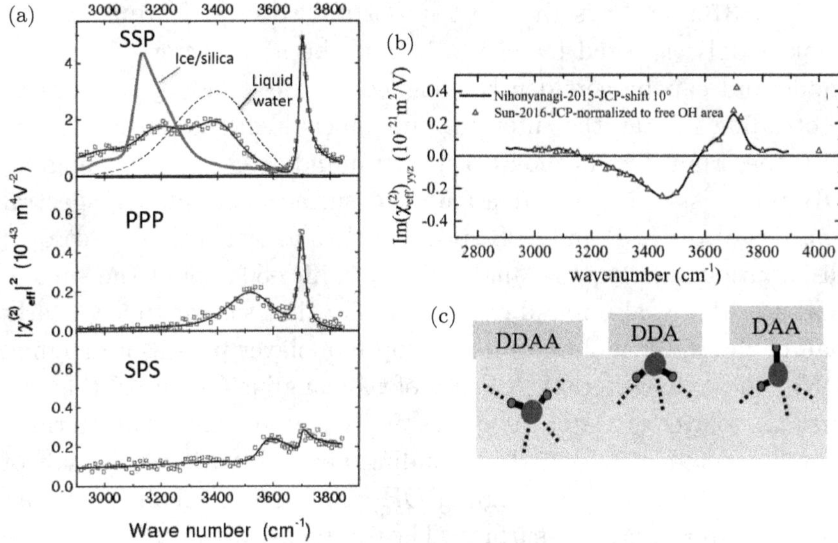

Fig. 8.1. (a) Reflected SF vibrational spectra, $|(\chi^{(2)}_{S,eff})_{yyz}|^2$, of the water/vapor interface in the OH stretching range taken with SSP, PPP, and SPS polarization combinations. IR absorption spectra of bulk water and ice are shown together with the SSP spectrum for comparison (after Ref. [2]). (b) $\mathrm{Im}(\chi^{(2)}_{S,eff})_{yyz}$ spectra of the water/vapor interface taken by Nihonyanagi *et al.* using broadband PS-SFVS (Ref. [4]) and Sun *et al.* (Ref. [5]) using narrow band PS-SFVS. The two spectra look identical, but the former were actually shifted up from the original spectrum by assuming a phase correction of $+10°$ on $(\chi^{(2)}_{S,eff})_{yyz}$ (after Ref. [5]). (c) Sketch describing water molecules with DDA, DAA, and DDAA hydrogen bonding.

to remove the Fresnel factors Fs in $(\chi^{(2)}_{S,eff})_{yyz}$. In the water/vapor case, it was found that the effect of the Fresnel factors is minor.[6] We can therefore simply refer to Fig. 8.1(b) to extract structural information about the water/vapor interface.

Intensive experimental and theoretical SFVS studies over the past two decades have brought us progressively a fair understanding of water/vapor interfacial structure, as described in a series of review articles. Actually, even without referring to any theoretical calculation, one can already have a qualitative picture on the structure from the $\mathrm{Im}(\chi^{(2)}_{S,eff})_{yyz}$ spectrum (Fig. 8.1(b)). The sharp peak at 3700 cm^{-1} is the most important spectral feature. As the peak is only detected by SFVS, it must be a characteristic surface mode and can be confidently assigned to the dangling OH stretch protruding out at the interface. Its much higher intensity in the SSP spectrum than in the SPS spectrum indicates that the dangling OH bond is well oriented about the surface normal. (A spectral analysis of the SSP and SPS spectra led to an azimuthally isotropic distribution with a polar angle spread of 40°–50° about the surface normal.) It was also found experimentally that similar to ice, nearly half of the water molecules in the top monolayer possess a dangling OH.[1] These characteristic features of the dangling OH reveal that the first monolayer of water molecules at the surface must have formed a well-ordered hydrogen (H-) bonding network with nearly half of the molecules having a dangling OH and the other half having an oxygen exposed at the surface. The subphase water molecules are H-bonded to molecules above and below, but bond ordering decays away rapidly toward the bulk liquid structure in about two to three monolayers. We note that bond ordering here should be considered in the statistical sense. As liquid, the H-bonding network must be dynamically varying with H bonds frequently attached and broken, but on average, interfacial water has a bond ordering better than bulk water, but worse than ice.

We can extend the description of the water/vapor interfacial structure further using our common knowledge about water molecules:[3] Each molecule with two OH extended at 104.5^o can have a maximum of four tetrahedral H bonds with neighbors, two donor (D) bonds from the molecule that donate a proton to bonding, and two acceptor (A) bonds that accept a proton from a neighbor in bonding. Depending on the relative geometry of the molecule with the bonding neighbor, the H-bonding strength varies and the OH stretch frequencies of the bonded molecules shift accordingly. Since H-bonding of water fluctuates widely, the vibrational spectrum of the bonded OH stretches is expected to be a broad band, as seen in Fig. 8.1. Thus, it is the H-bonded OHs that contribute to the negative broadband in the range of 3200 to 3700 cm^{-1} in Fig. 8.1(b). There are a few other tips: The OH stretch frequency of a water molecule is affected more by D-bonding than by A-bonding. The two OHs on each molecule with different OH bonding strengths with neighbors can be assumed to be independent; coupling of the two is not strong enough to cause significant change of the broad OH spectrum. Because the top monolayer is exposed to air, one or two of the H-bonds of each surface water molecule must be broken, so that only DAA and DDA, DD, AA, and DA molecules can exist in the monolayer. Here, DAA and DDA denote, respectively, water molecules connected to neighbors with (one D, two A) bonds and (two D, one A) bonds. (Fig. 8.1(c)) (DD, AA, and DA are special cases of DDA and DAA with one D or A having vanishingly small H-bonding strength.) Only DDAA molecules exist below the top monolayer assuming the probability of bond breaking is negligible.

We can see how different species contribute to the $\text{Im}(\chi^{(2)}_{S,eff})_{yyz}$ spectrum. First, the positive sharp peak must come from the dangling OH bond of DAA (including DA). (Intramolecular coupling of the free OH and the H-bonded OH may affect the peak width.) The AA species may also contribute, but its surface density is expected to be significantly lower, and on average, the two dangling OH bonds are oriented away from the surface normal, contributing weakly to the SSP SF signal. Furthermore, intramolecular coupling of the two free OHs of AA should split the two stretch modes into nondegenerate symmetric and antisymmetric stretches with a frequency difference of ~ 90 cm^{-1}. Indeed, in Fig. 8.1(b), the symmetric stretch can be seen as a weak shoulder at ~ 3620 cm^{-1} on the low-frequency side of the dangling OH peak, and the antisymmetric stretch is probably hidden in the high-frequency tail of the dangling OH peak. The shoulder disappeared in the vapor/HDO spectrum as expected because intramolecular coupling between OH and OD stretches is not appreciable.

Next, the negative broad band of $\text{Im}(\chi^{(2)}_{S,eff})_{yyz}$ in Fig. 8.1(b) must come from D-bonded OH with varying H-bonding strengths. There are roughly three times more D-bonded OHs than free OHs from DDA and DAA molecules in the top monolayer; essentially all of them point into the bulk water (although at a large angle from the surface normal), and contribute to the broad negative band from

~ 3100 to ~ 3600 cm^{-1}. They are the dominant species contributing to the spectral range of 3400 to 3600 cm^{-1}, as seen from the observation that this part of the spectrum barely changes when ions emerge at the interface; molecules in the top monolayer are hardly affected by the surface field created by emerging ions (to be described more in detail in Sec. 8.3). Underneath the top monolayer, all molecules are of the DDAA type. Again, owing to the widely varied H-bonding strength, they contribute to the broadband, but are red-shifted because on average, the 4-coordinated DDAA has weaker H-bonding than the 3-coordinated DDA and DAA. The DDAA molecules generally have equal number of D-bonded OH pointing up and down with opposite contributions to SFVS. However, because of the rapidly decreasing bond ordering away from the interface, the net orientation of OH of DDAA is up, contributing to a positive broad band, but the contribution is relatively weak compared to that from DDA and DAA. Partial cancellation of the positive and negative bands below 3500 cm^{-1} from DDAA and DDA/DAA leads to an overall negative band with a long tail toward lower frequency. It was suggested that the Fermi resonance between the overtone of the H-O-H bending vibration and bonded OH stretches could also contribute to the negative broad band, but theoretical support is not yet available.

Molecular dynamics (MD) simulations have been used to calculate the $\mathrm{Im}(\chi^{(2)}_{S,eff})_{yyz}$ spectrum since 1995.[7] Results from different models for MD all seem to be able to produce the main features of the $\mathrm{Im}(\chi^{(2)}_{S,eff})_{yyz}$ spectrum, i.e., the positive sharp dangling OH peak and the negative broad band. More sophisticated simulations in recent years have been successful in generating spectra that match well with the experimental one. They provide more quantitative description on the above qualitative interpretation of the $\mathrm{Im}(\chi^{(2)}_{S,eff})_{yyz}$ spectrum although details in the simulations still vary with models. Thus, the $\mathrm{Im}(\chi^{(2)}_{S,eff})_{yyz}$ spectrum of the neat water/vapor interface can now serve as a reference for all other water interfaces.

Observation of $\mathrm{Im}(\chi^{(2)}_{S,eff})_{yyz}$ spectra of water bending vibration at ~ 1650 cm^{-1} from water/air and water covered by charged surfactant monolayers has been reported by several groups.[8–12] Presumably the bending mode could yield more information about

H-bonding between water molecules because of its narrower band-width. There is however controversy on whether the observed spectrum comes from water molecules at the interface or from water bulk, or from both.[8–12] The argument mainly focuses on how the surface field from the charged surfactant monolayer could affect $\mathrm{Im}(\chi^{(2)}_{S,eff})_{yyz}$, but not much on how surface water molecules bond to surfactants. How the spectrum is related to the interfacial structure is also not clear. Obviously more work is needed.

8.3. Ions Emerging from Solutions at Water/Vapor Interfaces

Ions in water play an important role in atmospheric and environmental chemistry as ions emerging at a water/vapor interface can be directly involved in reactions with ambient gas molecules impinging on the interface. In early days, it was believed that ions would experience Coulomb repulsion from their own image charge and stay away from the interface.[13] Junwirth and Tobias, however, predicted by MD simulations in 2001 that polarizable ions actually like to appear at the interface and set up an electric double layer (EDL) with counter ions.[14] From free energy consideration, emergence of large polarizable ions is energetically favorable as it reduces their cavitation energy in bulk water together with only a relatively small increase of electrostatic energy due to deformation of the hydrated polarizable ions.[15]

Probing ions with low surface density at a water interface directly is challenging. We need surface-specific spectroscopic techniques with sufficiently high sensitivity. Section 7.4.1 described how SFVS can detect surface molecular ions (namely, HSO_4^- from H_2SO_4 solution) through their own vibrational spectra. It was also demonstrated that SHG can probe surface ions via their electronic resonances. However, the most sensitive technique of probing surface ions at a water interface is to monitor the effect on SHG/SFG from the electric double layer (EDL) in water set up by the surface ions (assuming ionic strength in solution is low so that the EDL is well defined). In fact, it has been noted that the SF vibrational spectra of different charged water interfaces can vary widely because of differences in EDL created by a variety of surface charges at interfaces as we shall see in what follows.

In the presence of an EDL, the effective surface nonlinear susceptibility, $(\overset{\leftrightarrow}{\chi}_{S,eff}^{(2)})_R$, measured by reflected phase-sensitive(PS) SHG/SFG has the expression[16]

$$(\overset{\leftrightarrow}{\chi}_{S,eff}^{(2)})_R = \overset{\leftrightarrow}{\chi}_S^{(2)} + \overset{\leftrightarrow}{\chi}_{EDL}^{(2)}$$

$$\overset{\leftrightarrow}{\chi}_{EDL}^{(2)} \equiv \int_{-\infty}^{0^+} \overset{\leftrightarrow}{\chi}_B^{(3)} \cdot \hat{z} E_0(z') e^{i\Delta k_{Rz} z'} dz' \tag{8.1}$$

$$\simeq \overset{\leftrightarrow}{\chi}_B^{(3)} \cdot \int_{-\infty}^{0^+} \hat{z} E_0(z') e^{i\Delta k_{Rz} z'} dz'$$

where $\overset{\leftrightarrow}{\chi}_S^{(2)}$ is the surface nonlinear susceptibility and $\overset{\leftrightarrow}{\chi}_{EDL}^{(2)}$ is the dc-field-induced second order nonlinear susceptibility from the EDL with $\overset{\leftrightarrow}{\chi}_B^{(3)}$ denoting the third-order electric dipole (ED) nonlinear susceptibility of bulk water, $\hat{z} E_0$ is the dc field in the EDL along the surface normal $\hat{z}$, and Δk_{Rz} is the phase mismatch of SHG/SFG in reflection in water. Negative surface charges lead to a positive $\hat{z} E_0$; accordingly, $\overset{\leftrightarrow}{\chi}_{EDL}^{(2)}$ bears the sign of surface charges. Bulk EQ contribution to SSP SFVS is negligible. When the surface ion density is low, its effect on $\overset{\leftrightarrow}{\chi}_S^{(2)}$ is negligible, and $\overset{\leftrightarrow}{\chi}_{EDL}^{(2)}$ can be deduced from the difference of measured $(\overset{\leftrightarrow}{\chi}_{S,eff}^{(2)})_R$ with and without surface charges. It turns out that a surface charge density of $\sim 10^{-4}$ C/m^2 (compared to 0.5 C/m^2 for a fully deprotonated fatty acid monolayer) with ionic strength of $\leq 10\,\mu$M in water can be detected this way. An example is shown in Fig. 8.2(a). It is seen that the $\mathrm{Im}(\chi_{S,eff}^{(2)})_{R,yyz}$ spectra of the water/vapor interfaces for neat water and 2.1M NaI solution taken by SSP SFVS are very different; I$^-$ ions emerging at the interface contribute to $\mathrm{Im}(\chi_{EDL}^{(2)})_{R,yyz}$, and hence the $\mathrm{Im}(\chi_{S,eff}^{(2)})_{yyz}$ spectrum, a positive bump extending from 3100 to 3450 cm^{-1}. Above 3450 cm^{-1}, the spectrum remains unchanged because it comes mainly from DDA and DAA water molecules in the top monolayer as mentioned in Sec. 8.2.1. A number of other

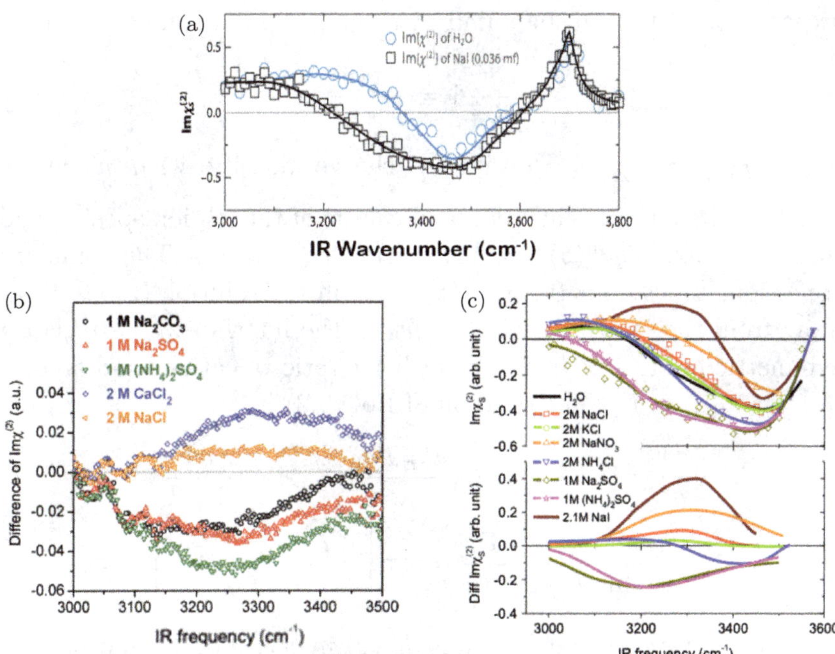

Fig. 8.2. (a) Comparison of $Im(\chi^{(2)}_{S,eff})_{yyz}$ spectra of water/air interfaces of neat water (filled squares) and water with 2.1 M NaI (open circles). Emergence of I^- at the interface leads to a more negative bonded OH stretch band (after Ref. [3]). (b) $Im(\chi^{(2)}_{S,eff})_{yyz}$ spectra of water/vapor interfaces for various salt solutions in comparison with that of the neat water/air interface (after Ref. [17]). (c) A similar set of $Im(\chi^{(2)}_{S,eff})_{yyz}$ spectra as (b) taken by Hua *et al.* The difference spectra in the lower frame are with reference to the neat water/air interface (after Ref. [18]).

salt solutions have also been measured. The spectral change of $Im(\chi^{(2)}_{S,eff})_{R,yyz}$ arising from $Im(\chi^{(2)}_{EDL})_{R,yyz}$ allows ranking of surface propensity of ions at the water/air interface, i.e., a Hofmeister series describing which ions are more likely to emerge at the interface[17,18] (Fig. 8.2(b,c)).

For quantitative determination of surface charge densities (σ) from SHG/SFG, we need to know the dependence of E_o in Eq. (8.1) on σ. For diluted ion solutions, the Guoy–Chapman (GC) theory has been proven to be successful in relating E_o and σ. It is based on the assumption that ions can be approximated as point charges, and the electrostatic potential $\Phi(z)$ and the dc $E_o(z) = -\partial\Phi(z)/\partial z$ they create in the EDL can be found by solving the Poisson equation

$$\frac{d^2}{dz^2}\Phi = -\frac{\rho(z)}{\varepsilon\varepsilon_0} \tag{8.2}$$

where $\rho(z) = \sum_i n_i(z)q_i$ is the charge density with n_i an q_i denoting the ion concentration and charge of the ith ion species, and $n_i(z) = n_{i0}\exp[-q_i\Phi(z)/k_BT]$ in thermal equilibrium. The boundary conditions are $\Phi(z) = 0$ and $n_i(z) = n_{i0}$ in the solution $(z >> 0)$ far away from the interface and $\Phi = \Phi_0$ at the interface $(z = 0)$. For a symmetric electrolytic solution with 1:1 ratio of cations and anions, $n_{0,+q} = n_{0,-q} \equiv n_0$. The solution of Eq. (8.2) is

$$\frac{d\Phi}{dz} = -E_0(z) = -\left(\frac{8k_BTn_0}{\varepsilon}\right)^{1/2}\sinh\left(\frac{q\Phi}{2k_BT}\right)$$

$$\frac{\tanh(q\Phi/4k_BT)}{\tanh(q\Phi_0/4k_BT)} = \exp\left[-\left(\frac{2n_0q^2}{\varepsilon k_BT}\right)^{1/2}z\right] \tag{8.3}$$

from which we can find $\Phi(z)$ and $E_0(z)$ in terms of Φ_0 and n_0, but σ is related to Φ_0 by $\sigma = \varepsilon\varepsilon_0(\frac{d\Phi}{dz})_{z=0} = -(8k_BTn_0\varepsilon)^{1/2}\sinh(\frac{q\Phi_0}{2k_BT})$. With $E_0(z)$ in terms of σ known, we now have, through Eq. (8.1), $\overset{\leftrightarrow}{\chi}^{(2)}_{EDL}$ in terms of σ if $\overset{\leftrightarrow}{\chi}^{(3)}_{B}$ is given. Conversely, we can find σ from the experimentally deduced $\overset{\leftrightarrow}{\chi}^{(2)}_{EDL}$.

For high ion concentrations, the ion size can no longer be neglected, but the modified Guoy–Chapman theory taking into account the ion size has been developed.[19] To distinguish different ion species, however, one has to resort to a more sophisticated theory:[20,21] An ion approaching an interface sees not only the electrostatic potential set up by surface charges, but also the image potential; moreover, it also experiences equivalent potentials due to its position-dependent cavitation energy for dissolving in water and electron redistribution in ions if they are polarizable. (If it is an interface of water with a material, the interaction potential of ions with the material must also be included.) Here, we will not go beyond the simple Gouy–Chapman model unless noted, believing that for individual ion species, the simple model should be able to provide relative surface ion densities quantitatively.

The bulk nonlinear susceptibility, $\overset{\leftrightarrow}{\chi}_B^{(3)}$, is a characteristic response coefficient of bulk water independent of water interfaces, and changes less than a few percent even if a few moles of ions are dissolved in water. The surface nonlinear susceptibility, $\overset{\leftrightarrow}{\chi}_S^{(2)}$, is also expected to remain approximately the same as that of pure water ($\overset{\leftrightarrow}{\chi}_{S0}^{(2)}$) even if the surface ion density reaches ~ 1 C/m^2, which is less than $\sim 5\%$ of the surface density of water molecules ($\sim 1.2 \times 10^{20}$/m^2). Yet, at $\sigma \sim 0.01$ C/m^2, $\overset{\leftrightarrow}{\chi}_{EDL}^{(2)}$ already can be readily detected. Thus, at low σ, we can obtain $\overset{\leftrightarrow}{\chi}_S^{(2)}(\sigma)$ from $\overset{\leftrightarrow}{\chi}_{S0}^{(2)}$ measured by PS-SFVS and deduce $\overset{\leftrightarrow}{\chi}_{EDL}^{(2)}(\sigma)$ from $\overset{\leftrightarrow}{\chi}_{EDL}^{(2)}(\sigma) = (\overset{\leftrightarrow}{\chi}_{S,eff}^{(2)})_R(\sigma) - \overset{\leftrightarrow}{\chi}_S^{(2)}(\sigma)$ from which we can find $\overset{\leftrightarrow}{\chi}_B^{(3)}$ if σ is known or, σ from $\overset{\leftrightarrow}{\chi}_{EDL}^{(2)}$ if $\overset{\leftrightarrow}{\chi}_B^{(3)}$ is known. This scheme works with both SHG and SFG, although the latter was first developed in conjunction with the goal to extract $\overset{\leftrightarrow}{\chi}_S^{(2)}$ spectra of a charged water interface as we shall discuss in Sec. 8.7.[16] It can be extended to high σ, but the modified GC theory may have to be used if the ionic strength in solution is high. In the following section, we describe how the scheme was used to find surface density of protons or hydroniums (H$^+$ or H$_3$O$^+$) at vapor/water interfaces of acid solutions.

8.4. Is the Pure Water/Vapor Interface Acidic or Basic?

There is a long standing controversy on the question whether the pure water/vapor interface is acidic or basic. This is a problem of great interest to researchers in atmospheric and environmental chemistry, but experimentally, reliable information on surface proton or hydroxyl ion affinity could only be obtained by SHG/SFG. Figure 8.3 shows the Im $(\chi_{S,eff}^{(2)})_R$ spectra of the water/air interfaces of three aqueous solutions, 1.2M HCl, 1.2M HI, and 1.2M NaOH.[22] In comparison with the spectrum of the neat water/vapor interface (also in Fig. 8.3), the spectra for the acid solutions appear more negative in the 3100–3500 cm^{-1} range (opposite to the case of NaI in Fig. 8.2(a)), indicating that the surface is positively charged owing to

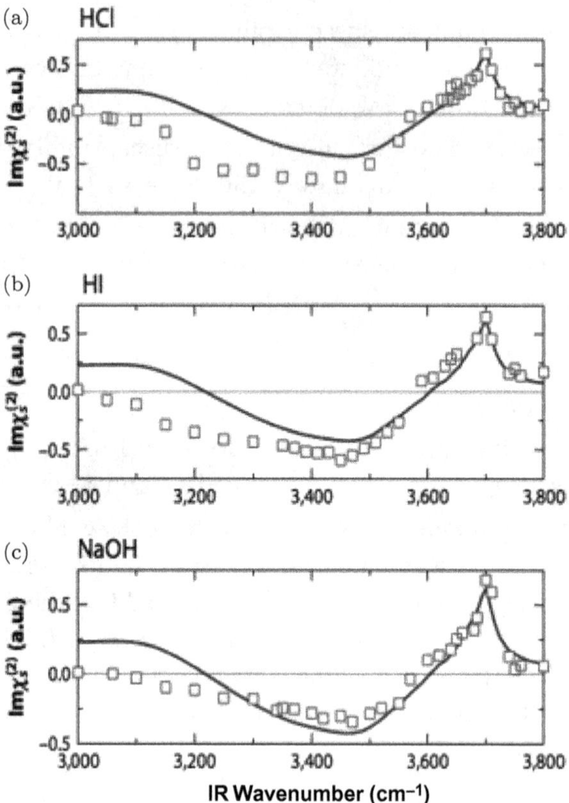

Fig. 8.3. $\mathrm{Im}(\chi^{(2)}_{S,eff})_{yyz}$ spectra of water/air interfaces in the OH stretch region for (a) 1.2M HCl solution, (b) 2M HI solution, and (c) 1.2M NaOH solution (open squares) in comparison with that for neat water (black curves) (after Ref. [22]). Because of an error in the phase measurement of $\mathrm{Im}(\chi^{(2)}_{S,eff})_{yyz}$, all spectra in the 3000–3300 range should be corrected by a down shift of 0.2 unit. See the $\mathrm{Im}(\chi^{(2)}_{S,eff})_{yyz}$ spectrum of the neat water/air interface in Fig. 8.1 for comparison.

surface excess of H^+ (or H_3O^+). The spectrum of the basic solution, however, is only slightly more positive in the 3200–3500 cm^{-1} range, suggesting a relatively small excess of OH^- at the surface. The more negative band between 3000 and 3200 cm^{-1} was assigned to the first solvation shell of the fully hydrated OH^- at the interface.[23] The results here seem to imply that for the same molar concentration of HCl, HI, and NaOH, protons or hydroniums emerge more readily at the water/vapor interface than OH^-, but this does not prove that the surface of neat water is acidic or basic. A better answer can be attained if the surface charge density as a function of bulk acid or basic concentration is known quantitatively and can be extrapolated to the neutral water case ($[H^+] = 10^{-7}$ or pH7) to see if the residue surface change density is positive, negative, or zero.

Quantitative measurement of surface densities of protons at water/vapor interfaces of hydrogen halide acidic solutions was performed in Ref. [24]. We take the case of HBr solution as an example. Figure 8.4A(a) displays the $Re(\chi^{(2)}_{S,eff})_{yyz}$ and $Im(\chi^{(2)}_{S,eff})_{yyz}$ spectra in the OH stretch region of the water/vapor interfaces measured by SSP PS-SFVS for a set of solutions with different HBr molar concentrations ([HBr]). In this range of [HBr], $\overset{\leftrightarrow}{\chi}^{(2)}_{S}$ is essentially independent of [HBr], as manifested by the observation of invariance of the $Im(\chi^{(2)}_{S,eff})_{yyz}$ spectrum in the 3500–3700 cm^{-1} section that originates from the top surface water monolayer (Sec. 8.3). We can therefore obtain the set of $Re\chi^{(2)}_{EDL,yyz}$ and $Im\ \chi^{(2)}_{EDL,yyz}$ spectra for different [HBr] from $\chi^{(2)}_{EDL,yyz} = (\chi^{(2)}_{S,eff})_{yyz} - \chi^{(2)}_{S,yyz}$ shown in Fig. 8.4A(b). Following the scheme discussed in Sec. 8.3, we can then find the surface charge density σ by fitting the $Im\ \chi^{(2)}_{EDL,yyz}$ spectrum with Eqs. (8.1) and (8.3) and the known spectrum of $\chi^{(3)}_{B,yyz}$ of bulk water, given in Fig. 8.4B, which was measured separately and consistently by SFVS on systems with different water interfaces.[16] The deduced σ versus [HBr] is plotted in Fig. 8.4C. Since we expect both H^+ and Br^- to appear at the interface, we have $\sigma = [\rho(H^+) - \rho(Br^-)]|e|$, where $\rho(x)$ denotes the surface density of x. It is possible to determine $\rho(H^+)$ [or $\rho(H_3O^+)$] and $\rho(Br^-)$ separately, knowing that in this range of [HBr], adsorption of ions at an interface can be regarded as independent of each other. By adding NaBr in an HBr solution to vary $[Br^-]$ (Na^+ is known to be repelled from the interface), measurement of σ allows us to find $\rho(Br^-)$ versus $[Br^-]$, and hence $\rho(Br^-)$ versus [HBr] in pure HBr solution plotted in Fig. 8.4C. Finally, we can obtain $\rho(H^+)$ from $\sigma/|e| + \rho(Br^-)$ for given $[H^+]$ or [HBr], also plotted in Fig. 8.4C. We expect the deduced $\rho(H^+)$ versus $[H^+]$ to be unchanged if Br is replaced by other halides if ion adsorption can be approximated as individual

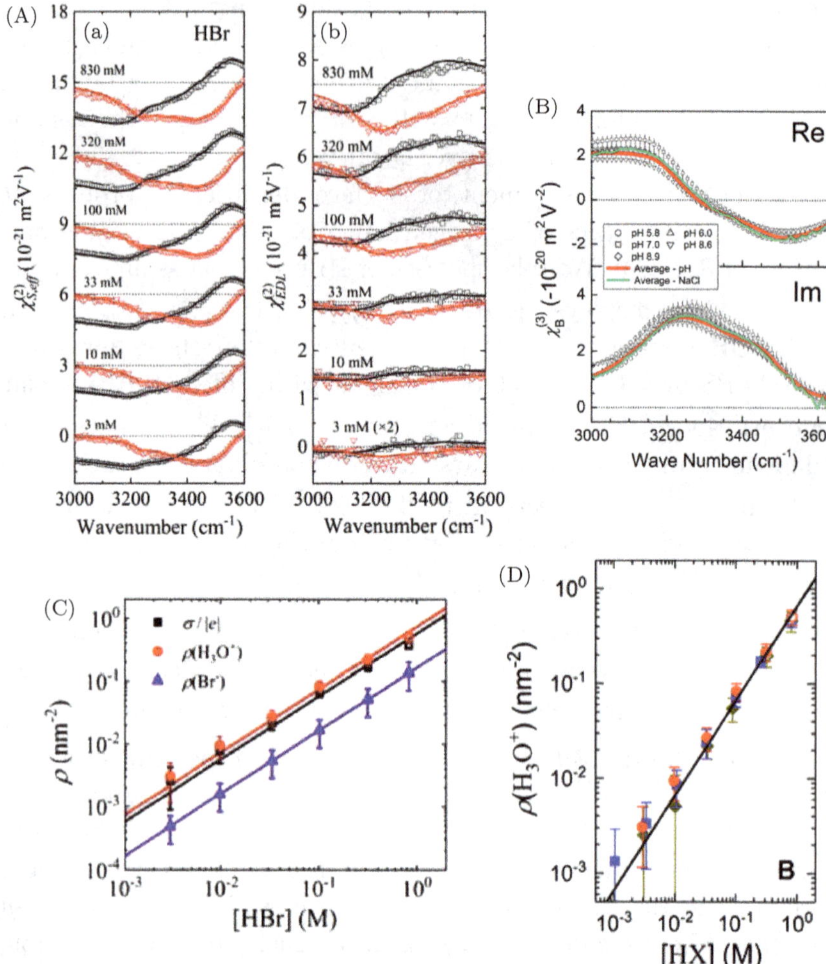

Fig. 8.4. (A) Frame (a): $Re(\chi^{(2)}_{S,eff})_{yyz}$ (black) and $Im(\chi^{(2)}_{S,eff})_{yyz}$ (red) spectra of water/air interfaces for solutions of different HCl molar concentrations. Frame (b): Corresponding spectra of $Re(\chi^{(2)}_{EDL})_{yyz}$ and $Im(\chi^{(2)}_{EDL})_{yyz}$ deduced from the spectra in Frame (a). (B) $Re(\chi^{(3)}_{B})_{yyz}$ and $Im(\chi^{(3)}_{B})_{yyz}$ of bulk water obtained by a series of measurements reported in Ref. [16]. Red and green spectra are average spectra from two different sets of measurements. (C) Deduced surface charge density σ, surface density of hydronium $\rho(H_3O^+)$ or $\rho(H^+)$, and surface density of bromine ions $\rho(Br^-)$ versus HBr concentration in water (see the text for details). (D) Deduced $\rho(H_3O^+)$ as a function of HX concentration in solution, where X denotes Cl, Br, or I, showing that the result is independent of halide ions (after Ref. [24]).

events. Similar measurements on HCl and HI solutions proved that this is indeed true as depicted in Fig. 8.4D. Extrapolation of the line of $\rho(H^+)$ versus $[H^+]$ in Figs. 8.4C and 8.4D yields, at bulk pH 7, a nonzero positive surface charge density with a corresponding surface $pH(=-\log_{10}[H^+]) = 6.35$, indicating that the neat water/air interface is weakly acidic. In the literature, a strong argument advocating that the neat water/air surface is basic came from the observation that water drops in oil appear to carry negative charges.[25] However, as we shall see in the next section, water/oil interface is different from water/air interface; negative OH^- ions from water are likely to emerge at a water interface covered by hydrophobic molecules.

The above scheme can be used to measure surface charge densities of other water interfaces, as we shall discuss further in the following sections. It is perhaps also obvious from the underlying principle that PS-SHG should be equally effective as PS-SFVS to measure σ.[26] A single input frequency with nonresonant SHG can effectively do the job, although multiple input frequencies can improve the accuracy. The experimental arrangement of phase-sensitive SHG is simpler. Being nonresonant, both $\overleftrightarrow{\chi}_S^{(2)}$ and $\overleftrightarrow{\chi}_B^{(3)}$ are real. We should have from Eq. (8.1) that $\mathrm{Im}\,\overleftrightarrow{\chi}_{S,eff}^{(2)} = \mathrm{Im}\,\overleftrightarrow{\chi}_{EDL}^{(2)}$. As in the SFG case, $\overleftrightarrow{\chi}_B^{(3)}$ of bulk water can be separately determined; it was found that for S-polarized input and P-polarized output at $2\omega \sim 15750$ cm^{-1}, $\chi_{B,zyy}^{(3)} = 9.6 \times 10^{-22}$ $\mathrm{m}^2/\mathrm{V}^2$. We can then find σ directly from the measured $\mathrm{Im}\,\overleftrightarrow{\chi}_{S,eff}^{(2)}$. We also see from Eq. (8.1) that at a given σ, change of the beam geometry leads to change of Δk, and hence change of $\overleftrightarrow{\chi}_{EDL}^{(2)}$ through the Gouy–Chapman theory, but not $\overleftrightarrow{\chi}_S^{(2)}$. Therefore, measurement of $\overleftrightarrow{\chi}_{S,eff}^{(2)}$ for two different Δks should allow us to find two unknowns, $\overleftrightarrow{\chi}_S^{(2)}$ and σ; measurement with more different Δks yields better accuracy. Actually, a closer look at the scheme can show that phase measurement of SHG is not even needed. Possible detection of a surface charge density of $\geq 10^{-4}\,\mathrm{C/m}^2$ has been demonstrated.[26] In data analysis, if the ionic strength of water is high and ion size is large, the modified Gouy–Chapman theory may have to be used.

8.5. Water Exposed to Insoluble Gas Atmosphere

Opposing media at an interface generally form a correlated interfacial system; their surface structures are affected by each other. In the case of an insoluble gas/water solution interface, the hydrophobic gas molecules can still adsorb on water via van der Waals interaction. Change of the water surface structure, e.g., by ions emerging at the interface can change the surface coverage of adsorbed molecules, and conversely, change of the surface coverage of adsorbates can change the surface water structure. SFVS being able to probe *in situ* both the adsorbates and the water interface is an effective tool to interrogate such interfaces. We use hexane(C_6H_{14})/water as an example.[27] The observed $|(\chi^{(2)}_{S,eff})_{yyz}|^2$ spectra in the CH and OH stretch range of hexane/water interfaces with different hexane gas pressures and two different pHs of water, taken by SSP SFVS, are shown in Fig. 8.5(a). The presence of the CH_3 symmetric and antisymmetric stretch modes of hexane at 2880 and 2960 cm^{-1} confirms that hexane molecules do adsorb on water. It was difficult to determine their surface coverage on water from the CH_3 modes because their orientation order is poor. However, it was noted that adsorbed hexane created a new dangling OH mode downshifted by 20 cm^{-1} from that of the neat water surface. This shift comes from the dangling OHs that are in close contact with adsorbed hexane molecules and experience the van der Waals interaction with hexane. As hexane gas pressure increased, the new mode grew while the old mode diminished; the transfer of strength was completed at 100 mmHg of hexane gas pressure, indicating that the hexane coverage on water had reached a full monolayer. Fitting of the dangling OH spectrum at a given hexane gas pressure by the two modes allowed deduction of the hexane surface coverage at that gas pressure. The deduced hexane surface coverage versus hexane gas pressure for the pH10 solution is plotted in the inset of Fig. 8.5(a).

The dependence of hexane surface coverage on hexane gas pressure was found to hardly vary with pH in water, but the average orientation of the adsorbed hexane molecules did vary. As seen in Fig. 8.5(b), the amplitude of the CH_3 symmetric stretch decreases

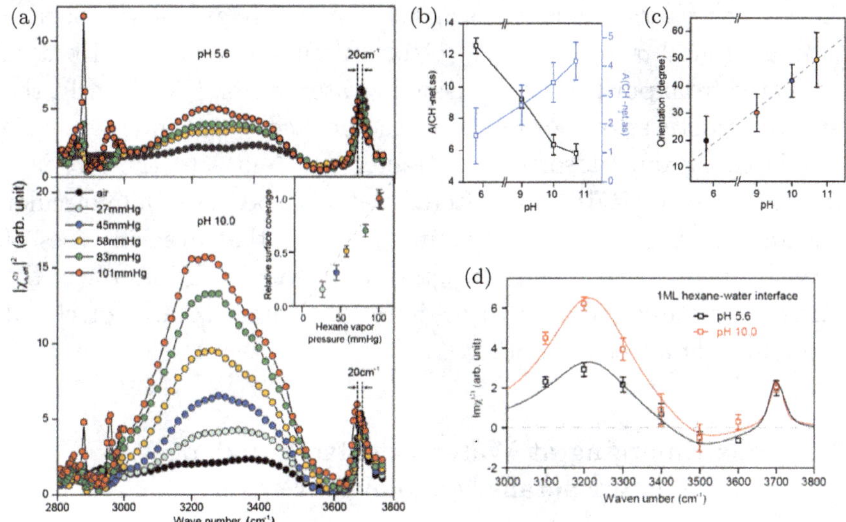

Fig. 8.5. (a) $|(\chi^{(2)}_{S,eff})_{yyz}|^2$ spectra of CH and OH stretches for water/hexane interfaces under different hexane gas pressures and two different pH values in water, pH 5.6 and pH 10.0. The fractional hexane coverage on water as a function of hexane gas pressure is plotted in the inset. (b) Amplitudes of the symmetric and antisymmetric CH_3 stretch modes, A(CH-ss) and A(CH-as), deduced from $|(\chi^{(2)}_{S,eff})_{yyz}|^2$ spectra with respect to pH in water. From the ratio of the two, the average orientation of adsorbed hexane molecules can be deduced as a function of pH, shown in (c). The $Im(\chi^{(2)}_{S,eff})_{yyz}$OH stretch spectra of water at pH 5.6 and 10.0 under a full monolayer of hexane are depicted in (d). The positive low-frequency broadband in each spectrum comes from EDL set up by OH^- emerging at the interface (after Ref. [27]).

and that of the antisymmetric stretch increases with increase of pH. From the ratio of the two, the average orientation of CH_3 can be deduced, as plotted in Fig. 8.5(c), suggesting that hexane chain must have inclined more toward the surface at higher pH. The effect of adsorbed hexane on water surface structure is more apparent. It is seen in Fig. 8.5(a) that the bonded OH stretch spectrum of water is stronger with higher hexane gas pressure corresponding to higher hexane coverage on water, more so at higher pH. Figure 8.5(d) shows that the $Im(\chi^{(2)}_{S,eff})_{yyz}$ spectrum is more positive at higher pH, indicating that more OH^- ions must have appeared at the water surface and set up a stronger EDL. Quantitative measurement of the surface charge density (Sec. 8.4) for a full monolayer hexane coverage found $\sigma \sim -8 \times 10^{-3}$ C/m^2, a value consistent with that estimated from purified oil emulsion in water in Ref. [25]. Clearly, the surface structures on both sides of the hexane/water interface were correlated: hexane adsorption attracted OH$^-$ ions to the interface and the emerging OH$^-$ ions altered the adsorbed hexane molecular orientation. This is a good example showing that even in cases of very weak interaction between opposing media at an interface, the surface structures of the two media at the interface are correlated and mutually affect each other.

8.6. Ions Emerging at Water Interfaces under Nonionic Surfactant Monolayears

Ions are expected to emerge at all types of water interfaces. As described in Sec. 6.4, impurites in a solution can adsorb at an interface if it is energetically favorable. Ions can be considered as impurities, having their Coulomb interactions with surroundings play an important role in their energetics. Because of their high reactivities, their appearance at water interfaces can strongly modify the water interfacial properties and is therefore of interest in many disciplines.

In Sec. 6.6, we described how SFVS can be used to probe microscopic structures of Langmuir monolayers on water without mentioning how a monolayer affects the surface water structure; this is what we shall now discuss. We first consider neutral surfactant monolayers on water, and take the fully packed hexadecanol (HD; CH$_3$(CH$_2$)$_{15}$OH) monolayer on water as a representative case. Interfaces of various electrolytic solutions covered by a HD full monolayer have been studied by SFVS in the CH and OH stretch ranges. The CH spectrum of hexadecanol was found to remain unchanged with different ion species and concentration up to 2M in solutions,[28] showing that the chain orientation and conformation of HD was not affected by possible change of water structure around the COH terminals as one would expect from the strong chain–chain interaction of fully packed HD. The interfacial OH stretch spectra,

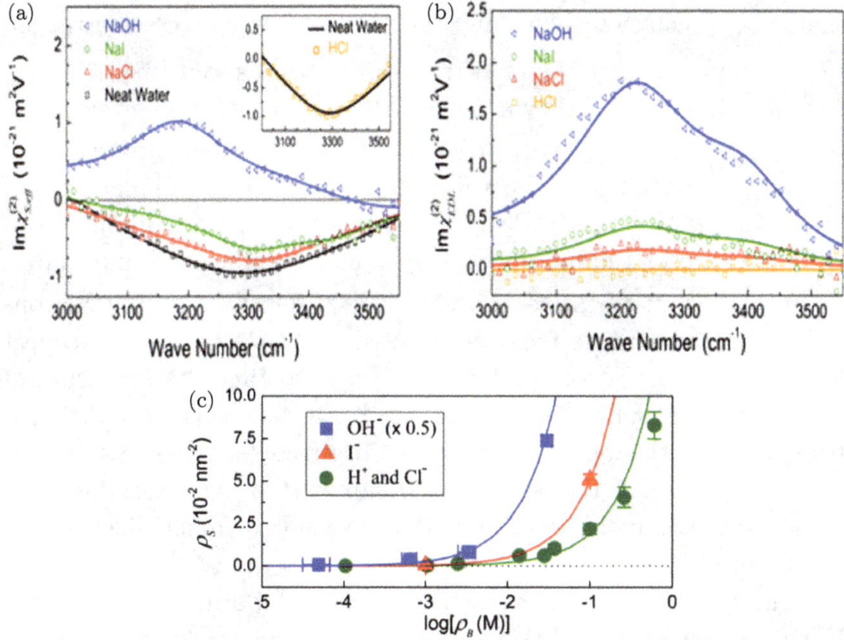

Fig. 8.6.   (a) $\mathrm{Im}(\chi^{(2)}_{S,eff})_{yyz}$ OH stretch spectra of hexadecanol (HD) mono-layer/water interfaces with 0.1M NaCl, 0.1M NaI, 0.1M HCl, and 0.03M NaOH in water in comparison with that of the HD monolayer/neat water interface. The inset shows that the spectra for the HD monolayer/HCl solution and HD monolayer/neat water interfaces are nearly the same. (b) Corresponding $\mathrm{Im}(\chi^{(2)}_{EDL})_{yyz}$ spectra deduced from the spectra in (a). (c) Surface densities of OH^-, I^-, H^-, and Cl^- extracted from the $\mathrm{Im}(\chi^{(2)}_{EDL})_{yyz}$ spectra in (b) with the help of the Gouy–Chapman theory (after Ref. [28]).

$\mathrm{Im}(\chi^{(2)}_{S,eff})_{yyz}$, of neat water (pH 5.7) and 0.1M solutions of HCl, NaOH, NaCl, and NaI are displayed in Fig. 8.6(a).[28] It is seen that all solutions except HCl appear to have a more positive spectrum than the neat water. The changes must have come from $\mathrm{Im}\, \chi^{(2)}_{EDL,yyz}$ of the EDL set up by negative ions in competition with H^+ emerging at the interface; Na^+ is known to be repelled from the interfaces by its own image. The spectra for the HCl solution and the neat water are almost identical. This is an indication that H^+ and Cl^- must have adsorbed at the interface with nearly the same surface density. Using the relation $\chi^{(2)}_{EDL,yyz} = (\chi^{(2)}_{S,eff})_{yyz} - \chi^{(2)}_{S,yyz}$ and knowing that $\chi^{(2)}_{S,yyz}$ can hardly change at such a low ion concentration, we can find the $\chi^{(2)}_{EDL,yyz}$ spectra from the measured $\mathrm{Im}(\chi^{(2)}_{S,eff})_{yyz}$ for all solutions, as shown in Fig. 8.6(b). From the strength of $\mathrm{Im}\,\chi^{(2)}_{EDL,yyz}$ in Fig. 8.6(b), we can see that the surface propensity of the various ions at the HD/water interface has the ranking order of $OH^- > I^- < H^+ \sim Cl^- > Na^+$. Analysis of the spectra of Fig. 8.6(b) with the help of the modified Gouy–Chapman theory[20] and known $\overleftrightarrow{\chi}^{(3)}_B$ provided quantitative values on surface densities of different ions at the HD/water interfaces, as plotted in Fig. 8.6(c) with respect to bulk ion concentration.[28] We should note that the spectrum of neat water under hexadecanol in Fig. 8.6(a) is appreciably different from that of the neat water/vapor interface (see Fig. 8.1(b)). It was believed that the former is dominated by the terminal OH of hexadecanol molecules because it looks very much like the IR spectrum of HD.

The result of the above work shows that both the hydrophilic HD/water interface and the hydrophobic hexane gas/water interface like to attract OH^- ions. It seems to suggest that surface ion propensity has little to do with hydrophilicity or hydrophobicity, but depends more on interaction of ions with adsorbed molecules. It would be interesting to conduct a similar study on water interfaces with a hydrophobic surfactant monolayer. Such a study was carried out on water in contact with alkyl chains of a fully packed octadecyltrichlorosilane (OTS) monolayer deposited on a silica substrate.[29] Figure 8.7(a) compares the SSP spectra of $|(\chi^{(2)}_{S,eff})_{yyz}|^2$ for water/OTS at pH 6.0 and water/vapor interfaces. As a signature of a hydrophobic interface, the dangling OH mode of water/OTS appears at 3680 cm^{-1}, red-shifted by 20 cm^{-1} from that of the neat water/air interface due to van der Waals interaction of the dangling OH with the methyl group of OTS. One would expect that the broad bonded OH band would also be similar to that of the neat water/air interface, but this is obviously not true. The $\mathrm{Im}(\chi^{(2)}_{S,eff})_{yyz}$ spectra of the bonded OH stretches for the water/OTS interface with different pHs (adjusted by dissolving HCl and NaOH in water) are displayed in Fig. 8.7(b). They show that at pH6 the broad band is positive

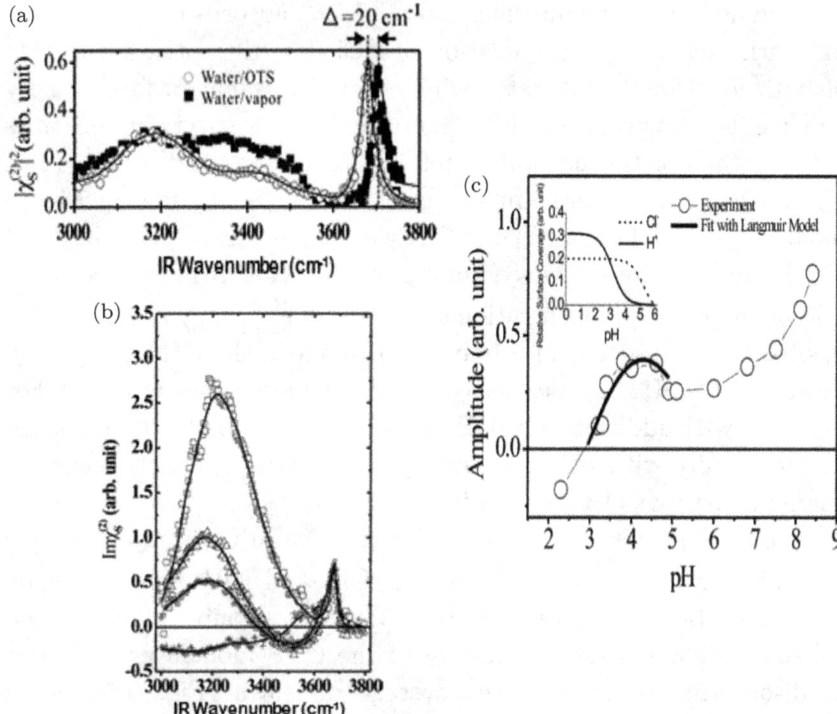

Fig. 8.7. (a) Comparison of $|(\chi^{(2)}_{S,eff})_{yyz}|^2$ spectra of OH stretches for the neat water/OTS/silica and neat water/vapor interfaces. (b) Im $\chi^{(2)}_{S.eff}$ spectra of the water/OTS/silica interfaces with the bulk pH at 11.0 (squares), 7.8 (triangles), 6.0 (stars), and 2.3 (crosses) set by solvated NaOH and HCl. (c) Peak value of Im $\chi^{(2)}_{S.eff}$ at 3200 cm^{-1} as a function of bulk pH. The thick solid curve is a fit assuming adsorption of Cl$^-$ and H$^+$ follows the simple Langmuir model. The adsorption isotherms of Cl$^-$ and H$^+$ separately measured are given in the inset (after Ref. [29]).

instead of negative as in the water/vapor case, and as pH increases, the positive band becomes much stronger while the dangling OH stretch remains the same. This change of the broad band must have come from the EDL set up by negative charges or OH$^-$ appearing at the interface. We take the peak value of the band at 3200 cm^{-1} to represent the spectral change, and plot it as a function of pH in Fig. 8.7(c). The result can be explained in terms of ion species at the interface. With only HCl in solution (pH < 5.7), both H$^+$ and

Cl^- can adsorb at the interface; initially Cl^- adsorbs more than H^+, but with increasing concentration of HCl (lowering pH), adsorption of both ions finally saturates with nearly the same surface density and yields a neutral interface; accordingly, the spectrum becomes close to that of the neutral water/air interface at pH $\sim$ 2.3. (The Cl^- adsorption isotherm in the inset of Fig. 8.7(c) was measured separately from the interface of OTS/water solutions with a range of NaCl concentrations.) As seen in Fig. 8.7(c), when pH decreases from 5.7 (by increasing concentration of HCl), $Im(\chi^{(2)}_{S,eff})_{yyz}$ first increases positively because Cl^- initially adsorbs more than H^+, begins to decrease at $\sim$pH4, crosses zero at $\sim$pH3, and becomes negative. For pH $>$ 5.7 with addition of NaOH in solution, the broad band increases monotonically with pH, manifesting an increase of negative surface charge density or OH^- at the interface.

Sum Frequency spectroscopic studies of the OTS/water system by measuring $|(\chi^{(2)}_{S,eff})_{yyz}|^2$ in the CH and OH stretch region have also been reported by several other groups. Tyrode and Liljeblad showed that the quality of the OTS monolayer, in terms of disordering or incomplete coverage on silica (Fig. 6.4), could affect the water surface spectrum: better quality led to a stronger dangling OH mode, but a weaker broad band.[30] They then suggested that the pH dependence of the OH spectrum could be due to deprotonation of the silica surface not covered by OTS, but they ignored possible ion adsorption at the interface. Unfortunately, for water interfaces, information provided by $|(\chi^{(2)}_{S,eff})_{yyz}|^2$ spectra are limited and could be misleading. Interpretation of the spectra based on presumption tends to be vague and confusing. In the present case, it was not possible to learn from the $|(\chi^{(2)}_{S,eff})_{yyz}|^2$ spectra whether the broad band described a net OH orientation pointing to or away from the interface and what the contribution from the EDL is. While it is possible that an imperfect OTS monolayer may lead to a small fraction of silica surface deprotonated, the observed broad band peaked at 3200 cm^{-1} in the $Im(\chi^{(2)}_{S,eff})_{yyz}$ spectrum (Fig. 8.7(b)) for the water/OTS/silica interface was nearly the same as that of the water/bare-silica interface at high pH. It was also found that adsorption of Cl^- from a 10 μM NaCl solution at the water/OTS/silica interface could readily be detected (inset of Fig. 8.7(a)) although it was known that 0.1M of NaCl in water could hardly affect the spectrum of the water/silica interface. These results, together with what we have learned from the water/hexane interface in Sec. 8.4, suggest that ions can readily adsorb on alkyl chains. Clearly, in order to firmly clarify the ambiguity, we need to repeat the PS-SFVS experiment on water/OTS/silica interfaces with both nearly perfect and less perfect monolayers.

8.7. Water Interfaces under Charged Surfactant Monolayers

Many surfactant monolayers can have their head groups in water protonated or deprotonated, or carrying cation, or anion, or both (zwitterionic). They are most relevant in practical applications such as cleaning, extraction of substances, etc., and in bioscience in connection with biomembranes. Water interfaces with surfactant monolayers that have their degree of protonation/deprotonation vary with pH in water are certainly more complicated than those with nonionic surfactant monolayers discussed in Sec. 8.6. The surface charge density results not only from ion adsorption at the interface, but also from the protonation/deprotonation process, and surface pH is generally different from bulk pH. Moreover, charged head groups tend to promote ion adsorption that could lead to structural change of both surfactants and interfacial water.

To see how SFVS can probe such interfaces, we first describe a general structural model for the interfaces. As mentioned in Sec. 8.1, the interfacial layer of two opposing media is generally composed of surface layers of both media with their structures affecting each other. In the case of a fully packed charged surfactant monolayer on water, ion adsorption at the interface can alter not only the structure of the surfactant head groups, but also the bonding structure of nearby water molecules. In addition, the surface charges create an EDL in which the bonding structure is essentially the same as in the bulk but molecules are appreciably reoriented by the field. We can divide the infacial layer into two regions (sketched in Fig. 8.8(a)): One

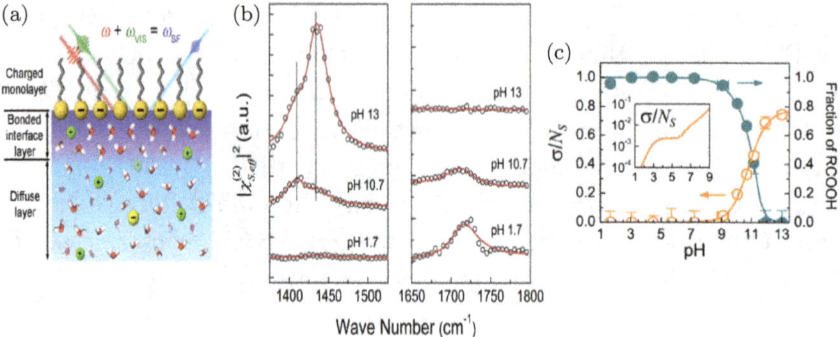

Fig. 8.8. (a) Cartoon showing that a surfactant monolayer interface can be divided into different regions: surfactant chain array, bonded interfacial layer (BIL) including surfactant head groups and an interfacial water layer that has a bonding structure appreciably different from bulk water, diffuse layer (DL) in which water molecules are appreciably reoriented by surface field, and bulk water. (b) $|(\chi^{(2)}_{S,eff})_{yyz}|^2$ spectra of the head groups of a lignoceric fatty acid monolayer on water with different pH values, displaying CO stretch modes of COOH, COO$^-$, and (COO$^-$)$_3$Na$^+$ at 1717, 1410, and 1434 cm^{-1}, respectively. (c) Surface charge density (σ) and fractional surface density of COOH versus bulk pH for the monolayer/water interface deduced from the measured CO stretch spectra. The charge density versus pH below pH9 in the inset was calculated from the deprotonation reaction (after Ref. [16]).

consists of the surfactant head groups (the surfactant chains are not likely to vary) and an atomically thin, vicinal water layer in which molecules are less mobile and their bonding structure appreciably different from that of the bulk and the other is the usually defined EDL. We call the former a bonded interfacial layer (BIL) and the latter a diffuse layer (DL) or EDL.[16] The BIL is most important because it directly governs the properties and functionalities of the interface. (Historically, description of a charged water interface always focused more on the water side. The water part of the BIL including the adsorbed ions was often called the Stern layer.) It is possible that the ion concentration in water is so high that the Debye screening length in water (hence the EDL thickness) becomes so small that it is comparable to a water monolayer. In that case, BIL and EDL can no longer be separated, and must be considered as a single layer. We show in the following how SFVS can be an effective tool to probe such charged water interfaces, and under certain circumstances, the vibrational spectra of BIL and EDL can be separately deduced. We take a lignoceric fatty acid (LA, $C_{23}H_{47}COOH$) monolayer on water as a representative case.[16]

We first discuss how to find surface charge density (σ) at the LA/water interface with given pH, adjusted by HCl and NaOH in water. The fatty acid head groups generally consist of three species: COOH, COO^-, and $(COO^-)_3Na^+$, identifiable by their CO stretches at 1717, 1410, and 1434 cm^{-1}, respectively[31] (Fig. 8.8(b)). Here, $(COO^-)_3Na^+$ refers to one Na^+ simultaneously interacting with three COO^- groups. We denote the surface fractions of the three species as X_{COOH}, X_{COO-}, and X_{COO-Na} normalized by $X_{COOH} + X_{COO-} + X_{COO-Na} = 1$. For pH > 9, fitting of the spectra by the three modes allow us to find X_i and hence the surface charge density $\sigma = |e|N_s(-X_{COO-} + X_{COO-Na}/3)$, where N_s is the surface density of LA. The result is plotted in Fig. 8.8(c). (As described in Sec. 8.4, σ could also be determined directly from SHG/SFG measurement with the help of the Gouy–Chapman theory.) For pH < 9, the $(COO^-)_3Na$ mode cannot be detected and the COO^- mode is too weak to be very quantitative; we have to rely on the reaction equation for deprotonation with $pK_a = 5.6$ (as compared to $pK_a = 5.25 \pm 0.35$ in bulk water) taken from Ref. [32] to find σ. It is known that

$$pK_a = pH_s - \log(X_{COO-}/X_{COOH})$$
$$X_{COO-} = -\sigma/|e|N_s \qquad (8.4)$$
$$pH_s = pH + e\Phi_0/2.3k_BT$$

where pH_s is the surface pH and Φ_0 is the surface potential related to σ by Eq. (8.3) from the Gouy–Chapman theory. Solution of the coupled Eqs. (8.3) and (8.4) with known pK_a determines σ and pH_s. The result is plotted in the inset of Fig. 8.8(c). It is seen that $\sigma \approx 0$ for pH < 2.5, $\sigma < 5\%$ for $2.5 <$ pH < 9, and $\sigma > 5\%$ for pH ≥ 9.

We discuss next how we can interrogate the interfacial structures of LA/water with different pH or σ using the $\text{Im}(\chi^{(2)}_{S,eff})_{yyz}$ spectra measured by PS-SFVS. Figure 8.9(a) displays a representative set of $\text{Im}(\chi^{(2)}_{S,eff})_{yyz}$ spectra for pH between 2.5 and 9. Below pH 2.5, the spectrum was found to remain unchanged, indicating that the

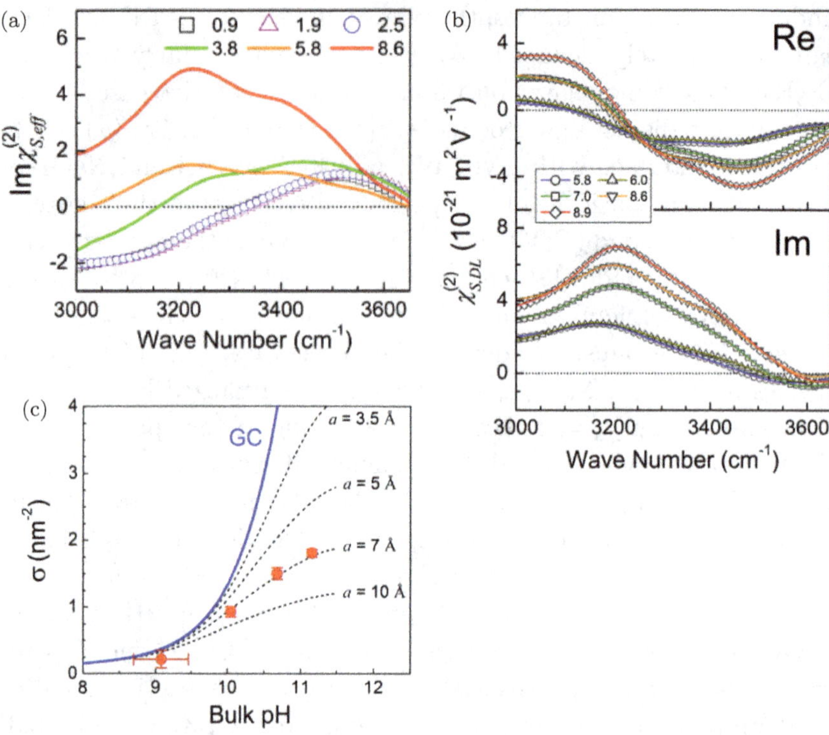

Fig. 8.9. (a) Measured $\mathrm{Im}(\chi^{(2)}_{S,eff})_{yyz}$ and (b) deduced $Re(\chi^{(2)}_{EDL})_{yyz}$ and $\mathrm{Im}(\chi^{(2)}_{EDL})_{yyz}$ OH stretch spectra of lignoceric fatty acid monolayer/water interfaces with a set of different bulk pH values from 0.9 to 8.6. More positive $\mathrm{Im}(\chi^{(2)}_{EDL})_{yyz}$ spectra at higher pH indicate more OH^- appearing at the interface. (c) σ versus pH calculated from $\mathrm{Im}(\chi^{(2)}_{EDL})_{yyz}$ spectra using the Gouy–Chapman theory (solid line) and modified Gouy–Chapman theory with different values of effective ion size (a) (dotted lines). The red dots are σ values given in Fig. 8.8(c) (after Ref. [16]).

interface is neutral with no surface charges from either deprotonation or ion adsorption and the spectrum must have come from the BIL: Since $(\overset{\leftrightarrow}{\chi}^{(2)}_{S,eff})_R = \overset{\leftrightarrow}{\chi}^{(2)}_{BIL} + \overset{\leftrightarrow}{\chi}^{(2)}_{EDL}$, we have $(\overset{\leftrightarrow}{\chi}^{(2)}_{S,eff})_R = \overset{\leftrightarrow}{\chi}^{(2)}_{BIL,0}$ in the absence of EDL. For $2.5 < \mathrm{pH} < 9$, σ is small; the BIL structure is hardly distorted and we should expect $\overset{\leftrightarrow}{\chi}^{(2)}_{BIL} \simeq \overset{\leftrightarrow}{\chi}^{(2)}_{BIL,0}$, but as discussed in Sec. 8.3, $\overset{\leftrightarrow}{\chi}^{(2)}_{EDL}$ for $\sigma > 0.1\%$ of a monolayer is readily detectable. We can find $\overset{\leftrightarrow}{\chi}^{(2)}_{EDL}(\sigma)$ by simply subtracting

$\overset{\leftrightarrow(2)}{\chi}_{BIL,0}$ off from the measured $(\overset{\leftrightarrow(2)}{\chi}_{S,eff})_R$. The $Re(\chi_{EDL}^{(2)})_{yyz}$ and $Im(\chi_{EDL}^{(2)})_{yyz}$ spectra deduced from Fig. 8.9(a) are presented in Fig. 8.9(b). Knowing $[\chi_{EDL}^{(2)}(\sigma)]_{yyz}$ and σ, we should then be able to find $Re\chi_{B,yyz}^{(3)}$ and $Im\,\chi_{B,yyz}^{(3)}$ spectra for bulk water from Eq. (8.1) with the help of the Gouy–Chapman theory. They are described in Fig. 8.4(c). Conversely, if $\overset{\leftrightarrow(3)}{\chi}_B$ is known, we can find σ from $\overset{\leftrightarrow(2)}{\chi}_{EDL}(\sigma)$. As expected, σ so obtained is consistent with $pK_a = 5.6$ assumed in the calculation. For $pH > 9$, we could again use Eq. (8.4) [with $\sigma = -N_S\,|e|\,X_{COO-} = -N_S\,|e|\,(1 - X_{COOH} + X_{COO\text{-}Na}/3)$] and the Gouy–Chapman theory to calculate σ for given pH and ionic concentration, but as seen in Fig. 8.9(c), they do not agree well with the experimentally deduced σ. Good agreement was however found using a modified Gouy–Chapman theory that assumed $pK_a = 5.6$ and an effective size of $a = 0.7$ nm for the counter ions (Fig. 8.9(c)). With $\sigma \geq 5\%$, the BIL structure can no longer be considered unchanged, and the difference between $\overset{\leftrightarrow(2)}{\chi}_{BIL}(\sigma)$ and $\overset{\leftrightarrow(2)}{\chi}_{BIL,0}$ can be appreciable. For given pH or σ, we can use the modified Gouy–Chapman model to find $\overset{\leftrightarrow(2)}{\chi}_{EDL}$ and then obtain $\overset{\leftrightarrow(2)}{\chi}_{BIL}(\sigma) = (\overset{\leftrightarrow(2)}{\chi}_{S,eff})_R - \overset{\leftrightarrow(2)}{\chi}_{EDL}(\sigma)$ from the measured $(\overset{\leftrightarrow(2)}{\chi}_{S,eff})_R$. The $\overset{\leftrightarrow(2)}{\chi}_{BIL}(\sigma)$ spectra at pH 10.6 and 12 so obtained are given in Fig. 8.10(a) in comparison with that of the neutral interface at pH 2.5. It is seen that the spectrum becomes more positive at higher pH where X_{COO-} and X_{COO-Na} of the fatty acid head groups are larger. This indicates that more water molecules in the BIL are reoriented to have their donor bonds better connected with the deprotonated head groups (sketched in Fig. 8.10(b)). More detailed information on structure dependence of the BIL on pH will have to come from MD simulations of the interface.

There have been extensive SFVS studies on ionic surfactant monolayers on water by many groups. Most of them were carried out by SF intensity $(|\chi_{S,eff}^{(2)}|^2)$ spectroscopy displaying CH stretch modes from hydrocarbon chains of surfactants and a broad OH stretch band mainly from water molecules in the EDL set up by the high surface charge density associated with surfactant head

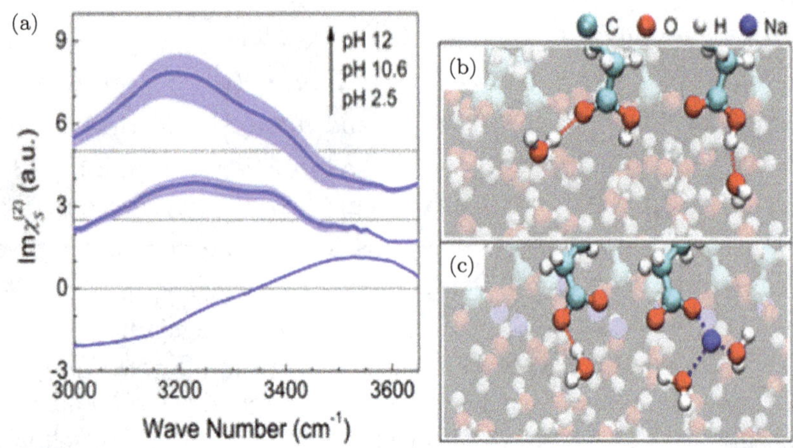

Fig. 8.10. (a) Deduced $\mathrm{Im}(\chi_{BIL}^{(2)})_{yyz}$ (labelled as $\mathrm{Im}\overset{\leftrightarrow}{\chi}_S^{(2)}$ in the figure) spectra of the BIL of lignoceric monolayer/water interfaces with bulk pH values at 2.5, 10.6, and 12. (b) and (c) Cartoons describing possible water bonding structures in the BIL of a neutral and fully deprotonated monolayer, respectively (after Ref. [16]).

groups. A few of them reported $\mathrm{Im}(\chi_{S,eff}^{(2)})_{yyz}$ spectra that reveal broad positive and negative OH stretch bands in response to negative and positive surface charges, respectively. As examples, the $\mathrm{Im}(\chi_{S,eff}^{(2)})_{yyz}$ OH stretch spectra of water surfaces covered by three model ionic surfactant monolayers are presented in Fig. 8.11:[33] positively charged DPTAP (1,2-dipalmitoyl-3-trimethylammonium propane), negatively charged DOPEG (1,2-dioleoyl-sn-glycero-3-phosphoethyleneglycol), and zwitter ionic DMPS (1,2-dimyristol-sn-glycero-3-phospho-L-serine) on isotopically diluted water $(H_2O:HOD:D_2O = 1:8:16)$. The cationic DPTAP and the anionic DOPEG induce, respectively, the negative and positive bands. The zwitter-ionic DMPS also generates a positive band, indicating that it carries effectively a net negative surface charge density. Increase of ionic concentration in water would enhance field screening and reduce the width of the EDL and decrease the band strength. This was experimentally confirmed by adding NaCl in water. Because of the overwhelming contribution of $\overset{\leftrightarrow}{\chi}_{EDL}^{(2)}(\sigma)$ to SF spectra in these cases, deduction of $\overset{\leftrightarrow}{\chi}_{BIL}^{(2)}(\sigma)$ from SF measurement would be very difficult.

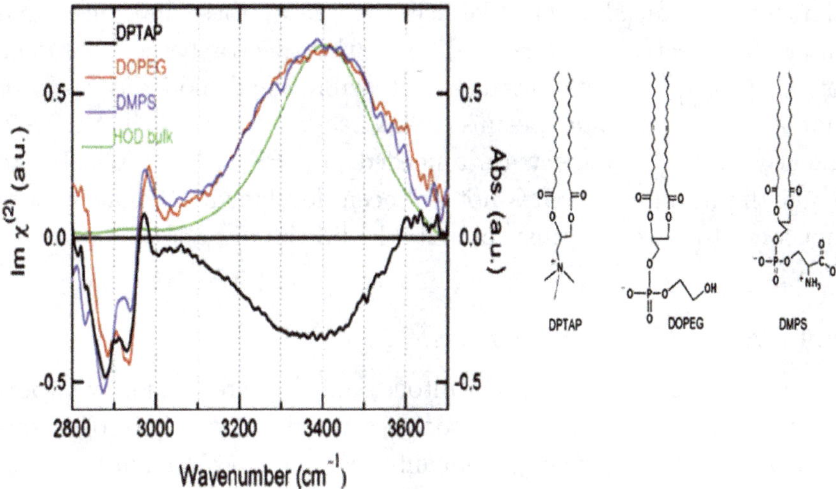

Fig. 8.11. Reflected SSP Im $\chi^{(2)}_{S.eff}$ spectra of isotopically diluted interfacial water (H_2O:HOD:D_2O = 1:8:16) under DPTAP, DOPEG, and DMPS monolayers in comparison with the IR absorption spectrum of bulk HOD in the OH stretch region. The molecular structures of the three surfactants are depicted on the right. (after Ref. [33]).

8.8. Water/Liquid Interfaces

SF spectroscopy is a unique tool to study liquid/liquid, and particularly water/liquid, interfaces. To probe the microscopic structure of such buried interfaces, however, is still quite challenging. So far, only SFVS $|\chi^{(2)}_{S,eff}|^2$ measurements of water/liquid interfaces have been reported. The first study was on a water/hexane interface; the observed $|(\chi^{(2)}_{S,eff})_{yyz}|^2$ spectrum was similar to that of the water/vapor interface with a dangling OH mode signifying the hydrophobic nature of the interface.[34] Richmond and coworkers later studied a number of water interfaces with polar and nonpolar liquids.[35] In all cases, the broad bonded OH stretch band dominated the spectrum. Water interfaces with nonpolar liquids showed a spectrum very similar to that of the hexane/water interface. Those with polar liquids had the dangling OH mode strongly suppressed and the broad band reduced in strength presumably because interaction of water and polar liquid molecules had led to a more diffusive interfacial layer. Ions can emerge at these interfaces and alter the spectrum. Interpretation of the spectra based on fitting of the $|\chi^{(2)}_{S,eff}|^2$ spectra with discrete vibrational modes is however not reliable and could be misleading, as we discussed in Sec. 8.2. Clearly, $\mathrm{Im}(\chi^{(2)}_{S,eff})$ spectra are needed, but PS-SFVS applicable to liquid/liquid interfaces has not yet been developed although it does not seem to have any unsurmountable difficulty.

8.9. Water/Oxide Interfaces

Water/oxide interfaces are ubiquitous, and play a role in many important natural and industrial processes ranging from heterogeneous catalysis, chemical sensing, biomembrane chemistry to environmental chemistry and geochemistry.[36] Protonation/deprotonation of silicate minerals, for example, is the key to their weathering process and serves as a major dioxide sink in the global carbon cycle.[37] Surface charging including ion adsorption is essential in most of these processes. Yet, despite extensive studies since Helmholtz's days more than 160 years ago, the microscopic understanding of such interfaces is still very poor. Lack of effective experimental tools is the reason. Recently, SFVS has made a giant step forward to change the situation as we shall discuss in what follows with water/silica interfaces taken as a representative case.

Simple oxide interfaces exposed to air are usually passivated by hydrogen. In water, they can be deprotonated (protonated) if pH of water is sufficiently high (low). Ions in water can adsorb at an interface to alter its surface charge density (σ). As described in Sec. 8.7, we can divide the interface, from the structural point of view, into two regions, BIL and DL (EDL). In the present case, the BIL includes the surface oxide layer and surface water molecules that have their H-bonding structure correlated with the oxide surface. The main interest on water/oxide interfaces is in finding, for a given system, the surface charge density, the surface potential distribution, the structure of EDL, and most importantly, the structure of BIL. The latter is unfortunately most difficult to glean, but as we discussed in Sec. 8.7, it is now possible to use SFVS with the help of the

Gouy–Chapman (or modified Gouy–Chapman) theory to find the vibrational spectra of $\chi^{(2)}_{S,eff}$ and $\chi^{(2)}_{BIL}$ separately, from which we can learn about the structure of BIL.

Water/silica interfaces have been studied extensively with different techniques. Phase-sensitive SFVS has been most successful in providing detailed structural information. It is known that if pH > 2, a silica/water interface would be negatively charged by the deprotonation process, $SiOH + OH^- \rightarrow SiO^- + H_2O$. Eisenthal and coworkers first used SHG to probe the EDL of the water/silica interface set up by deprotonated SiOH.[38] Their result with different pHs in water suggested the existence of two pK_a values, 4.5 and 8.5, for deprotonation of SiOH presumably at two different silanol sites (Fig. 8.12(a)). Soon after, surface SFVS measuring $|\chi^{(2)}_{S,eff}|^2$ was applied to the interface with a few different pHs and salt concentrations to demonstrate efficacy of the technique to probe buried interfaces.[39] Such measurements were also performed by several other research groups, and the observed spectra are fairly consistent. However, as we mentioned in Secs. 8.2 and 8.6, interpretations of the broad OH stretch band of $|\chi^{(2)}_{S,eff}|^2$ were necessarily vague, often confusing and untrustworthy. The Im $\chi^{(2)}_{S,eff}$ spectra from PS-SFVS are needed. Two sets of spectra, one on water/crystalline quartz and the other on water/silica, are shown in Figs. 8.12(b) and 8.12(c) with different pH values.[40,41] The two sets are similar except for a shift in frequency. A close examination of the spectra in Fig. 8.12(c) suggests that each spectrum could be separated into a high frequency and a low frequency band, the former increasing with pH in the pH 2–6 range and the latter in the pH 6–10 range, qualitatively reflecting the two pK_a values revealed in the SHG measurement. Even so, the information that can be extracted from the Im $\chi^{(2)}_{S,eff}$ spectra is very limited if their $\chi^{(2)}_{BIL}$ and $\chi^{(2)}_{EDL}$ spectral components are not known separately.

With the scheme described in Sec. 8.7, it should now be possible to separately deduce $\chi^{(2)}_{BIL}$ and $\chi^{(2)}_{EDL}$ from reflected PS-SFVS that measures $\overleftrightarrow{\chi}^{(2)}_{S,eff} = \overleftrightarrow{\chi}^{(2)}_{BIL} + \overleftrightarrow{\chi}^{(2)}_{EDL}$: For a neutral interface, $\overleftrightarrow{\chi}^{(2)}_{EDL} = 0$ and $\overleftrightarrow{\chi}^{(2)}_{S,eff} = \overleftrightarrow{\chi}^{(2)}_{BIL}(\sigma = 0) \equiv \overleftrightarrow{\chi}^{(2)}_{BIL,0}$; when the surface charge

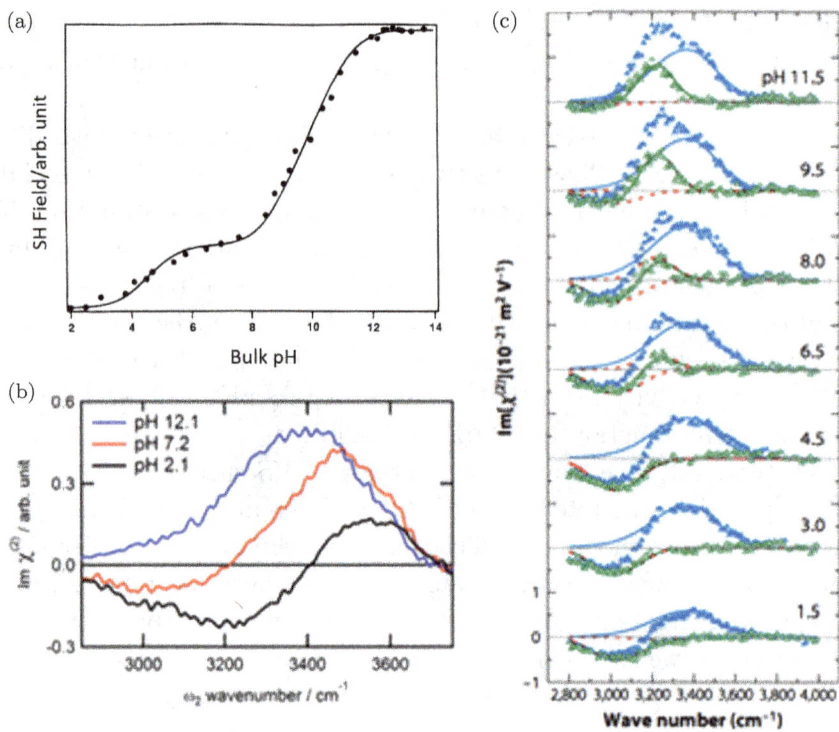

Fig. 8.12. (a) Reflected SH field amplitude as a function of bulk pH from the water/silica interface showing a two-step variation. The dots are experimental data and the solid curve is a theoretical fit (after Ref. [38]). (b) Reflected SSP Im $\chi^{(2)}_{S,eff}$ spectra of the water/silica interface at three different bulk pH values (after Ref. [41]). (c) Reflected SSP Im $\chi^{(2)}_{S,eff}$ spectra of the water/α-quartz interface at different bulk pH values. Each spectrum can be fit by a high-frequency band and a low-frequency band that have different strength variations with pH (after Ref. [40]).

density is small ($\sigma < 5\%$), we have $\overset{\leftrightarrow}{\chi}^{(2)}_{BIL}(\sigma) \simeq \overset{\leftrightarrow}{\chi}^{(2)}_{BIL,0}$ independent of σ, and hence $\overset{\leftrightarrow}{\chi}^{(2)}_{EDL}(\sigma) = \overset{\leftrightarrow}{\chi}^{(2)}_{S,eff} - \overset{\leftrightarrow}{\chi}^{(2)}_{BIL,0}$. With the help of the Gouy–Chapman model, we can then find σ from the deduced $\overset{\leftrightarrow}{\chi}^{(2)}_{EDL}(\sigma)$. We can also find σ by using two or more different beam geometries of SHG/SFG as described in Sec. 8.4; the scheme is applicable to all cases with $\overset{\leftrightarrow}{\chi}^{(2)}_{BIL}(\sigma) \simeq \overset{\leftrightarrow}{\chi}^{(2)}_{BIL,0}$ or not. Knowing σ and the corresponding $\overset{\leftrightarrow}{\chi}{}^{(2)}_{EDL}(\sigma)$, we can obtain $\overset{\leftrightarrow}{\chi}{}^{(2)}_{BIL}(\sigma) = (\overset{\leftrightarrow}{\chi}{}^{(2)}_{S,eff})_R - \overset{\leftrightarrow}{\chi}{}^{(2)}_{EDL}(\sigma)$. Such an experiment allows finding of σ, surface potential distribution and $\overset{\leftrightarrow}{\chi}{}^{(2)}_{EDL}(\sigma)$ and $\overset{\leftrightarrow}{\chi}{}^{(2)}_{BIL}(\sigma)$ for water/silica interfaces over a wide range of pH and salt concentrations in water. This scheme is being implemented to interrogate silica/ and other oxide/water interfaces.[42] However, when the ion concentration in water is very high, the EDL (or the Debye screening length) may become so narrow that it can no longer be separated from the BIL. In such cases, we must consider BIL and EDL as a combined single layer. This is the case of water/silica interfaces with bulk pH $\sim$ 12 recently investigated by Tahara and coworkers.[43]

Microscopic structure of interfacial water can be learned from molecular dynamics simulation of the observed interfacial water spectrum. In such simulations on water/oxide interfaces, the oxide surface structure is usually assumed given, e.g., a bulk terminated surface structure, but the assumption may not be correct. In fact, it is common that a solid surface structure tends to reconstruct in response to environmental change. To fully understand properties and functions of oxide/water interfaces, we need to know the surface structures of both oxides and water and their possible changes in response to change of water solution. This is feasible with SFVS.

We described in Sec. 7.2.2 how we could use SFVS to probe surface phonons (vibrations) of solids and presented quartz (silica) in air as an example. We can apply the same technique to probe surface vibrations of water/silica interface and learn about the surface structural change of silica when pH and ion concentration in water are varied. This has been worked out by W.T. Liu and coworkers.[44] The spectrum of Si-OH and Si-O- stretch vibrations and its changes with pH varying from 2 to 12 were observed. Spectral analysis with the help of MD simulations was able to identify a surface reconstruction process of silica upon deprotonation of SiOH. The discovery confirms the need of characterizing the surface structures of both oxide and water in characterization of the interfacial structure of an oxide/water interface, and generally, the same is expected for all interfaces.

Water/oxide interfaces can be charged by electrical bias that adds another independent controlling parameter to vary the charge state of the interfaces. W.T. Liu and coworkers have come up with a promising bias scheme that greatly extends the range of the surface charge states that can be explored by SFVS or other techniques.[45] It employs a three-layer structure like the Si/SiO_2/water system described in the last paragraph. As sketched in Fig. 8.13(a), the bias is across the three-layer junction. Such a junction device is actually well known as an electrolyte-insulator-semiconductor (EIS),

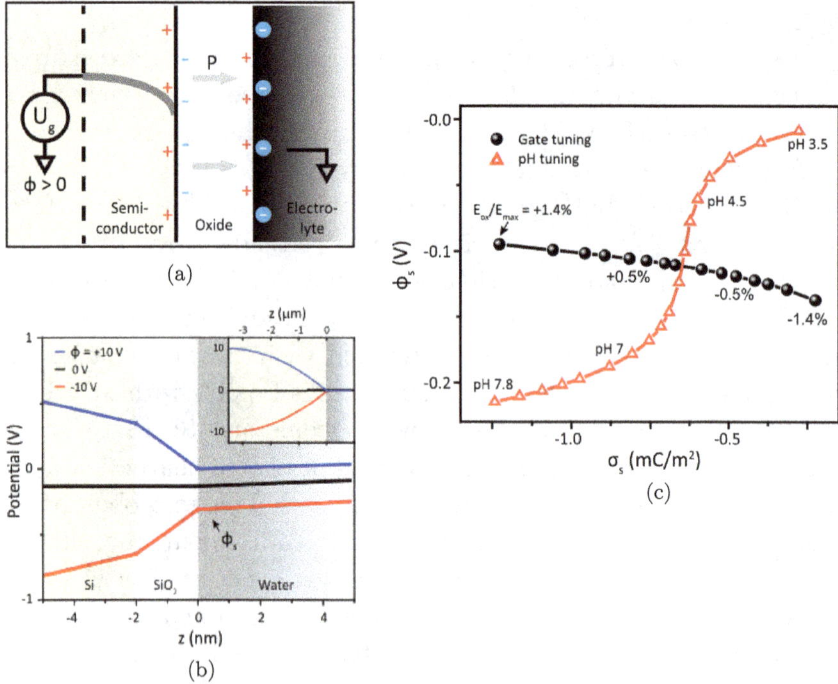

Fig. 8.13. (a) Schematic representation of a semiconductor/oxide/electrolyte EIS junction under a positive electrical bias. (b) Calculated potential distributions (ϕ) with positive, zero, and negative overall potentials for the device in (a). The inset shows the distributions over a larger range. (c) Calculated potential variation at the SiO_2/water interface (ϕ_s) as a function of surface charge density (σ_s) for the EIS junction in comparison with that of a simple bulk silica/water interface. In the same range of σ_s, the change of ϕ_s in the EIS case is much less than in the bulk silica/water case (after Ref. [45]).

ion-sensitive, field-effect transistor, and has long been widely used as ion and biomolecule sensor in biomedical applications. The underlying theory is well developed, allowing finding of potential distribution in the three layers for a given bias on such a device.[46] An example is shown in Fig. 8.13(b) for an $Si/SiO_2/water$ EIS with an overall potential drop of ϕ across the junction.[45] Qualitatively, the operation principle is as follows: At $\phi = 0$, the potential distribution is flat. The potential variation with finite ϕ mainly occurs in the Si and silica layers. As $\phi > 0$ becomes more positive, the field at the $SiO_2/water$ interface attracts more anions from water to the interface, and through deprotonation and ion adsorption, makes the surface charge density (σ) at the interface more negative. This change of surface charge density $(\Delta\sigma)$ creates a potential that opposes the change of the applied bias $(\Delta\phi)$, resulting in a change of surface potential, $\Delta\phi_s$, at the interface much less than $\Delta\phi$. Consequently, even if ϕ is varied to change σ over a wide range, ϕ_s remains quite small, resulting in a weak and little changed EDL whose contribution to SFVS can be more easily accounted for. Similarly, for $\phi < 0$, more negative ϕ leads to more positive σ. In both cases, the tuning range of σ is limited by the maximum $|\phi|$ that can be applied to the EIS junction before electric breakdown of the oxide layer occurs. It is possible to use this scheme to cover the whole range of σ that is normally covered by pH tuning, but variation of $\Delta\phi_s$ is much less, as seen in Fig. 8.13(c). Together with tuning of pH, the electric bias can cover a wide range of surface charge states, possibly beyond the normally explored region, and SFVS can be used to interrogate the interfacial structure over the whole range. This scheme has already been successfully demonstrated on the $Si/SiO_2/water$ EIS junction.[45] It promises exciting results in a new regime and could pave the way for studies of surface charging and electrochemistry of different water/oxide interfaces.

8.10. Electrochemical Interfaces of Water

Electrochemistry (EC) is one of the most important branches of chemistry. It appears ubiquitously in many areas of modern science and technology. The familiar EC processes are electrolysis, corrosion, electroplating, charging and discharging of batteries, artificial photosynthesis for renewable energy, and ion transportation through membranes, just to name a few. Research on EC aims at understanding and controlling EC reactions. Identification of atomic/molecular species appearing at electrode/electrolyte interfaces and characterization of their behavior in response to varying potential are often the focus. However, despite its long history, EC still remains a field short of advanced research tool. Cyclic voltammetry (CV) measuring charge transfer and current as functions of applied potential that cycles back and forth is the commonly employed technique for EC studies. Since EC interfaces are buried, they are not accessible by many surface probes. Infrared reflection absorption spectroscopy (IRRAS) and X-ray spectroscopy have been developed to study EC with good success, but they suffer to some extent from not being sufficiently surface specific. One would expect SHG/SFG could be a viable tool. Indeed, experiments have shown that the technique is viable, but removal of a few obstacles is needed.

In fact, it was an EC interface first used to demonstrate that SHG/SFG can be an effective tool to probe surfaces and interfaces.[47] In that experiment, SHG from a silver electrode in a 0.1M KCl water solution of an EC cell was detected to track formation and dissociation of AgCl on silver during a CV cycle. The surface of Ag was roughened for the experiment. As shown in Fig. 8.14(a), when the potential of Ag with reference to the saturated calomel electrode was switched from −0.30 v to 0.08 v, a negative current began to flow, indicating formation of AgCl on the Ag electrode. Correspondingly, the SHG signal rose rapidly, but soon appeared to saturate. The fast rise of the signal was in response to the formation of the first AgCl monolayer, and the saturated signal suggested multilayer growth of AgCl because SHG is only sensitive to surface change. With the potential switched to 0.02 v, the current reversed direction and AgCl was reduced back to Ag, but the SHG signal only decreased sharply when the last monolayer of AgCl started to disappear as seen from vanishing of the current. It was also found that if 0.05M of pyridine

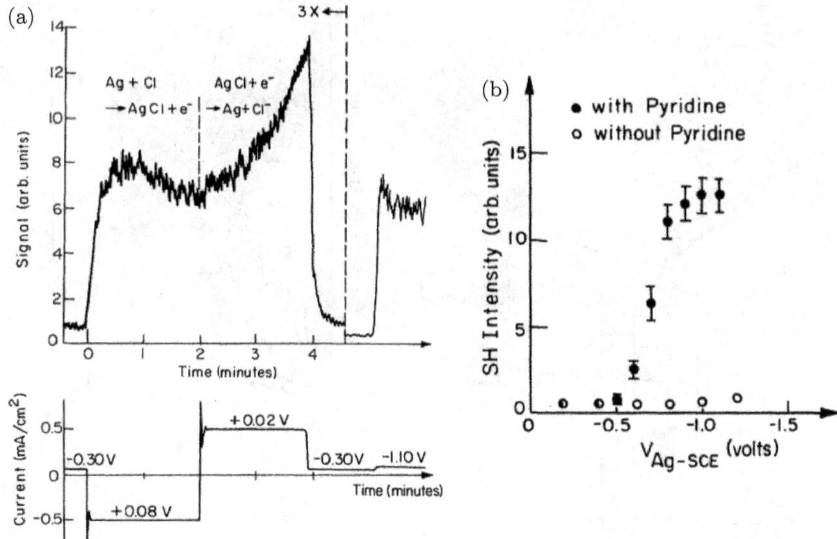

Fig. 8.14. (a) Time-varying EC current (lower curve) and second harmonic generation from the Ag electrode (upper curve) in a 0.1M KCl solution as functions of time during an EC cycle. The voltages listed on the lower curve are the Ag electrode potentials, V_{Ag-SCE}, with respect to the saturated calomel reference electrode in the EC cell. (b) Second harmonic signal versus V_{Ag-SCE} when pyridine was added to the electrolytic solution following completion of the EC cycle (after Ref. [47]).

was mixed into the electrolyte, the SHG signal shot up when the bias potential increased more negatively beyond –0.5 v, signifying adsorption of pyridine on Ag (Fig. 8.14(b)). The signal strength from the roughened Ag surface was very strong. This was because SHG experienced a $\sim 10^4$ local field enhancement from the local plasmon resonance of the roughened Ag. Soon afterwards, it was realized that SHG from an EC interface is often so strong that local field enhancement is not needed for SHG to probe EC processes.

Spectroscopic information is however necessary to identify and characterize species in EC studies. Guyot-Sionnest and Tedjeddine first applied SFVS to monitor CO and CN adsorption on a Pt electrode in an EC cell.[48] In order for the IR input to access the water/electrode interface, they devised a cell with the entrance

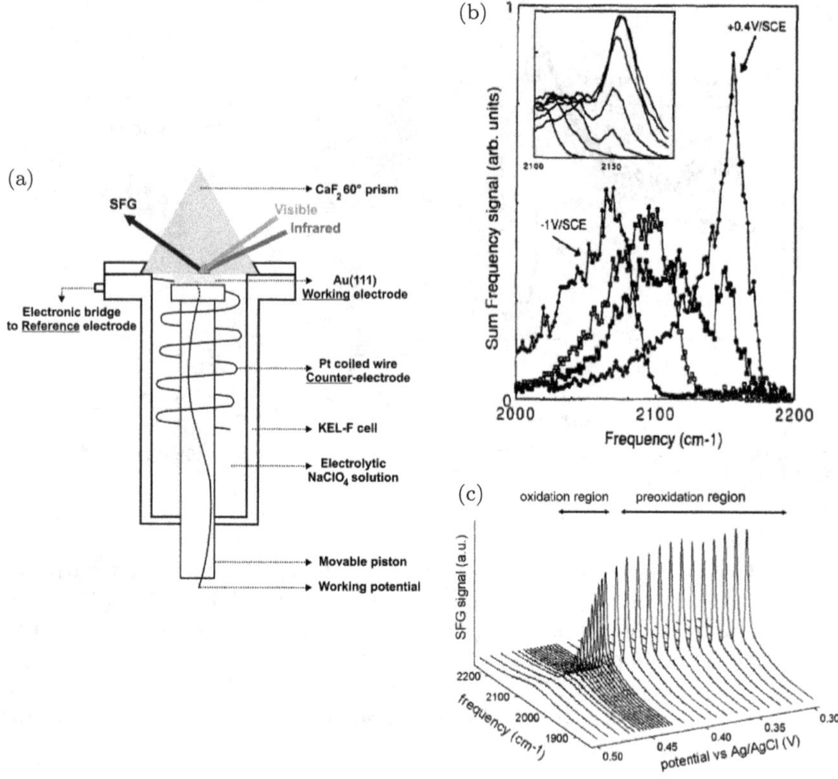

Fig. 8.15. (a) Sketch of an EC cell for SF vibration spectroscopy and the beam geometry. (b) SF vibrational spectra of CN^- adsorbed on a polycrystalline platinum electrode in a 0.1 M $NaClO_4$, and 0.025 M KCN solution with the electrode potential at -1v (filled diamonds), -0.6 v (open squares), -0.2 v (filled squares), and $+0.4$ v (open diamonds), respectively. The inset shows the growth of the 2150 cm^{-1} band when the potential increases from -0.8 v to $+0.4$ v in 200 mV steps (after Ref. [48]). (c) Sum frequency CO stretch spectra taken with increasing potential of Pt for a CO monolayer initially adsorbed on Pt in a CO-free 0.1M H_2SO_4 electrolytic solution. Decay of the CO mode is due to CO oxidation (after Ref. [49]).

window or prism pressed against Pt to limit the water film thickness to less than 10 μm so that attenuation of the IR input in passing through the water film to reach the interface was tolerable (Fig. 8.15(a)). This is the scheme later adopted by others for SFVS studies of EC. We show two examples here. One reported by Guyot-Sionnest and Tedjeddine is on CN adsoption from a 0.1M $NaClO_4$

and 0.025M KCN solution on a Pt electrode with different bias voltages. The observed PPP $|\chi^{(2)}_{S,eff}|^2$ stretch spectra of adsorbed CN are reproduced in Fig. 8.15(b). At -1v bias, an asymmetric band peaked at 2070 cm^{-1} is present. It blue shifts as the bias potential increases; and then a new band peaked at 2150 cm^{-1} begins to emerge at -0.2 v and grows rapidly with increase of bias. The two phases were attributed to the presence of two CN species with opposite orientations. The low-frequency band was from CN with N bonded to Pt and the high-frequency band from CN with C bonded to Pt. Another example is on oxidation of a CO monolayer adsorbed on a Pt electrode in a 0.1M H_2SO_4 CO-free electrolytic solution[49] (Fig. 8.15(c)). The series of PPP spectra of the CO stretch show that as the bias potential increases, CO starts to disappear by oxidation to CO_2 , very rapidly in the 0.43–0.47 v range.

Over the past decades, SFVS has been used to study different EC processes: corrosion, metal deposition, decomposition of alcohols, CO oxidation, hydrogen evolution reaction, intermediates of reactions, monolayer phase transition, and so on with their dependences on temperature and electrolyte composition, etc.[50,51] Through vibration resonances, the technique allows identification and tracking of adsorbed species, their absence and presence at the electrode, their surface density and adsorption geometry, and their interaction with the surroundings, providing valuable information about EC processes at the molecular level. Essentially all studies so far have been focused on adsorbed species at electrodes.

There are still obvious shortcomings for SFVS to become a prime tool for EC studies: To have more complete knowledge about EC at an interface, one would like to know not only what appears at the interface, but also the water structure at the interface. It is known that Im $\chi^{(2)}_{S.eff}$ spectra are more informative, but so far, only $|\chi^{(2)}_{S,eff}|^2$ spectra have been taken for EC interfaces. Because of high reflectivity of metal electrodes, only SFVS with PPP polarization can be readily detected. In order to measure spectra of other polarizations and weaker vibrational modes, improvement of SFVS sensitivity is needed. Removal of some of the above deficiencies should be possible through modification of the EC cell and beam geometry.

The narrow water gap between the electrode and optical window of the usual EC cell constructed for SFVS could cause major difficulty and is a possible source of worry. It is not clear whether EC would be different if the gap is too small. Although attenuation of the IR input in the path reaching the EC interface is reduced by reducing the gap, transmitted SFG through the water film may still contribute an unwanted signal that interferes with the desired signal from the interface. The surface signal generated at the interface between the cell window and water solution may also be a concern. It is possible to make sure if the SF signal is from the EC interface if the gap thickness can be varied, but this is cumbersome. The problem could be solved if we could have the IR input incident from the electrode side, but the input attenuation would be worse even if thin metal films are used. It is however possible to have strong IR field on a metal film through excitation of surface plasmons from the backside.[52] The experimental arrangement is sketched in Fig. 8.16(a). As a demonstration, the scheme was employed to monitor desorption and readsorption of a self-assembled thiol monolayer (1-octadecanethiol, $CH_3(CH_2)_{16}CH_2SH$) on a gold electrode in a 10 mM KOH solution. The cyclic voltammogram (CV) with a thiol monolayer prepared on the gold electrode is described in Fig. 8.16(b). The CV scan starts at the bias voltage of +1.2 v. The observed $|\chi^{(2)}_{S,eff}|^2$ vibrational spectrum (with Fresnel coefficients properly removed) of the EC interface at 1.2 v in Fig. 8.16(c) shows a clear CH_3 symmetric stretch mode at 2870 cm^{-1}, indicating the presence of an ordered all-trans thiol monolayer. A cathodic scan downward to –0.5 v should desorb the monolayer completely as evidenced by the dip around –0.2 v in the CV scan of Fig. 8.16(a), but the observed spectrum at –0.5 v still exhibits a clear CH stretch band that can be attributed to a chain-ordered monolayer with some gauche defects. An anodic scan back to +1.2 v recovers the original spectrum as seen in Fig. 8.16(c). The result is a conformation of earlier studies that the desorbed thiol monolayer may have remained as an integrated monolayer in the neighborhood of the Au electrode ready to be readsorbed.

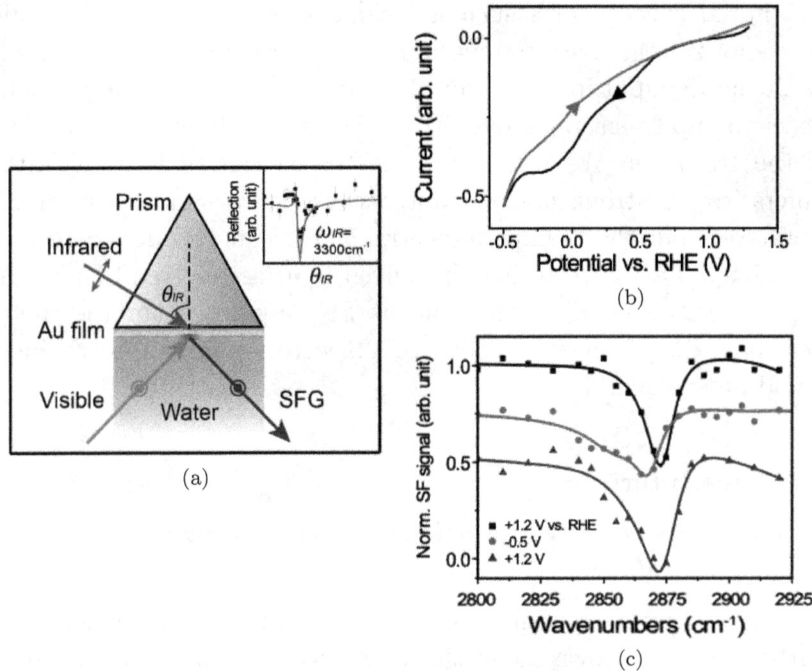

Fig. 8.16. (a) Schematic representation of the electrochemical cell and the beam geometry for SFVS with IR plasmon excitation. Inset: Reflection of IR input at 3300 cm^{-1} from the prism/gold/air interface as a function of incident angle; the dip indicates surface plasmon excitation. Dots are measured data, and the solid line is from calculation (after Ref. [52]). (b) CV plot of a thiol-covered gold electrode in 10 mM KOH solution. The potential of the gold electrode is with reference to a reverse hydrogen electrode in the cell. The cycle began with the cathodic scan (arrow down) and was followed by the anodic scan (arrow up). (c) SF vibrational spectra (with Fresnel coefficients removed) of CH stretches from the thiol monolayer taken at 1.2 v, and −0.5 v at the beginning and end of the cathodic scan, respectively, and finally back to 1.2v at the end of the anodic scan (after Ref. [52]).

The above surface plasmon scheme avoids the need of a narrow water gap and enhances the IR field at the EC interface. It may also be possible to enhance the input fields of different polarizations and improve the sensitivity of SFVS by a proper design of 2D metamaterials as metal electrodes. W.T. Liu and coworkers have had initial success with such a scheme showing that SFVS could have enough sensitivity to measure water spectra at EC interfaces.[53] With the IR input incident from the electrode side, it would also be easier to implement phase-sensitive SFVS through phase adjustment of the IR beam. We realize that SFVS measuring $|\chi^{(2)}_{S,eff}|^2$ often suffers from a strong nonresonant background originating from the metal electrode that hampers spectral analysis and reduces detection sensitivity. The problem may be solved by time-resolved SFVS that suppresses the nearly instantaneous SF response from the metal electrode, but if phase-sensitive SFVS were available, the problem would be avoided.

8.11. Ice Interfaces

8.11.1. *Reflected SF vibrational spectra from ice interfaces*

Ice interfaces play a pivotal role in nature and in our modern life, but still little is known about them. One would think SFVS could be an effective means to probe ice interfaces. Wei *et al.* first applied the technique to the ice/vapor, ice/OTS/silica, and ice/silica interfaces with focus on the basal (0001) plane of the hexagonal Ih ice.[54] Figure 8.17(a) displays the reflected SSP and PPP $|\chi^{(2)}_{S,eff}|^2$ spectra in the OH stretch range for the ice/vapor interface at different temperatures. The SPS spectra are weak and barely detectable. The dangling OH stretch is prominent in both SSP and PPP; it grows stronger and narrower, and red-shifts somewhat, as temperature decreases. The bonded OH stretch band of SSP is significantly narrower and stronger (by more than 30 times) than that of the water/vapor interface; the peak at $\sim$3180 cm^{-1} is enhanced and red-shifted with decrease of temperature. The same band in PPP is not as strong, but appears to have a component at higher frequencies that decreases with temperature. The spectra were reproduced later by other groups, although the details appeared to vary.[55] Shultz and coworkers extended the measurement to the prism planes of hexagonal ice and observed spectra similar to those of the basal

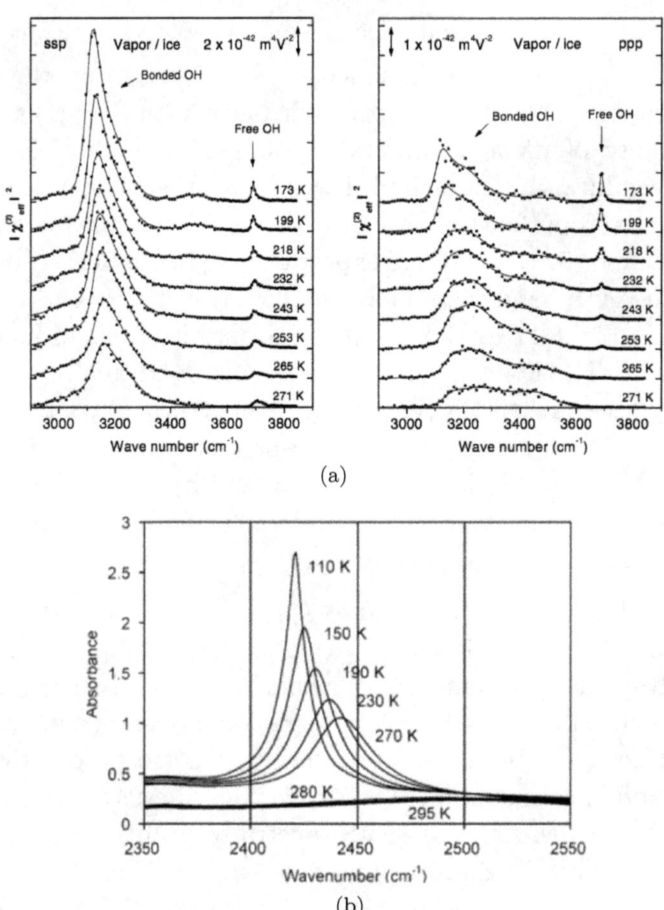

Fig. 8.17.   (a) Reflected SSP and PPP SF $|\chi^{(2)}_{S,eff}|^2$ spectra of OH stretches from a basal-face ice/vapor interface at various temperatures (after Ref. [54]). (b) IR absorption spectra of D_2O ice showing temperature dependence similar to that of the bonded OH peak in $|\chi^{(2)}_{S,eff}|^2$ (after Ref. [59]).

plane.[56] Yamaguchi and coworkers used PS-SFVS to probe different Ih-ice/vapor interfaces,[57] but their $\mathrm{Im}(\chi^{(2)}_{S,eff})_{yyz}$ spectrum for the basal face of ice appears to have a phase difference from that obtained by Smit *et al.*[58] In all reports, it is clear that the dangling OH peak is from ice surface, but unclear on whether the broad bonded OH band is also surface-dominated. In fact, there are a number of reasons to believe that electric–quadrupole (EQ) contribution (i.e., the $-\hat{z} \cdot \overset{\leftrightarrow}{\chi}^{(2)}_{q3}$ term discussed in Sec. 7.4.2) from the bulk is not negligible. Consider the spectrum taken with SSP polarization. (1) Because of its more ordered crystalline structure, ice should have a significantly larger bulk EQ contribution than liquid water. (2) Judging from the water/vapor case, if the spectrum is dominated by the ice interface, one would expect that the dangling OH peak and the bonded OH band should have similar strength per cm^{-1}, but the spectra in Fig. 8.17(a) show that the former is orders of magnitude weaker. (3) The bonded OH stretch bands observed from different ice interfaces are similar, suggesting that surface contribution is not likely dominant. (4) The IR absorption spectra of ice in the bonded OH stretch region, given in Fig. 8.17(b), display temperature dependence similar to the SF spectra.[59] (5) Transmitted SFVS measurement yielded a value of $|\hat{z} \cdot \overset{\leftrightarrow}{\chi}^{(2)}_{q3}|$ for EQ bulk contribution close to that of the measured $|\chi^{(2)}_{S,eff}|$. The above list does not mean that the EQ bulk contribution should dominate in reflected SFVS from ice. In some experiments, spectral changes resulting from perturbation of ice interfaces were actually observed, indicating that surface contribution was significant. It is most likely that both surface and bulk contribute to SFVS from ice interfaces. Temperature dependences of the IR and SFVS spectra seem to suggest that bulk contribution may be more significant around 3100 cm^{-1}.

There are many reports in the literature on SFVS of ice interfaces, attempting to extract interfacial ice structure from observed spectra.[55] In most of them, interpretations of experimental results are based on the assumption that the spectra come only from interfaces. They have created a great deal of confusion and controversy. Without justification of bulk contribution being negligible, such interpretations of the observed spectra are not trustworthy. Theoretical calculations of SFVS on ice interfaces also have not taken bulk contribution into account. Thus, it is fair to say that we still do not have good knowledge on interfacial structure of ice. Development of techniques that can separately probe surface and bulk is needed.

8.11.2. *Surface melting of ice*

Surface melting of ice was first proposed by Faraday.[60] He suggested that the reason ice could be sintered with little pressure was due to the presence of a liquid layer on ice surfaces even below the bulk ice melting temperature. His idea was confirmed in the past half century by various modern techniques although quantitative results vary. One would think that SFVS could also be employed to probe surface melting of ice and the surface liquid layer. It is indeed possible to detect spectral change of the bonded OH stretches in $\overleftrightarrow{\chi}^{(2)}_{S,eff}$ of ice as temperature decreases below the bulk melting temperature. Unfortunately, as mentioned in the last section, their bulk contribution to SFVS is not likely negligible and attempts to learn interfacial structural change of ice as a function of temperature from the temperature-dependent spectrum of $\overleftrightarrow{\chi}^{(2)}_{S,eff}$ are not conclusive.

The dangling OH peak at ~ 3700 cm^{-1} is the only spectral feature characteristic of a hydrophobic ice interface. Presented in Fig. 8.18(a) are the expanded basal ice/vapor spectra of the dangling OH mode taken at different temperatures. It is seen that the SSP and PPP spectra have different temperature dependence, indicating that the dangling OH must have its orientation, or orientation distribution, vary with temperature. If the basal ice surface is solid with a bulk-terminated structure, we expect that the dangling OH should be fixed along the surface normal. Increase of temperature may cause the dangling OH to librate, resulting in an increasingly broad orientation distribution. As described in Sec. 6.4, we can use the ratio of the PPP and SSP amplitudes of the dangling OH mode, A_{PPP}/A_{SSP}, to determine the orientation spread. Figure 8.18(b) shows A_{PPP}/A_{SSP} as a function of temperature deduced from the spectra in Fig. 8.18(a).[61] With the assumption of an azimuthally isotropic orientation distribution of $f(\theta, \phi) = $ constant for $\theta < \theta_m$ and 0 otherwise, it was found that θ_m increases from 0 at ~ 200K to $\sim 60°$ at 273K, as illustrated in Fig. 8.18(c) (Sun *et al.* suggested from MD simulations that $f(\theta, \phi)$ could be an exponential decay function, but any single-parameter $f(\theta, \phi)$ is not expected to change the above result significantly[62]).

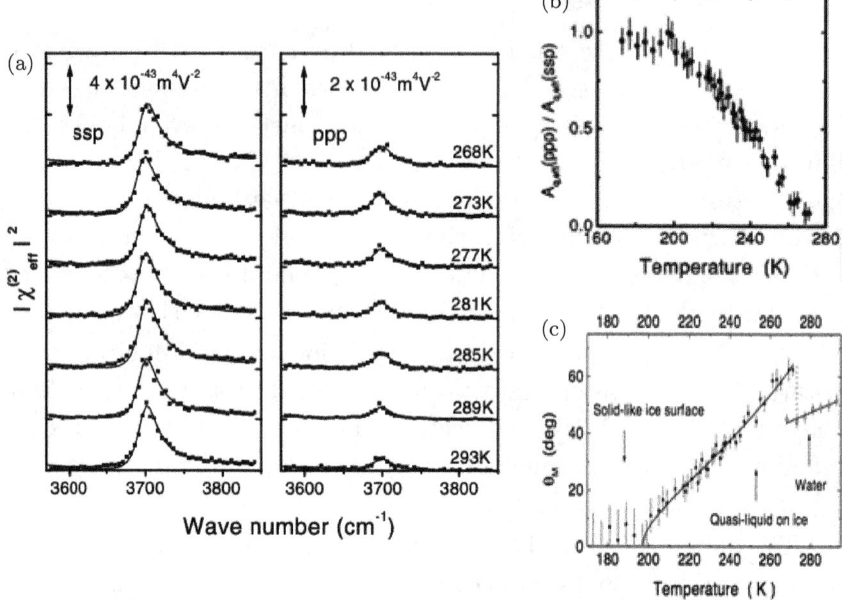

Fig. 8.18. (a) Expanded version of the dangling OH mode in the spectra of Fig. 8.17(a). (b) Temperature dependence of the ratio of the PPP and SSP amplitudes of the dangling OH mode deduced from (a). (c) Temperature dependence of the polar angle spread (θ_M) of the dangling OH due to libration. Near 273K, ice has a larger θ_M than liquid water (after Ref. [61]).

The plot in Fig. 8.18(c) indicates that the dangling OH starts to librate at 200K. This however does not mean that surface melting of ice starts at 200K if melting refers to lattice deformation in the usual definition. Surface lattice deformation of ice is likely to occur at a significantly higher temperature as was observed by other techniques such as X-ray diffraction and ellipsometry. A surprising result of the SFVS study is that the orientation spread of the dangling OH near the bulk melting temperature appears larger than that of the supercooled water at the same temperature and exhibits some hysteresis effect upon temperature cycling. This is an indication that the liquid layer from surface melting, often called a quasi-liquid layer, is structurally different from water of the liquid phase and surface melting of ice is incomplete, i.e., the quasi-liquid layer does not continuously expand and merge into bulk liquid with increase of temperature toward bulk melting.

8.12. Ferroelectric Ice Films

The unique and peculiar properties of water are known to originate from the hydrogen bonding characteristics of water molecules: the oxygen of each molecule can have two donor bonds donating two protons to neighbors and two acceptor bonds accepting two protons from neighbors. For a perfect hexagonal Ih ice crystal, the four hydrogen bonds should connect with four equally distant oxygens of neighboring water molecules with one and only one proton in each bond, forming a periodic lattice. This is the well-known Bernal–Fowler–Pauling ice rule in ice physics. Pauling showed that under this rule there exist many possible ways to orient water molecules in the ice lattice, leading to a fairly large residual entropy and impossibility to have ferroelectric order even at 0 K.[63] If, however, one could arrange to have a surface molecular monolayer of ice fully polar oriented, then the ice rule would require the subsequent monolayer by monolayer all fully polar oriented, as sketched in Fig. 8.19(a). Such a ferroelectric order can only be broken if defects come in to violate the ice rule. This is the case of Ih ice grown on Pt(111).

The hexagonal surface of Ih ice has a lattice constant close to that of the hexagonal surface of Pt(111), as illustrated in Fig. 8.19(b). Therefore, epitaxial growth of Ih ice on Pt(111) is feasible. Water molecules adsorbed on Pt(111) preferentially with either H or O facing Pt (an ambiguity yet to be settled) would form a polar-oriented monolayer, and the subsequent epitaxially grown water layers would adopt the same polar ordering. Obviously, SHG/SFG is an effective tool to detect such polar ordering.

Su *et al.* first reported SFVS studies of epitaxially grown ice with different layer thicknesses on Pt(111) at 137K.[64] The observed OH stretch spectra are presented in Fig. 8.19(c). The dangling OH mode at 3700 cm^{-1} appears in all spectra and remains unchanged

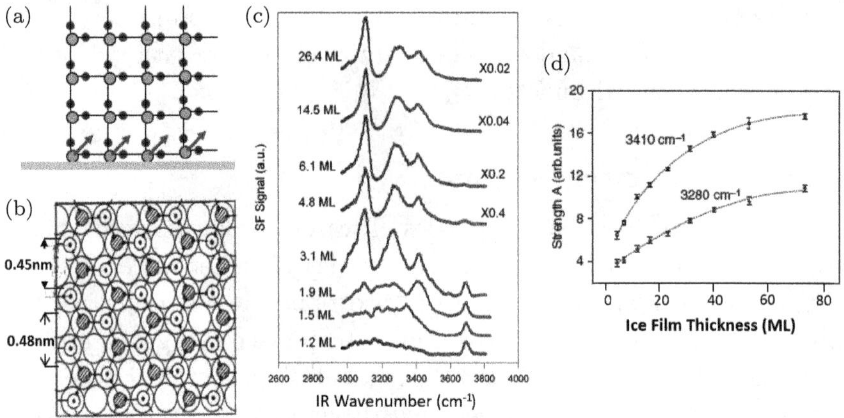

Fig. 8.19. (a) Two-dimensional sketch of polar ordering of water monolayers adsorbed on a surface following the ice rules. (b) Drawing showing that the lattice of the basal plane of hexagonal ice matches well with that of the Pt(111) surface. (c) SSP SF vibrational spectra in the OH stretching region for a set of ice films of different thicknesses grown on Pt(111). (d) Amplitudes of the 3280 cm^{-1} and 3410 cm^{-1} modes of the ice film versus film thickness. The 3100 cm^{-1} peak has a similar dependence (after Ref. [64]).

with layer thickness as one would expect from a surface vibrational mode. Three characteristic peaks of bonded OH stretches of Ih ice at 3100, 3280, and 3410 cm^{-1} show up distinctly after the ice layer thickness goes over 2 ML, and grow nonlinearly with thickness. The resonant amplitudes of the different modes deduced from the spectra as functions of thickness are plotted in Fig. 8.19(d). All of them increase sub-linearly with thickness and toward saturation at large thickness. The result indicates that the ice film must have grown polar-oriented layer by layer, i.e., growth of a ferroelectric-ordered film. If the ice film were paraelectric, these modes would remain unchanged like the dangling OH peak; the EQ bulk contribution to SFVS from a thin film should be negligible. The sub-linear, instead of linear, growth of the modes was due to defects created by thermal fluctuations that broke the ice rule and impeded the polar-ordered growth.

Matsumoto and coworkers have repeated the measurement and confirmed the ferroelectric ice growth on Pt(111).[65] They extended the ice layer thickness to more than 200 ML and observed that the resonant amplitude increased as square root of layer thickness beyond 20 ML. They also made PS-SFVS measurement on the ice films on Pt(111) and found that the bonded OH peaks are negative with reference to the positive dangling OH peak,[66] indicating that the ferroelectric-ordered OH should be pointing downward toward Pt. This is obviously in conflict with the result that the dangling OH is pointing upward; according to the ice rule, the ferroelectric-ordered OH should be pointing in the same direction as the dangling OH. This conflict is yet to be resolved. Matsumoto and coworkers devised a sophisticated model to explain their observation on the bonded OH modes. Unfortunately, it is not clear how the spectra were analyzed in detail. For example, whether IR absorption and phase mismatch of SFG in thick ice films were taken into account in the analysis was not mentioned.

8.13. Summary and Prospects

Sum frequency vibrational spectroscopy (SFVS) is a unique tool for studies of water interfaces. It has provided most of current information about the molecular structure of water interfaces. Applications to studies of ice interfaces, however, still have a few difficulties to iron out. We summarize the main points discussed in this chapter as follows.

- SFVS can effectively probe OH stretch vibrations of water interfaces, but surface specificity can only be assured with the SSP polarization.
- An interfacial water spectrum generally shows a broad bonded-OH stretch band; there is a dangling OH peak if the interface is hydrophobic.
- Interpretation of SFVS intensity ($|\chi^{(2)}_{S,eff}|^2$) spectra on the bonded OH stretches of water interfaces is often confusing and misleading. $\mathrm{Im}(\chi^{(2)}_{S,eff})$ spectra from phase-sensitive (PS) SFVS directly reflecting the H-bonded continuum resonances are needed.

- Good understanding of the water/vapor interfacial structure has been achieved from its $\mathrm{Im}(\chi^{(2)}_{S,eff})$ spectrum with the help of MD simulations.

- Emergence of ions at vapor/water interfaces interrogated by SFVS is apprehended. The pure water/vapor interface is believed to be weakly acidic from a SFVS study.

- PS-SFVS has also been successfully applied to interfaces of water with gases, surfactant monolayers, and oxides.

- With ions appearing at water interfaces, an electric double layer (EDL) set up by the surface ions can strongly modify the bonded OH stretch band. The $\mathrm{Im}(\chi^{(2)}_{S,EDL})$ spectrum for the EDL extracted from PS-SFVS allows quantitative evaluation of surface ion densities with the help of the Gouy–Chapman theory.

- The $\mathrm{Im}(\chi^{(2)}_S)$ spectrum of a charged water interface characterizes not only the bonding structure of water monolayers at the interface, but also the surface structure of the opposing medium; interfacial structural changes induced by alteration of water solution can be monitored by $\mathrm{Im}(\chi^{(2)}_S)$ as well.

- Applications of SFVS to oxide/water and electrochemical interfaces are promising. New schemes are being developed that would greatly enhance the capability of the technique.

- SFVS studies of ice interfaces have not been conclusive because bulk contribution to SFG may not be negligible.

- On surface melting of ice, libration of the dangling OH bond can be detected to start at $\sim 200\mathrm{K}$ by SFVS, but lattice deformation requires knowledge of the surface spectrum of bonded OH stretches.

- Growth of ferroelectric ice films on Pt(111) can be monitored by SFVS, but the ordering direction is still a mystery.

SFVS is expected to remain a most viable technique for studies of water interfaces for a long time, but PS-SFVS for certain types of water interfaces is yet to be developed. With SFVS being the only viable technique, studies of water/solid and electrochemical interfaces would be most interesting and also relevant in the near future. Current development of the technique to facilitate such studies is very encouraging and could provide great opportunities for new research in these fields. Theoretical help to interpret the observed spectra will be very much needed.

There exist a large number of review articles on SFVS studies of water interfaces; many of them are listed in the following list of References. From what we have discussed in this chapter, however, they should be taken with a grain of salt.

References

1. Du, Q.; Superfine, R.; Freysz, E.; Shen, Y. R.: Vibrational Spectroscopy of Water at the Vapor Water Interface. *Phys. Rev. Lett.* 1993, 70, 2313–2316.
2. Wei, X.; Shen, Y. R.: Motional Effect in Surface Sum-Frequency Vibrational Spectroscopy. *Phys. Rev. Lett.* 2001, 86, 4799–4802.
3. Ji, N.; Ostroverkhov, V.; Tian, C. S.; Shen, Y. R.: Characterization of Vibrational Resonances of Water-Vapor Interfaces by Phase-Sensitive Sum-Frequency Spectroscopy. *Phys. Rev. Lett.* 2008, 100, 096102.
4. Nihonyanagi, S.; Kusaka, R.; Inoue, K. I.; Adhikari, A.; Yamaguchi, S.; Tahara, T.: Accurate Determination of Complex $X^{(2)}$ Spectrum of the Air/Water Interface. *J. Chem. Phys.* 2015, 143, 124704.
5. Sun, S.; Liang, R.; Xu, X.; Zhu, H.; Shen, Y. R.; Tian C. S.: Phase Reference in Phase-Sensitive Sum-Frequency Vibrational Spectrosopy. *J. Chem. Phys.* 2016, 144, 244711.
6. Tian, C. S.; Shen, Y. R.: Sum Frequency Vibrational Spectroscopy Studies of Water/Vapor Interfaces. *Chem. Phys. Lett.* 2009, 470, 1–6.
7. Morita, A.; Hynes, J. T.: A Theoretical Analysis of the Sum Frequency Generation Spectrum of the Water Surface. II. Time-dependent Approach. *J. Phys. Chem. B* 2002, 106, 673–685.
8. Kundu, A.; Tanaka, S.; Ishiyama, T.; Ahmed, M.; Inoue, K.; Nihonyanagi, S.; Sawai, H.; Yamaguchi, S.; Morita, A.; Tahara, T.: Bend Vibration of Surface Water Investigated by Heterodyne-Detected Sum Frequency Generation and Theoretical Study: Dominant Role of Quadrupole. *J. Phys. Chem. Lett.* 2016, 7(13), 2597–2601.
9. Seki, T.; Chiang, K. Y.; Yu, C. C.; Yu, X.; Okuno, M.; Hunger, J.; Nagata, Y.; Bonn, M.: The Bending Mode of Water: A Powerful Probe for Hydrogen Bond Structure of Aqueous Systems. *J. Phys. Chem. Lett.* 2020, 11, 8459–8469.
10. Moll, C.; Versluis, J.; Bakker, H.: Direct Evidence for a Surface and Bulk Specific Response in the Sum-Frequency Generation Spectrum of the Water Bending Vibration. *Phys. Rev. Lett.* 2021, 127, 116001.

11. Seki, T.; Yu, C. C.; Chiang, K. Y.; Tan, J.; Sun, S.; Ye, S.; Bonn, B.; Nagata, Y.: Disentangling Sum-Frequency Generation Spectra of the Water Bending Mode at Charged Aqueous Interfaces. *J. Phys. Chem. B* 2021, 125, 7060–7067.

12. Almed, M.; Nihonyanagi, S.; Kundu, A.; Yanaguchi, S.; Tahara, T.: Resolving the Controversy over Dipole versus Quadrupole Mechanism of Bend Vibration of Water in Vibrational Sum Frequency Generation Spectra. *J. Phys. Chem. Lett.* 2020, 11, 9123–9130.

13. Onsager, L.; Samaras, N. N. T.: The Surface Tension of Debye-Huckel Electrolytes. *J. Chem. Phys.* 1934, 2, 528–536.

14. Jungwirth, P.; Tobias, D. J.: Molecular Structure of Salt Solutions: A New View of the Interface with Implications for Heterogeneous Atmospheric Chemistry. *J. Phys. Chem. B* 2001, 105, 10468–10472.

15. Netz, R. R.; Horinek, D.: Progress in Modeling of Ion Effects at the Vapor/Water Interface. *Ann. Rev. Phys. Chem.* 2012, 63, 401–418.

16. Wen, Y. C.; Zha, S.; Yang, S. S.; Tian, C. S.; Shen, Y. R.: Unveiling Microscopic Structures of Charged Water Interfaces by Surface-Specific Vibrational Spectroscopy. *Phys. Rev. Lett.* 2016, 116, 016101.

17. Tian, C. S.; Byrnes, S. J.; Han, H. L.; Shen, Y. R.: Surface Propensities of Atmospherically Relevant Ions in Salt Solutions Revealed by Phase-Sensitive Sum Frequency Vibrational Spectroscopy. *J. Phys. Chem. Lett.* 2011, 2, 1946–1949.

18. Hua, W.; Jubb, A. M.; Allen, H. C.: Electric Field Reversal of Na_2SO_4, $(NH_4)_2SO_4$, and Na_2CO_3 Relative to $CaCl_2$ and $NaCl$ at the Air/Aqueous Interface Revealed by Heterodyne Detected Phase-Sensitive Sum Frequency. *J. Phys. Chem. Lett.* 2011, 2, 2515–2520.

19. Borukhov, I.; Andelman, D.; Orland, H.: Steric Effects in Electrolytes: A Modified Poisson-Boltzmann Equation. *Phys. Rev. Lett.* 1997, 79, 435–438.

20. Levin, Y.: Polarizable Ions at Interfaces. *Phys. Rev. Lett.* 2009, 102, 147803.

21. Levin, Y.; dos Snatos, A. P.: Ions at Hydrophobic Interfaces. *J. Phys. Cond. Matter* 2014, 26, 203101.

22. Tian, C. S.; Ji, N.; Waychunas, G. A.; Shen, Y. R.: Interfacial Structures of Acidic and Basic Aqueous Solutions. *J. Am. Chem. Soc.* 2008, 130, 13033–13039.

23. Imamura, T.; Ishiyama, T.; Morita, A.: Molecular Dynamics Analysis of NaOH Aqueous Solution Surface and the Sum Frequency Generation Spectra: Is Surface OH– Detected by SFG Spectroscopy? *J. Phys. Chem. C,* 2014, 118, 29017–29027.

24. Chiang, K. Y.; Dalstein, L.; Wen, Y. C.: Affinity of Hydrated Protons at Intrinsic Water/Vapor Interface Revealed by Ion-Induced Water Alignment. *J. Phys. Chem. Lett.* 2020, 11, 696–701.

25. Beattie, J. K.; Djerdjev, M.: The Pristine Oil/Water Interface: Surfectant-Free Hydroxide-Charged Emusions. *Angew. Chem. Int. Ed. Engl.* 2004, 43, 3568–3571.

26. Dalstein, L.; Chiang, K. Y.; Wen, Y. C.: Direct Quantification of Water Surface Charge by Phase-Sensitive Second Harmonic Spectroscopy. *J. Chem. Phys. Lett.* 2019, 10, 5200–5205.

27. Yang, S.; Chen, M.; Su, Y.; Xu, J.; Wu, X.; Tian, C. S.: Stabilization of Hydrooxide Ions at the Interface of a Hydrophobic Monolayer on Water via Reduced Proton Transfer. *Phys. Rev. Lett.* 2020, 125, 156803.

28. Wen, Y.; Zha, S.; Tian, C. S.; Shen, Y. R.: Surface pH and ion Affinity at the Alcohol Monolayer/Water Interface Studied by Sum-Frequency Spectroscopy. *J. Phys. Chem. C* 2016, 120, 15224–15229.

29. Tian, C. S.; Shen, Y. R.: Structure and Charging of Hydrophobic Material/Water Interfaces Studied by Phase-Semsitive Sum-Frequency Vibrational Spectroscopy. *Proc. Nat. Acad. Sci.* 2009, 106, 15148–15153.

30. Tyrode, E.; Liljeblad, J. F. D.: Water Structure Nest to Ordered and Disordered Hydrophobic Silane Monolayers: A Vibrational Sum Frequency Spectroscopy Study. *J. Phys. Chem. C* 2013, 117, 1780–1790.

31. Tang, C. Y.; Allen, H. C.: Ionic Binding of Na+ versus K+ to the Carboxylic Acid Head group of Palmitic Acid Monolayers Studied by Vibrational Sum Frequency Generation Spectroscopy. *J. Phys. Chem. A* 2009, 113, 7383–7393.

32. Le Calvez, E.; Blaudez, D.; Buffeteau, T.; Desbat, B.: Effect of Cations on the Dissociation of Arachidic Acid Monolayers on Water Studied by Polarization-Modultaed infrared Reflection-Absorption Spectroscopy. *Langmuir* 2001, 17, 670–674.

33. Mondal, J. A.; Nihonyanagi, S.; Yamaguchi, S.; Tahara, T.: Structure and Orientation of Water at Charged Lipid Monolayer/Water Interfaces Probed by Heterodyne-Detected Vibrational Sum Frequency Generation Spectroscopy. *J. Am. Chem. Soc.* 2010, 132, 10656–10657.

34. Du, Q.; Freysz, E.; Shen, Y. R.: Surface Vibrational Spectroscopic Studies of Hydrogen Bonding and Hydrophobicity. *Science* 1994, 264, 826–828.

35. Moore, F. G.; Richmond, G. L.: Integration or Segregation: How Do Molecules Behave at Oil/Water Interfaces? *Acc. Chem. Res.* 2008, 41, 739–748.

36. Putnis, A.: Why Mineral Interfaces Matter. *Science* 2014, 343, 1441–1442.

37. Stumm, W.; Wollast, R.: Coordination Chemistry of Weathering. Kinetics of the Surface-Controlle Dissolution of Oxide Minerals. *Rev. Geophys.* 1990, 28, 53–69.

38. Ong, S. W.; Zhao, X. L.; Eisenthal, K. B.: Polarization of Water-Molecules at a Charged Interface – Second Harmonic Studies of the Silica Water Interface. *Chem. Phys. Lett.* 1992, 191, 327–335.

39. Du, Q.; Freysz, E.; Shen, Y. R.: Vibrational-Spectra of Water-Molecules at Quartz Water Interfaces. *Phys. Rev. Lett.* 1994, 72, 238–241.

40. Ostroverkhov, V.; Waychunas, G. A.; Shen, Y. R.: New Information on Water Interfacial Structure Revealed by Phase-Sensitive Surface Spectroscopy. *Phys. Rev. Lett.* 2005, 94, 046102.

41. Myalitsin, A.; Urashima, S.; Nihonyanagi, S.; Yamaguchi, S.; Tahara, T.: Water Structure at the Buried Silica/Aqueous Interface Studied by Heterodyne-Detected Vibrational Sum-Frequency Generation. *J. Chem. Phys. C* 2016, 120, 9357–9363.

42. Liu, W. T.; Wen, Y. C. and coworkers (Work in progress).

43. Urashima, S. B.; Myalitsin, A.; Nihonyanagi, S.; Tahara. T.: The Topmost Water Structure at a Charged Silica/Aqueous Interface Revealed by Heterodyne-Detected Vibrational Sum Frequency Generation Spectroscopy. *J. Phys. Chem. Lett.* 2018, 9, 4109–4114.

44. Li, X.; Brigiano, F. S.; Pezzotti, S.; Liu, X.; Chen, H.; Li, Y.; Li, H.; Shen, Y. R.; Gaigeot, M.-P.; Liu, W. T.: Unveiling Structural Evolution of Oxide Surface in Liquid Water (To be published).

45. Wang, H.; Xu, Q.; Liu, Z.; Tang, Y.; Wei, G.; Shen, Y. R.; Liu, W. T.: Gate-Controlled Sum-Frequency Vibrational Spectroscopy for Probing Charged Oxide/Water Interfaces. *J. Phys. Chem. Lett.* 2019, 10, 5943–5948.

46. Vlasov, Y. G.; Tarantov, Y. A.; Bobrov, P. V.: Characteristics and Sensitivity Mechanisms of Electrolyte-Insulator-Semiconductor System-Based Chemical Sensors: A Critical Review. *Anal. Bioanal. Chem.* 2003, 376, 788–796.

47. Chen, C. K.; Heinz, T. F.; Ricard, D.; Shen, Y. R.: Detection of Molecular Monolayers by Optical Second Harmonic Generation. *Phys. Rev. Lett.* 1981, 46, 1010–1012.

48. Guyot-Sionnest, P.; Tadjeddine, A.: Spectroscopic Investigations of Adsorbates at the Metal Electrolyte Interface Using Sum Frequency Generation. *Chem. Phys. Lett.* 1990, 172, 341–345.

49. Lu, G. Q.; Lagutchev, A.; Dlott, D. D.; Wieckowski, A.: Quantitative Vibrational Sum-Frequency Generation Spectroscopy of Thin Layer Electrochemistry: CO on a Pt Electrode. *Surf. Sci.* 2005, 585, 3–16.

50. Humbert, C.; Busson, B.; Six, C.; Gayral, A.; Gruselle, M.; Villain, F.; Tadjeddine, A.: Sum-Frequency Generation as a Vibrational and Electronic Probe of the Electrochemical Interface and Thin Films. *J. Electroanal. Chem.* 2008, 621, 314–321.

51. Gardner, A. M.; Saeed, K. H.; Cowan, A. J.: Vibrational Sum-Frequency Generation Spectroscopy of Electrode Surfaces: Studying the Mechanisms of Sustainable Fuel Generation and Utilization. *Phys. Chem. Chem. Phys.* 2019, 21, 12067–12086.

52. Liu, W. T.; Shen, Y. R.: In Situ Sum-Frequency Vibrational Spectroscopy of Electrochemical Interfaces with Surface Plasmon Resonance. *Proc. Nat. Acad. Sci.* 2014, 111, 1293–1297.

53. Liu, Z.; Li, Y.; Xu, Q.; Wang, H.; Liu, W. T.: Coherent Vibrational Spectroscopy of Electrochemical Interfaces with Plasmonic Nanograting. *J. Phys. Chem. Lett.* 2020, 153, 080903.

54. Wei, X.; Miranda, P. B.; Zhang, C.; Shen, Y. R.: Sum-Frequency Spectroscopic Studies of Ice Interfaces. *Phys. Rev. B* 2002, 66, 085401.

55. Yamaguchi, S.; Suzuki, Y.; Nojina, Y.; Otosu, T.: Perspective on Sum Frequency Generation Spectroscopy of Ice Surfaces and Interfaces. *Chem. Phys.* 2019, 522, 199–210.

56. Bisson, P. J.; Shultz, M. J.: Hydrogen Bonding in the Prism Face of Ice Ih via Sum Frequency Vibrational Spectroscopy. *J. Phys. Chem.* A 2013, 117, 6116–6125.

57. Nojima, Y.; Suzuki, Y.; Takahashi, M.; Yamaguchi, S.: Proton Order toward the Surface of Ice Ih Revealed by Heterodyne-Detected Sum Frequency Generation Spectroscopy. *J. Phys. Chem. Lett.* 2017, 19, 5031–5034.

58. Smit, W. J.; Tang, F.; Nagata, Y.; Sanchez, M. A.; Hasegawa, T.; Backus, H. G.; Bonn, M.; Bakker, H. J.: Observation and Identification of a New OH Stretch Vibrational Band at the Surface of Ice. *J. Phys. Chem. Lett.* 2017, 8, 3656–3660.

59. Zelent, B.; Nucci, N. V.; Vanderkool, J. M.: Liquid and Ice Water and Glycerol/Water Glasses Compared by Infrared Spectroscopy from 295 to 12K. *J. Phys. Chem.* A 2004, 108, 11141–11150.

60. Faraday, M.: Lecture Before the Royal Institution Reported in the Athenaeum no. 1181. 840 (1850).

61. Wei, X.; Miranda, P. B.; Shen, Y. R.: Surface Vibrational Spectroscopic Study of Surface Melting of Ice. *Phys. Rev. Lett.* 2001, 86, 1554–1557.

62. Sun, S.; Tang, F.; Imoto, S.; Moberg, D. R.; Ohto, T.; Paesani, F.; Bonn, M.; Backus, E. H. G.; Nagata, Y.: Orientation Distribution of Free OH Groups of Interfacial Water is Exponential. *Phys. Rev. Lett.* 2018, 121, 246101.

63. Pauling, L.: The Structure and Entropy of Ice and of Other Crystals with Some Randomness of Atomic Arrangement. *J. Am. Chem. Soc.* 1935, 57, 2680–2684.

64. Su, X. C.; Lianos, L.; Shen, Y. R.; Somorjai, G. A.: Surface-Induced Ferroelectric Ice on Pt(111). *Phys. Rev. Lett.* 1998, 80, 1533–1536.

65. Sugimoto, T.; Aiga, N.; Otsuki, Y.; Watanabe, K.; Matsumoto, Y.: Emergent High-T_c Ferroelectric ordering of Strongly Correlated and Frustrated Protons in a Heteroepitaxial Ice Film. *Nat. Phys.* 2016, 12, 1063–1068.

66. Aiga, N.; Sugimoto, T.; Otsuki, Y.; Watanabe, K.; Matsumoto, Y.: Origin of Emergent High-Tc Ferroelectric Ordering in Heteroepitaxial Ice Films: Sum Frequency Generation Vibrational Spectroscopy of H_2O and D_2O Ice Films on Pt(111). *Phys. Rev. B* 2018, 97, 075410.

Recent Review Articles

- Jubb, A. M.; Hua, W.; Allen, H. C.: Environmental Chemistry at Vapor/Water Interfaces: Insights from Vibrational Sum Frequency Generation Spectroscopy. *Ann. Rev. Phys. Chem.* 2012, 63, 107–130.

- Nihonyanagi, S.; Mondal, J. A.; Yamaguchi, S.; Tahara, T.: Structure and Dynamics of Interfacial Water Studied by Heterodyne-Detected Vibrational Sum-Frequency Generation. *Ann. Rev. Phys. Chem.* 2013, 63, 579–603.

- Ishiyama, T.; Imamura, T.; Morita, A.: Theoretical Studies of Structures and Vibrational Sum Frequency Generation Spectra at Aqueous Interfaces. *Chem. Rev.* 2014, 114, 8447–8470.

- Bjorneholm, O.; Hansen, M. H.; Hodgeon, A.; Liu, L. M.; Limmer, D. T.; Michaelides, A.; Pedevilla, P.; Rossmeisl, J.; Shen, H.; Tocci, G.; Tyrode, E.; Walz, M. M.; Werner, J.; Bluhm, H.: Water at Interfaces. *Chem. Rev.* 2016, 116, 7698–7726.

- Han, H. L.; Horowitz, Y.; Somorjai, G. A.: A Review on *In Situ* Sum Frequency Generation Vibrational Spectroscopy Studies of Liquid–Solid Interfaces in Electrochemical Systems. In K. Wandelt (ed.), *Encyclopedia of Interfacial Chemistry: Surface Science and Electrochemistry*, Vol. 1.1, pp. 1–12. Eleseiver: Amsterdam, 2018.

- Yamaguchi, S.; Suzuki, Y.; Nojima, Y.; Otosu, T.: Perspective on Sum Frequency Generation Spectroscopy of Ice Surfaces and Interfaces. *Chem. Phys.* 2019, 522, 199–210.
- Backus, E. H. G.; Schaefer, J.; Bonn, M.: Probing the Mineral-Water Interface with Nonlinear Optical Spectroscopy. *Angew. Chem. Int. Ed. Engl.* 2020, 60, 10482–10501.
- Ge, A.; Inoue, K.; Ye, S.: Probing the Electrode-Solution Interfaces in Rechargable Batteries by Sum-Frequency Generation Spectroscopy. *J. Chem. Phys.* 2020, 153, 170902.
- Gonnela, G.; Backus, E. H. G.; Nagata, Y.; Bonthuis, D. J.; Loche, P.; Schlaich, A.; Netz, R. R.; Kuhnle, A.; McCrum, I. T.; Koper, M. T. M.; Wolf, M.; Winter, B.; Mejer, G.; Campen, R. K.; Bonn, M.: Water at Charged Interfaces. *Nature Rev. Chem.* 2021, 5, 466–485.
- Chowdhury, A. U.; Muralidharan, N.; Daniel, C.; Amin, R.; Belharouak, I.: Probing the Electrolyte/Electrode Interface with Vibrational Sum Frequency Generation Spectroscopy: A Review. *J. Power Sources* 2022, 506 (To be published June 20, 2022).

Chapter 9

Polymer Interfaces

Polymers are the most wonderous materials ever discovered by mankind. They appear everywhere and have countless applications in our modern world. Organic chemists like magicians seem to be able to synthesize any polymers for any applications one desires. Interest in bulk polymers often focuses on their mechanical properties, although more recently it has been extended to electrical and optical properties. Analysis of bulk composition and structure can be readily carried out. Many functions and applications of polymers, however, rely on their surface structure and properties. For example, polymers have been extensively used to repair or replace body parts, but the applicability relies on biocompatibility of the polymers in bodies. Currently, techniques available for *in situ* probing of polymer surfaces and interfaces are quite limited. Again, SFG spectroscopy has found a niche here. It has been developed into a most effective tool for probing polymer surfaces, especially buried polymer interfaces. We present in this chapter a wide variety of cases to show how SFG spectroscopy has impacted polymer science and technology.

9.1. General Considerations

Polymers are composed of repeated molecular units covalently linked to form long chains or networks of cross-linked chains. Each chain appears to have a backbone with repeated molecular side groups or chains attached to it and the two ends capped by molecular groups or chains. The backbone dictates the mechanical properties, but the side groups may emerge at surfaces and define the surface functionality and chemistry. In some cases, the end groups and plasticizers (additives in polymers for modification of their bulk mechanical properties) may also come prominently to surfaces. In design of polymers, chemists know how to synthesize the backbone to provide desired bulk properties. They also know from basic principles and intuition the likely surface structure and properties of polymers, and how slight modification of molecular structure could lead to drastically different surface properties without changing much of the bulk properties. They even know how the polymer surface structure would vary in different environments. The problem is that despite their success in predicting surface structures and properties, they would still need experimental confirmation of their prediction. Generally, to get a rough idea about the surface structure of a polymer, one needs to know the molecular species appearing at the surface and their orientation. Infrared and Raman spectroscopies are not surface-specific even with the total reflection beam geometry. X-ray photoemission spectroscopy (XPS) and near edge X-ray absorption fine structure spectroscopy (NEXAFS) can help, but the polymer sample has to be placed in vacuum, and surface specificity of the techniques is often not limited to top monolayers. Scanning microscopy can yield information of surface molecular arrangement, but not much about the details of molecular units; it also has difficulties to probe polymer surfaces in real environments and monitor their structural changes. Surface sum frequency vibrational spectroscopy is obviously a much more versatile and effective tool. Through observed vibrational spectra, SFVS can provide the information needed: identification of molecular groups appearing at a polymer surface or interface and determination of their orientations from the spectral dependence on input/output polarizations. For instance, appearance of well-ordered methyl and/or methylene groups at a polymer surface guarantees the surface's hydrophobicity, whereas hydroxyl and carbonyl groups make it hydrophilic. Specific functionality of a polymer surface requires specific molecular groups emerging at the surface.

SHG instead of SFG was first employed to study rapid polymerization of monolayers on water when irradiated with UV light and learn about their polymerization kinetics and molecular orientation.[1] It was also used to monitor the very different photo-isomerization kinetics of polymer monolayers and multilayers on water.[2] For most surface studies of polymers, however, SFVS is needed to provide microscopic structural information. Since 1997,[3] SFVS has been increasingly adopted for surface characterization of polymers. On the one hand, information gleaned from SFVS studies leads to better understanding of how specific surface molecular structure can result from given bulk structure, selected environment, and particular surface treatment. On the other hand, SFVS studies provide help on molecular-based design of polymers for particular applications.

Sum frequency spectroscopy as a surface analytical tool is particularly suitable for polymers because essentially all bulk polymers (with a few exceptions) have inversion symmetry that assures surface specificity of SFVS. Most polymers are composed of light atoms (H, C, N, O, F, etc.) so that their signature vibrational modes appear in the frequency range currently reachable by SFVS. Unlike the liquid case, the problem of EQ bulk contribution to SFVS is not so worrisome. Molecular groups appearing at a polymer surface tend to be polar-oriented; they often produce a characteristic SF spectrum with negligible EQ bulk contribution. It is also possible to prepare polymer films with variable thickness down to $<1\,\mu$m; whether the EQ bulk contribution is negligible or not can be seen from independence or dependence of SFVS on film thickness. Surface specificity of SFVS on polymers can also be judged from spectral changes when polymer surfaces are perturbed. Nearly all polymer interfaces are accessible by SFVS, and *in situ* probing allows monitoring of interfacial structural changes in real environment and real time. As mentioned in Sec. 8.5, the surfaces of two media forming an interface generally have their surface structures affecting each other, and SFVS can probe both. This is important for studies of buried polymer interfaces.

Many polymer applications involve buried polymer interfaces. Compared to other buried interfaces, such as liquid/solid, they are relatively easy to probe because the polymer film thickness can be varied. Consider a polymer film on a substrate in air. If the substrate is transparent and the polymer film is thick, reflected SFVS from the air side and the substrate side can probe the polymer/air and polymer/substrate interfaces, respectively. If the film is thin so that beam attenuation through the film is tolerable, and the substrate has a significantly higher refractive index than the polymer, total beam refraction at critical angles can be used to selectively enhance reflected SFVS from one or the other interface and separately probe one or the other.[4] If the substrate is non-transparent, it is possible for the input to access the polymer/substrate interface from the air side as long as the film is sufficiently thin. The reflected SFVS signal has contributions from both polymer/air and polymer/substrate interfaces, but from the signal dependence on the film thickness, the two contributions can be separated, i.e., separate determination of the surface nonlinear susceptibilities ($\chi_{S,a/p}^{(2)}$ and $\chi_{S,p/s}^{(2)}$) for the two interfaces.

Consider more generally a polymer sandwiched between media 1 and 2, one of which can be non-transparent. We can use the common three-layer model to calculate the Fresnel coefficients at the two interfaces and find the SF output as a function of film thickness and surface nonlinear susceptibilities ($\chi_{S,1/p}^{(2)}$ and $\chi_{S,p/2}^{(2)}$) of the two interfaces, as well as the EQ bulk nonlinear susceptibility ($\chi_{BQ}^{(2)}$) of the polymer if significant.[5] Measurement of SFVS dependence on film thickness then allows separate determination of $\chi_{S,a/p}^{(2)}$, $\chi_{S,p/s}^{(2)}$, and $\chi_{BQ}^{(2)}$. The EQ bulk contribution is usually negligible in such measurement if the film thickness is less than 1 μm. The scheme can, in general, be extended to multilayer systems.

So far, most SFVS studies on polymers have focused on $|\chi_{S,eff}^{(2)}|^2$ measurement, and analysis of the result relies on fitting of the observed $|\chi_{S,eff}^{(2)}|^2$ spectra. Phase-sensitive SFVS permits direct measurement of the complex $\chi_{S,eff}^{(2)}$, but has not yet been widely applied to polymers. Its implementation to yield Im $\chi_{S,eff}^{(2)}$ helps not only in asserting polar orientation of surface molecular groups, but also in better resolving overlapping vibrational modes in spectra. Mode overlapping is a common difficulty in spectral analysis of complex polymers. Separation and assignment of vibrational modes from different, but very similar, molecular groups on a polymer chain can be very challenging. This problem may become less severe once we have more spectral data on closely related polymers.

In the following sections, we shall present representative cases for different polymer interfaces to illustrate the wide range of issues that SFVS is capable of addressing.

9.2. Polymer-Air Interfaces

Polymers may have similar bulk structures but very different surface properties. They may also have different bulk structures but similar surface properties. This is because surface properties of a polymer are mainly governed by surface molecular units appearing at the polymer surface with little dependence on bulk composition. In this section, we describe a number of cases to illustrate how SFVS is uniquely suited to identify and probe surface molecular groups emerging at polymer/air interfaces, expected or unexpected. We show in Fig. 9.1 the SF vibrational spectra of air/polymer interfaces of a set of polymers that have the same $-C-C-$ backbone chain but different pendants attached to the chain: polyethylene [PE, $(CH_2\text{-}CH_2)_n$], polyvinyl chloride [PVC, $(CH_2\text{-}CHCl)_n$], polyvinyl alcohol [PVA, $(CH_2\text{-}CHOH)_n$], polymethyl acrylate [PMA, $(CH_2\text{-}CH\ COOCH_3)_n$], polymethyl methacrylate [PMMA, $(CH_2\text{-}CCH_3COOCH_3)_n$], polypropylene, [PP, $(CH_2\text{-}CCH_3)_n$], and polystyrene [PS, $(CHC_6H_5\text{-}CH_2)_n$]. (See the insets of Fig. 9.1(A)). These are all relatively common polymers known for their wide applications. Since air can be considered as hydrophobic, we expect hydrophobic molecular groups likely to emerge at polymer/air interfaces. In the case of PE, it is the hydrophobic CH_2 groups pointing into air at the surface that have broken inversion symmetry and contribute to SFVS. Indeed, the spectra (Fig. 9.1(a)) show two peaks at 2851 and 2926 cm^{-1} that can be attributed to the symmetric and antisymmetric stretch modes of CH_2, respectively.[3] Appearance of both stretch modes and their polarization dependence indicate that on average, the symmetric axis of CH_2 tilts away from the surface normal. For PVA and PVC, we

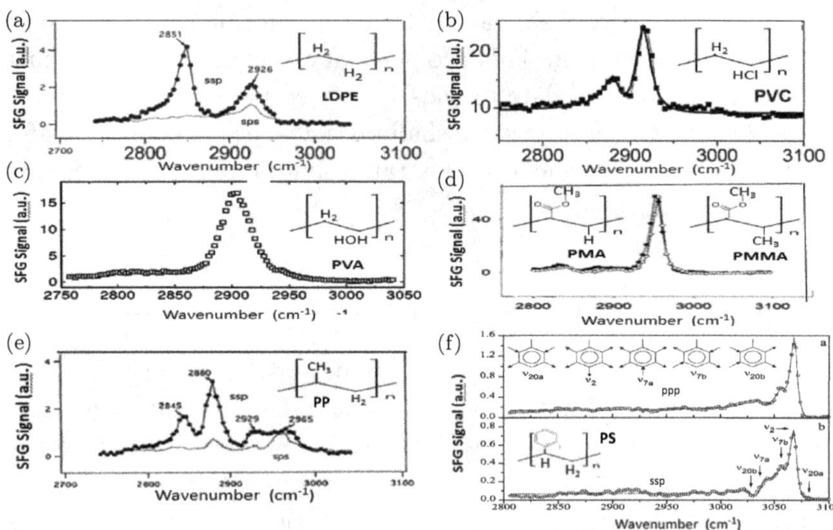

Fig. 9.1. SF vibrational spectra in the CH stretch range of various polymers in air: (a) low density PE (after Ref. [3]); (b) PVC (after Ref. [7]); (c) PVA (after Ref. [6]); (d) PMA (open circles) and PMMA (filled circles) (after Ref. [8]); (e) PP (after Ref. [3]); and (f) PS (after Ref. [4]). All spectra were taken with SSP polarizations except the ones labeled with SPS in (a) and (e) and PPP in (f).

should also expect CH_2 groups at air/polymer interfaces, which is confirmed by the SSP spectra in Figs. 9.1(b) and 9.1(c).[6,7] Both spectra display a strong peak at $\sim$2905 cm^{-1} that can be assigned to the symmetric stretch mode of CH_2. The weak mode at $\sim$2860 cm^{-1} in the PVC spectrum was believed to come from methyl end groups capping the chains. Absence of the antisymmetric stretch in the spectra indicates that the CH_2 symmetric axis is nearly along the surface normal, and correspondingly, the backbones lie in the surface plane. For PMA and PMMA, both have CO_2CH_3 groups on one side of the backbone, but on the opposite side, one has CH and the other has CH_3. The SSP SF spectra of PMA and PMMA in air are essentially identical showing a single sharp peak at 2955 cm^{-1} (Fig. 9.1(d)), which can be assigned to the symmetric CH_3 stretch of CO_2CH_3. This indicates that in both cases, the CO_2CH_3 groups protrude out of the surface with the symmetric axis of CH_3 close to the surface normal and the backbone lying along the surface (as in the case of PVA).[8]

The isotactic polypropylene (IPP) has an SSP spectrum (Fig. 9.1(e)) originated from both CH_2 and CH_3 side groups: symmetric and antisymmetric stretches of CH_2 at 2845 and 2929 cm^{-1}, and symmetric and antisymmetric stretches of CH_3 at 2880 and 2965 cm^{-1}.[3] The spectrum seems to indicate that both CH_2 and CH_3 groups should have protruded at the surface. Actually, it is well known in polymer science that the side groups of IPP are arranged in staggered positions that generate a helical chain conformation with successive methyl groups twisted by 120°.[9] With the backbones lying along the surface, both CH_2 and CH_3 have 2/3 of their groups point out at the surface with an average orientation of $\sim$60° from the surface normal.[3] Finally, for PS, the hydrophobic styrene rings should be pointing into air; the SSP spectrum (Fig. 9.1(f)) indeed displays a set of ring modes (described in the inset of Fig. 9.1(f)).[4] We note that in many reports in the literature, SFVS measurements on polymers were taken with SSP, SPS, and PPP polarizations and the results provided quantitative information on polar orientation or/and orientation distribution of selected surface groups (Sec. 6.3).[10]

As mentioned above in the case of PVC, end molecular groups of polymer chains could also appear at the surface and be detected in surface SF spectra despite their small volume density in the bulk. For better illustration, we consider here three polyethylene glycols [PEG, R-$(OCH_2CH_2)_n$-R'] with different end groups R and R': PEG diol with R=R'=OH, PEG methyl ether with R=OH and R'=OCH_3, and PEG dimethyl ether with R =R'=OCH_3, all having a molecular weight of $\sim$2000.[11] Figure 9.2 shows IR absorption spectra of the bulks on the left and SSP SF vibrational spectra of the polymer/air interfaces on the right for the three PEGs. The IR spectra are identical because the end groups hardly affect the bulk structure, but the surface SF spectra are clearly different. The O-CH_2 symmetric stretch of the main chains at 2865 cm^{-1} dominates the SF spectrum of the PEG-diol/air interface (Fig. 9.2(a)); the hydrophilic OH end groups do not like to come to the surface. For the PEG-methyl-ether/air interface, however, both O-CH_2 from the main chains and

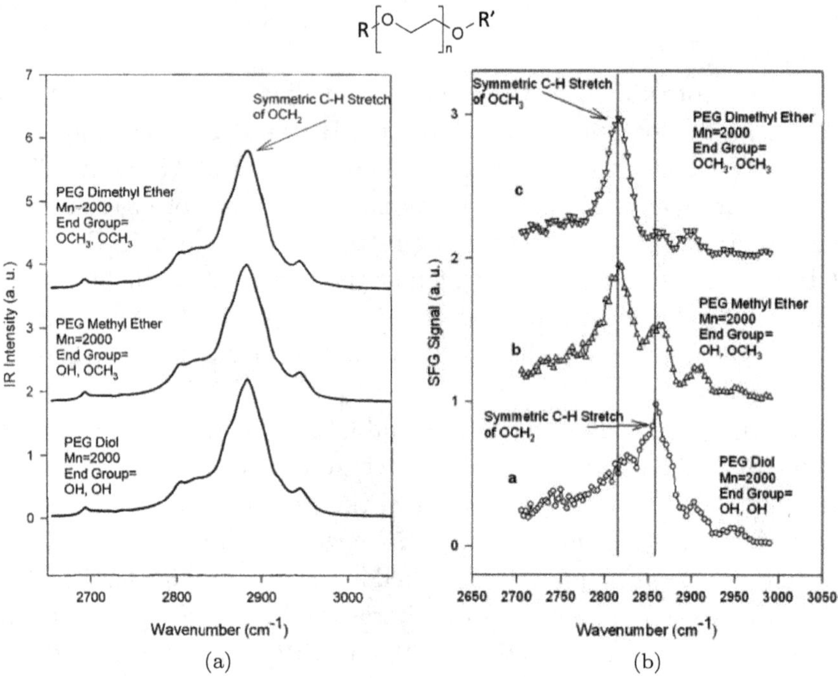

Fig. 9.2. (a) IR spectra and (b) SF vibrational spectra in the CH stretch range for PEG polymers of molecular weight Mn = 2000 with three different end groups: PEG diol, PEG methyl ether, and PEG dimethyl ether. The molecular structure of PEG is shown at the top. The IR spectra for the three polymers are essentially the same, but the SFVS spectra are clearly different (after Ref. [11]).

O-CH$_3$ symmetric stretches from the end groups (at 2865 and 2820 cm^{-1}, respectively) are notable in the SF spectrum (Fig. 9.2(b)); apparently the OCH$_3$ end groups have covered part of the surface. For the PEG-dimethyl-ether/air interface, only the O-CH$_3$ symmetric stretch is prominent, indicating that OCH$_3$ end groups must have covered a large fraction of the polymer surface.[11] This is an example showing that even the end groups of polymer chains can dominate at air/polymer interfaces if they are more hydrophobic than the main chains.

A polymer may have different bulk structures depending on the chain configuration, and its surface structure varies accordingly. We take the case of PEs with two different molecular weights as an

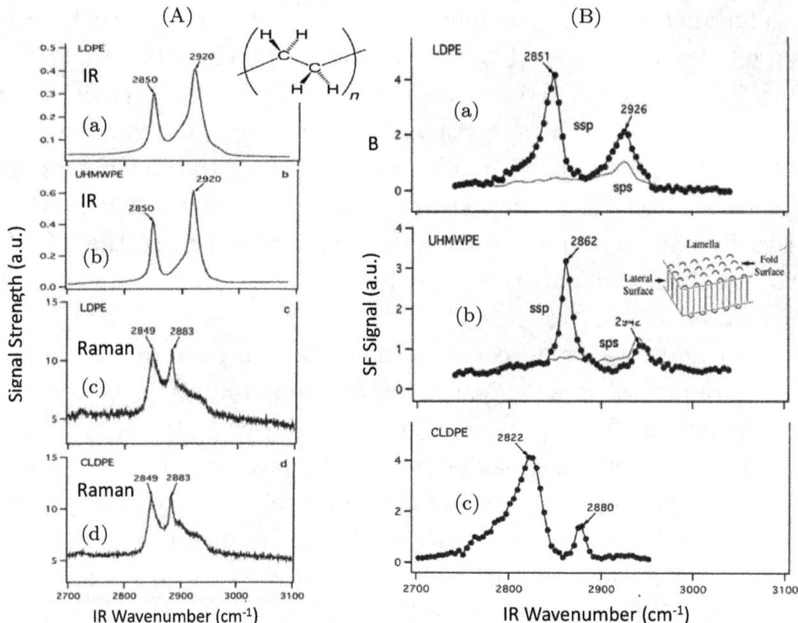

Fig. 9.3. A. (a) and (b) IR spectra of LDPE and UHMWPE, respectively, and (c) and (d) Raman spectra of LDPE and CLDPE, respectively, in the CH stretch range. B. SF vibrational spectra with SSP (filled circles) and SPS (thin lines) polarizations for three PE samples: (a) LDPE, (b) UHMWPE, and (c) CLDPE in the CH stretch range. The molecular structure of a repeating unit of PE is shown at the top, and the surface structure of UHMWPE is sketched in the inset of B(b) (after Ref. [3]).

example,[3] a low-density PE (LDPE) with MW 3×10^4 and an ultra-high molecular weight PE (UHMWPE) with MW 10^6. As seen in Figs. 9.3(Aa), 9.3(Ab), 9.3(Ba), and 9.3(Bb), the IR absorption spectra of the two PE bulks are nearly the same, but the SF spectra of the two polymer/air interfaces are different. The two peaks from CH_2 symmetric and antisymmetric stretches in the SF spectra for UHMWPE appear narrower and blue-shifted in comparison with those of LDPE. The difference was believed to come from the more orderly lamellar structure of UHMWPE near the surface as sketched in the inset of Fig. 9.3(Bb):[3] chain bending at the surface makes neighboring surface CH_2 groups fan out, reduces their interactions, and leads to the narrower and blue-shifted spectral peaks of CH_2.

The spectra of a commercial low-density PE (CLDPE) were also displayed in Fig. 9.3 for comparison. The Raman spectra (Figs. 9.3(Ac) and (Ad)) for the bulks of LDPE and CLDPE appear the same, but the SF spectra are significantly different. The two peaks in the spectrum of CLDPE seem to belong to the CH stretches of methoxy, instead of methylene groups. It is known that methoxy derivatives are common additives in commercial PE. In the CLDPE sample tested, the additives must have diffused preferentially to the polymer surface in air.

That additives could come to the surface of polymers is actually an environmental issue. Plasticizers are generally used as additives to improve flexibility, transparency, and durability of polymers, but they often get segregated to the surface and leach to air or water. We show here the case of PVC plasticized by bis-2-ethylhexyl phthalate [DEHP, molecular formula in the inset of Fig. 9.4(a)] that softens PVC as an example. Figure 9.4 presents the spectra of PVC mixed with different amounts of DEHP.[12] The spectrum of coherent antiStokes Raman spectroscopy (CARS) probing the bulk does not change until the wt% of DEHP reaches ~25 (Fig. 9.4(a)), but the corresponding SF vibrational spectrum for the polymer/air interface exhibits strong characteristic feature of DEHP even before the wt% of DEHP reaches 5 (Fig. 9.4(b)). Obviously, DEHP likes to appear at

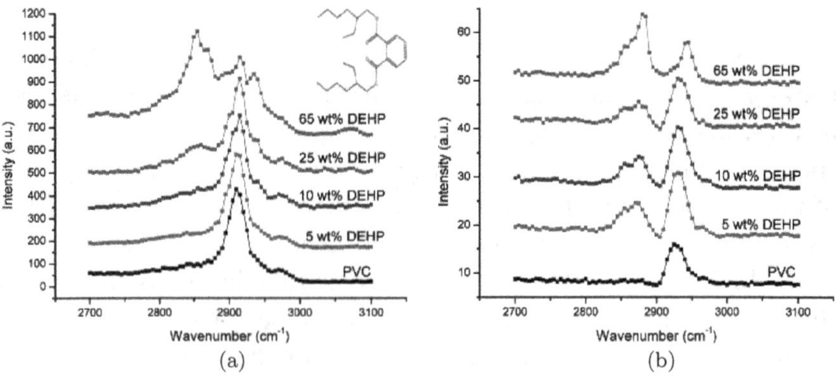

Fig. 9.4. (a) CARS spectra and (b) SSP SF vibrational spectra in the CH stretch range of annealed PVC films with different wt% of DHEP additives. The molecular structure of DHEP is shown in the inset of (a) (after Ref. [12]).

the surface. In a separate experiment,[13] diethyl phthalate (DEP) and dibutyl phthalate (DBP) were used as plasticizers for PVC. Again, SFVS found that both DEP and DBP were segregated to the surface, but DBP/PVC was more stable than DEP/PVC in terms of leaching into air.

Polymer blends are composite polymers designed to provide polymers with desired surface and bulk properties. The minority component is often so chosen that it prefers to emerge at an interface; variation of its concentration changes the surface structure and properties, but does not affect the bulk properties. We take the biopolymer blend of PHE (phenoxy) and BSP (biospan-SP, which are polyurethanes with polydimethylsiloxane (PDMS) as end groups) as an example; the chemical formulae of the various components are given in Fig. 9.5(a).[14] The SSP SF vibrational spectra for the polymer blend with different concentrations of the minority component, BS, are presented in Fig. 9.5(b).[15] The surface spectrum of pure PHE

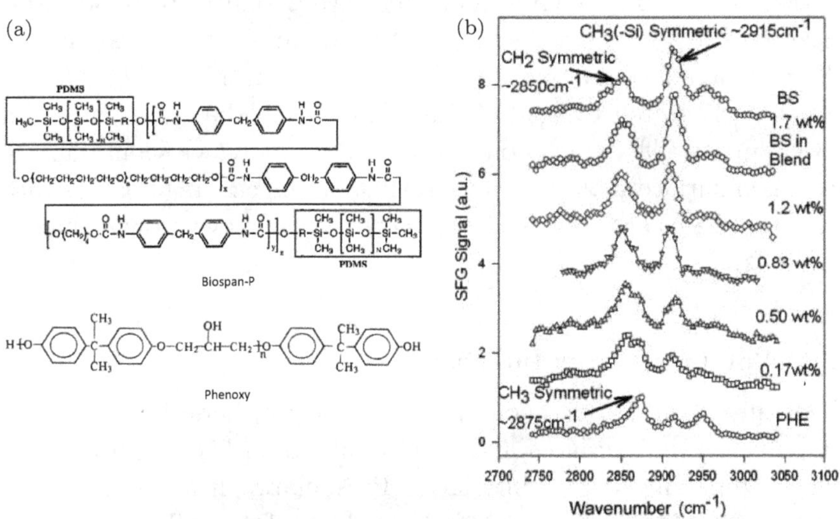

Fig. 9.5.   (a) Molecular structures of PDMS-grafted biospan (BS) and phenoxy (after Ref. [14]). (b) SSP SF vibrational spectra for a series of PHE/BS polymer blends with different BS wt%, showing that the characteristic CH_2 stretch modes of the PDMS end groups of BS emerge at the surface even before the BS concentration reaches 0.17% (after Ref. [15]).

exhibits a prominent CH_3 symmetric stretch peak at 2875 cm^{-1}, but with increasing BSP concentration in the blend, the spectrum rapidly changes to that of BSP, dominated by the CH_2 and CH_3-Si symmetric stretch modes at 2850 and 2915 cm^{-1} from the biospan and its PDMS end groups, respectively. Apparently, BSP is segregated to the air/polymer interface with its hydrophobic PDMS end groups largely covering the surface. This is an example that even the end groups of a long-chain minority polymer component can appear dominating at a surface.

The glass transition temperature (T_g) of a polymer describes the "melting" temperature of the polymer above which the polymer bulk is soft and rubbery and below which it is hard and brittle. One may wonder if the polymer surface and bulk have different T_g. Knowing that at polymer/air interfaces, surface molecular units of polymers have free space on one side to move around, one would think that a polymer surface exposed to air would have a lower T_g than the bulk. However, searches for a surface T_g different from that of the bulk in PP,[16] PVA,[17] and PS[18] have all failed. In these cases, according to the SFVS results mentioned earlier, the backbones of the polymers lie close to the surface and their rather rigid structure is not likely to relax ahead of the backbones in the bulk. To have a lower surface glass transition temperature, a polymer should have a less rigid surface molecular structure or arrangement to assure that the long chains at the surface can more readily move around than in the bulk.

9.3. Polymer-Water Interfaces

Generally, hydrophobic molecular units of polymers like to emerge at polymer/air interfaces and hydrophilic units like to go into the water at polymer/water interfaces. Presumably, if a polymer were polymerized in water, its surface would preferentially be covered by hydrophilic groups. Although backbones of polymers can hardly move below the glass transition temperature, the side groups protruding out at the surface are still mobile to some extent in response to environmental changes. When a dry polymer is in contact with water, hydrophilic groups would tend to come to the surface, and when an initially wet polymer is dried, hydrophobic groups would come to the surface. Many polymer applications involve polymers immersed in water, and it is important to know how their surface structure changes accordingly. To monitor such changes *in situ*, SFVS is currently the only effective means. We discuss here a few examples of surface reconstruction of polymers when immersed in water.

We consider first a group of water-insoluble polymers with the same backbone, but different side groups: PVC, PMMA, PBMA (polybutyl methacrylate), POMA (polyoctyl methacrylate), and PS. As seen in Fig. 9.1(b), the SSP SF vibrational spectrum of the PVC/air interface is dominated by the symmetric CH_2 stretch, indicating that the CH_2 groups are protruding out at the surface. With PVC immersed in water, it was found that the CH_2 mode in the SSP spectrum decreased gradually in time, but simultaneously a new peak associated with CHCl appeared at 2970 cm^{-1} in the PPP spectrum, suggesting that the surface CH_2 groups were reoriented (or disordered) toward the surface and the CHCl groups toward water.[7] The spectrum was only partially recovered when the PVC was dried because water had disordered the surface layer of the film irreversibly. Polymethacrylates of different alkyl chain lengths on the ester (-COO-) side groups restructure differently in water.[19] Figure 9.6 presents the surface CH stretch spectra of PMMA, PBMA, and POMA in air, in contact with water, and back in air and dried. (Contribution of Fresnel coefficients has not been removed from the spectra.) For PMMA, when the SSP and SPS $|\chi^{(2)}_{S,eff}|^2$ spectra had their Fresnel coefficients removed, they appeared nearly unchanged at all stages, suggesting little surface structural variation upon hydration/dehydration.[19] The appearance of the CH_3 symmetric and antisymmetric stretches of CO_2CH_3 at 2955 cm^{-1} in the SSP spectra and at 2990 cm^{-1} in the SPS spectra, respectively, in Fig. 9.6(a) implies that the methyl groups were oriented close to the surface normal with little angular spread. For PBMA (Fig. 9.1(b)), the spectra suggest that the methyl groups at the end of the ester butyl chains in air were tilted away from the surface normal. The spectral changes upon contact with water were appreciable, but were

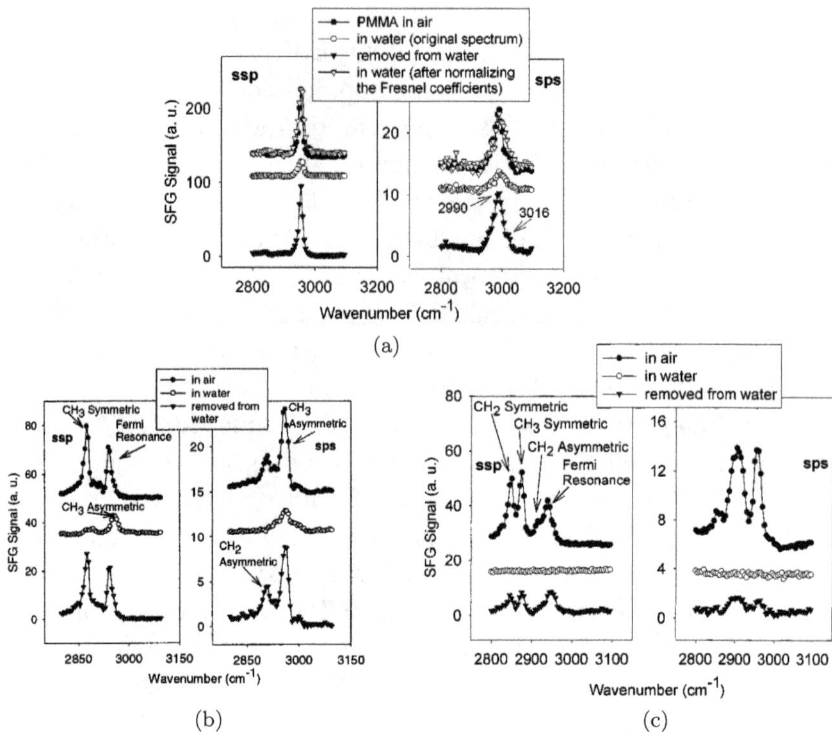

Fig. 9.6. SSP SF vibrational spectra in the CH stretch range of different polymers in air, in contact with water, and back in air and dried: (a) PMMA, (b) PBMA, and (c) POMA (after Ref. [19]).

reversible when dried. Detailed spectral analysis revealed that the methyl groups bent more toward the surface in water. For POMA (Fig. 9.1(c)), the methyl groups of the octyl chains in air tilted even more from the surface normal. The spectrum disappeared when POMA was in water and barely recovered when dried. It was believed that since the sample was measured at temperature above T_g, the backbones of POMA might have been disordered in contact with water and the surface structure distorted randomly in an irreversible way. We note that in SFVS studies of the PMAs, the C=O stretch was also observed, showing that COO^- emerged at the surface both in air and in water.[20] In the case of PS, it was also seen that the SSP spectrum of the ring modes decreased in intensity upon contact with water, signifying reorientation of the phenyl rings toward the surface.[21]

Poly 2-hydroxylethyl methacrylate (PHEMA) is another polymer with the same -C-C- backbone as PMAs. It is structurally similar to PMMA, but the side groups of CH_3 and $COOCH_3$ of PMMA are replaced by CH_3 and the more flexible $COOC_2H_4OH$, one hydrophobic and the other hydrophilic. If the side groups were able to rotate freely, one could expect that the CH_3 groups would like to stick out at the surface in air, but the C_2H_4OH groups like to stay in water. In reality, the side groups could only rotate to some extent. Even so, when PHEMA is in contact with water, the C_2H_4OH groups could partly emerge at the surface rendering the surface partially hydrophilic, as was detected by SFVS.[22] This characteristic feature makes PHEMA a good material for contact lenses: lens surface is hydrophilic when in contact with eye and hydrophobic when exposed to air. Figure 9.7 shows the SF vibrational spectra of the PHEMA in air and water. In air, the spectrum (Fig. 9.7(a)) is dominated by the symmetric stretch and Fermi resonance of CH_3 at 2880 and 2945 cm^{-1} indicating that the methyl groups at the surface are nearly along the surface normal toward air. In water, the symmetric stretch of CH_2 at 2854 cm^{-1} appears strong in the spectrum (Fig. 9.7(b)) signifying emergence of C_2H_4OH groups at the surface. With PHEMA out of water and dried, the CH_2 mode gradually disappears (Fig. 9.7(c), about 15 minutes out of water) and the spectrum reverses back to almost that of the original dry PHEMA (Fig. 9.7(a)). Similar behavior could be observed on more complex polymers. The PDMS-grafted biospan polymer (denoted as BSP with molecular structure given in Fig. 9.5(a)) is an example.[14] As seen in Fig. 9.8, the SSP SF spectrum of BSP in air is dominated by the CH_3 stretch modes of PDMS at 2919 and 2963 cm^{-1} and the CH_2 stretches of biospan at 2851 and 2785 cm^{-1}, indicating that the surface is covered by both biospan and PDMS, but after being immersed in water, it shows gradual increase of the CH_2 stretch modes and corresponding decrease of the CH_3 symmetric stretch at 2919 cm^{-1}. The spectral change describes a surface rearrangement of having the more hydrophobic PDMS end groups hide away from water and the

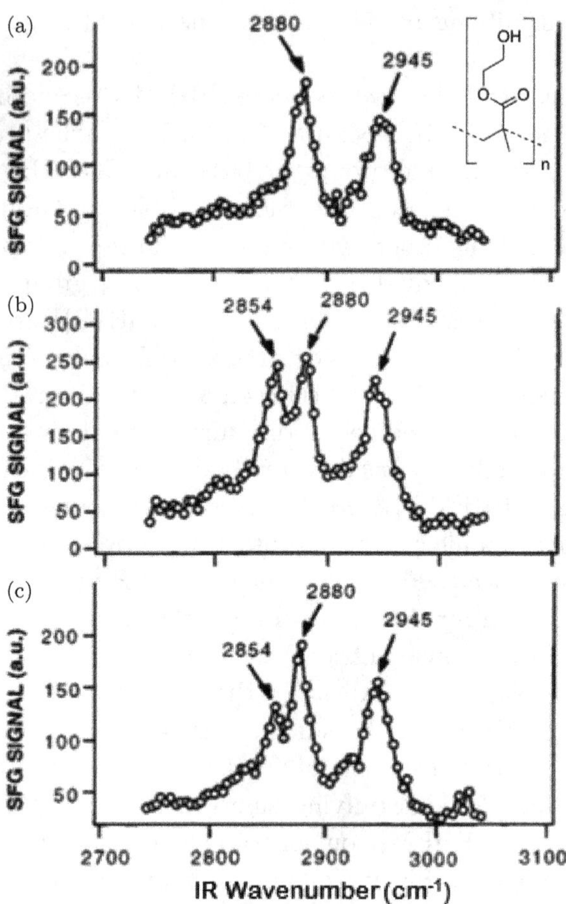

Fig. 9.7.   SSP SF vibrational spectra in the CH stretch range of various PHEMA interfaces. (a) Dry PHEMA/air interface. (b) PHEMA/water interface with PHEMA in water. (c) Hydrated PHEMA surface after being dried in air for ~15 minutes. Molecular formula of PHEMA is shown in (a) (after Ref. [22]).

less hydrophobic BS face water. Because the molecular fragments are large, the surface reconstruction takes hours to reach a steady state. Biocompatibility is a major concern of polymer applications in biomedical science and technology. The above examples show that SFVS can help in providing clues for design of biopolymers with proper surface structures and properties for selective applications.

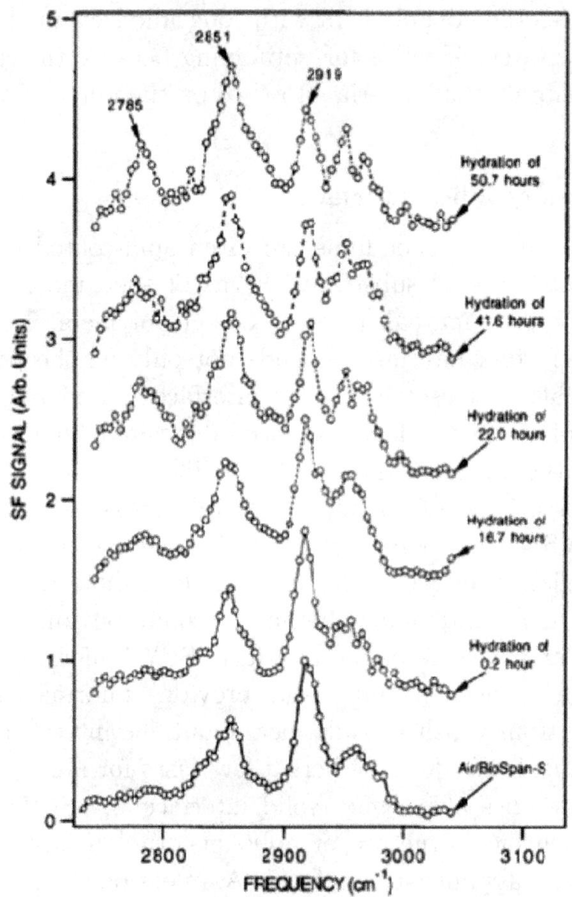

Fig. 9.8.   SSP SF vibrational spectra of PDMS-grafted biospan (BSP), taken at different times after being immersed in water, showing surface structural change of BSP with time from more hydrophobic to partially hydrophilic (after Ref. [14]).

Additives in polymers may preferentially reside at the surface, and subsequently leach to the surrounding. Leaching can be serious if the additives like to interact with water. As demonstrated by Z. Chen and coworkers, SFVS is an effective tool for in situ monitoring of the plasticized polymer/water interfaces and possible leaching effect.[7,23] Biofouling is a serious problem in marine science and technology that SFVS can help to investigate. Z. Chen's group has studied a variety of polymers that show anti-fouling effect, and found, for instance, that biocide-tethered polymers with long alkyl chains sticking out at the surface were effective for antifouling because the chains could disrupt membranes of bacteria adsorbed on the polymer surface.[24,25]

9.4. Polymer/Solid Interfaces

For applications, polymer films are often spin-coated or deposited from solution on solid substrates. Films of solid materials such as metals and oxides may also be coated on polymers. The structure of a polymer/solid interface depends not only on the polymer and solid materials, but also on how the interface is prepared. Interfacial structures of spin-coated and solution-deposited films of the same polymer on the same substrate could be different, and so could be the interfacial structures of polymer on metal versus metal on polymer. Again, SFVS is the only tool to probe the molecular structures of such buried interfaces. The main issue is how well a polymer is bonded to a solid, and whether it could be understood from the molecular structure of the interface. SFVS allows identification of interfacial molecular units and provides information on their orientation, from which a crude idea about the interfacial structure can be attained. We describe here a few cases for illustration.

So far, studies of polymer/solid interfaces by SFVS have been focused on surface structure of polymers; the surface structure of solids is often assumed to be fixed. As mentioned in the previous sections, since surface molecular units of polymers are relatively mobile, they can adjust to the environment they experience in forming an interface: hydrophobic (hydrophilic) molecular groups tend to emerge at a polymer surface if the opposite medium is hydrophobic (hydrophilic). Entropy may also play a role in the case of complex polymer chain conformation and arrangement at an interface.

Polymer coating on oxides (or transparent solids) for surface protection or change of surface functionality has found wide applications. Fused silica (SiO_2) and sapphire (Al_2O_3) are representative oxides often used to study polymer/oxide interfacial structures. Common oxide surface is hydrophilic, likely to repel hydrophoblic side groups

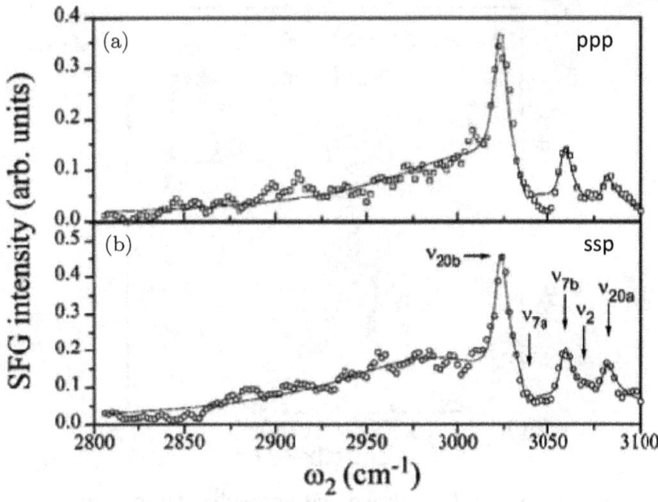

Fig. 9.9.   SF vibrational spectra of a PS/sapphire interface with (a) PPP and (b) SSP polarizations. The ν_{20b} mode at 3023 cm^{-1} is most prominent (after Ref. [4]).

at a polymer surface. The first SFVS experiment on a polymer/oxide interface was on an air/polystyrene (PS)/sapphire film using the total reflection beam geometry that facilitates measurement of both PS/air and PS/sapphire interfaces[4] (Sec. 9.1). The SSP and SPS spectra for the PS/air interface were presented in Fig. 9.1(f), and those for the PS/sapphire interface are displayed in Fig. 9.9. The two sets of spectra are obviously very different, with the latter being significantly weaker. Both can be fit by the five breathing modes of the phenyl ring described in the inset of Fig. 9.1(f), but their relative mode strengths differ drastically. For instance, the ν_2 mode at 3067 cm^{-1} is most prominent in the PS/air spectra, but the ν_{20b} mode at 3023 cm^{-1} is most prominent in the PS/sapphire spectra. The polarization dependence of the spectra allows determination of the average orientation of the phenyl ring; it was found that the phenyl rings protruding out of the surface have their symmetric axes tilted at 20° from the surface normal at the polymer/air interface and at 70° at the polymer/sapphire interface. This indicates that indeed the hydrophobic phenyl rings are repelled by the sapphire surface. As mentioned earlier, the interfacial structure of a polymer film could

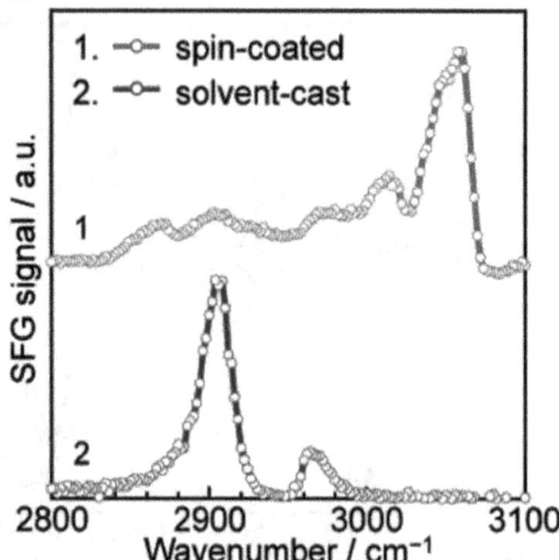

Fig. 9.10. SSP SF spectra for spin-coated and solvent-casted polystyrene films, both annealed at 393°C for 24 hours before measurement. The former dominated by a phenyl ring breathing mode at 3067 cm^{-1} shows that the phenyl rings are oriented close to the surface normal, but the latter has no trace of the ring breathing modes, suggesting that the surface molecular arrangement of PS is highly disordered (after Ref. [26]).

depend on how the film is prepared. Tsuruta *et al.* performed an SFVS measurement on an air/PS/quartz system with the PS film on quartz prepared by two different methods, spin coating and solvent casting[26] (Fig. 9.10). They used a beam geometry that collected the SSP SF signal mainly from the PS/sapphire interface. For the spin-coated PS, the spectrum is dominated by the phenyl ring modes, fairly similar to that of Fig. 9.1(f) observed by Gautem *et al.*,[4] but for the solvent casted PS, it is dominated by the CH_2 stretch modes from the methylene groups of the PS backbones. The result implies that the spin-coated PS has the backbones lie nearly along the surface caused by the centrifugal shear force on them during coating, but the solvent-casted PS has disordered backbones resulting from random structural formation upon solvent evaporation.

The case of a PBMA/silica interface is similar. The SSP and SPS spectra for the interface deduced from SFVS measurement are given in Fig. 9.11.[27] It is seen that they closely resemble those of the PBMA/water interface, but differ from those of the PBMA/air

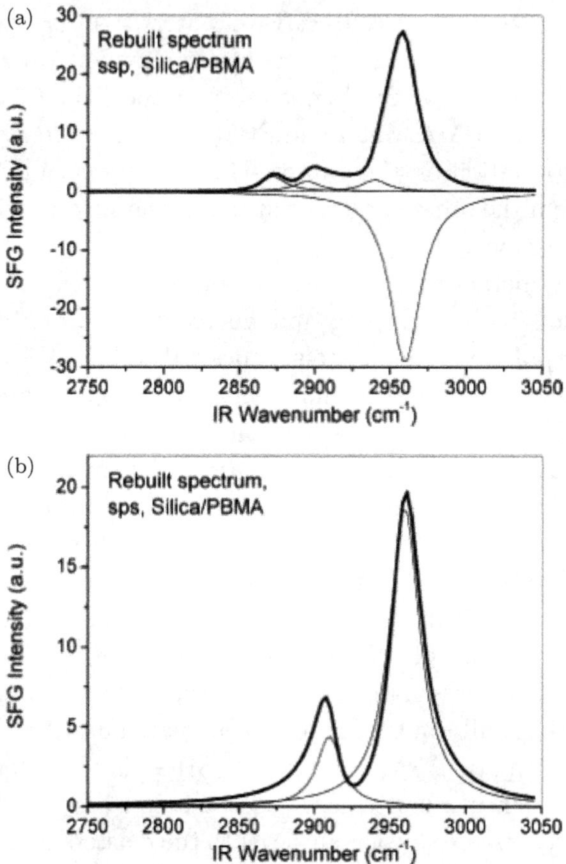

Fig. 9.11. SF vibrational spectra of the PMMA/silica interface deduced from fitting of the observed spectra with (a) SSP and (b) SPS polarizations. The mode components of the spectra are plotted with thin lines. The peak at 2960 cm^{-1} is attributed to the CH$_3$ antisymmetric stretch of the CO$_2$CH$_3$ side groups and has negative amplitude due to CH$_3$ orientation (after Ref. [27]).

interface in both strength and spectral profile (Fig. 9.6(b)). As described in Sec. 9.3, the symmetric stretch mode of the methyl group on the butyl side chain is prominent in the SSP spectrum of the PBMA/air interface, but greatly diminishes in the spectrum of the PBMA/water and now in the spectrum of PBMA/silica as well. The result suggests that the hydrophobic methyl group tilts away from the surface normal in air, but bends further toward the polymer surface when in contact with water or silica.[27] Analysis of the polarization dependence of the spectra found that the average tilt angle of the methyl group was $\sim 40^o$ at the PBMA/air interface and $\sim 70°$ at the PBMA/water and PBMA/silica interfaces.[27,28] More detailed information on these interfacial structures can probably be obtained from the other spectral features of the spectra with the help of MD simulations.

Polymer/metal interfaces are ubiquitous in modern industry in connection with packaging and microelectronics. Robust stable bonding of polymers and metals is desired and SFVS can provide useful molecular structural information for proper design of such interfaces. Second harmonic generation (SHG) was first used to probe metal on polymer, namely, the formation of copper/polyimide (PI) interfaces that are crucial for compact multilayer microelectronic circuit structures. It is known that Cu does not wet PI. Evaporated Cu would diffuse on the surface and form clusters if the temperature is below the glass transition temperature (T_g), but could disappear into the subsurface of PI by diffusion if $T > T_g$. In an early experiment, SHG was used to monitor the process and learn when and how a thin uniform Cu film could be coated on PI.[29] At $T < Tg$, Cu cluster formation was detected by SHG that experienced local plasmon resonance enhancement on Cu clusters as they grew in size, and later decreased to a constant level as the clusters coalesced into a continuous film. At $T > Tg$, disappearance of Cu particulates into PI during and after evaporation could be seen from the decrease of SHG. After they saturated the PI subsurface, SHG stayed at a constant level signifying the presence of a continuous Cu film; apparently when the subsurface was saturated by Cu particulates, Cu could coat the PI surface without having to go through the stage of cluster formation. SHG also detected growth of continuous Cu film on a PI surface covered by a monolayer of Ti that acted as a surfactant layer binding PI on one side and Cu on the other side.[29]

Polymer films on metals have been studied by SFVS to gather information on molecular structures of polymer at polymer/metal interfaces. Both air/polymer/metal and silica/polymer/metal systems have been used.[30,31] Consider the latter case with input from the silica side. The reflected SFG signal from the polymer/metal interface is expected to be overwhelmingly strong because the local field at the polymer/metal interface is so large that the weak signal from the silica/polymer interface is negligible. This was confirmed experimentally by showing that the SF spectrum was independent of the polymer film thickness using the silica/PMMA/Ag layer system as an example.[31] The SSP spectrum for the PMMA/Ag interface was dominated by a single sharp peak at 2955 cm^{-1} similar to the PMMA/air spectrum (Fig. 9.1(d)), which was assigned to the symmetric CH_3 stretch of the $COOCH_3$ side groups attached to the main chains. Analysis of the observed SSP and PPP spectra suggested that the CH_3 groups pointed away from Ag and tilted more from the surface normal than at the PMMA/air interface because silver likes to repel hydrophobic molecular groups. Generally, for design of polymers to form a robust interface with metals, we need to know how polymer molecular units establish bonds with metal; strong bonding requires establishment of chemical bonds between the two. Unfortunately, this is still a challenge for SFVS since the vibrational frequency of relevant modes is out of the range of currently available IR sources.

Bond formation between polymers and oxides at polymer/oxide interfaces has been detected by SFVS. The interface of TiO_2 and polyethylene terephthalate ($[CO_2(C_6H_4)CO_2(C_2H_4)]_n$, PET) provides an example.[32] The SPS spectrum of a PET/air interface was dominated by the 1710 cm^{-1} stretch mode of C=O that lies close to the surface (Fig. 9.12(a)). When a TiO_2 film was deposited on PET, an additional peak appeared at lower frequency that could be assigned to the stretch mode of C=O bonded to Ti with a tilt from the surface (Fig. 9.12(b)). Similar spectroscopic change was observed when an Al_2O_3 film was deposited on PET.

Polymer/polymer interfaces are also of interest in many applications. Again, strong polymer/polymer bonding requires formation of chemical bonds at the interface, and SFVS can be used to probe chemical reaction that establishes the bonds. We consider the work of Chen *et al.* on bonding of poly(4-aminomethyl-p-xylylene-co-p-xylylene) (polymer 1) and poly(4-formyl-p-xylylene-co-p-xylylene)

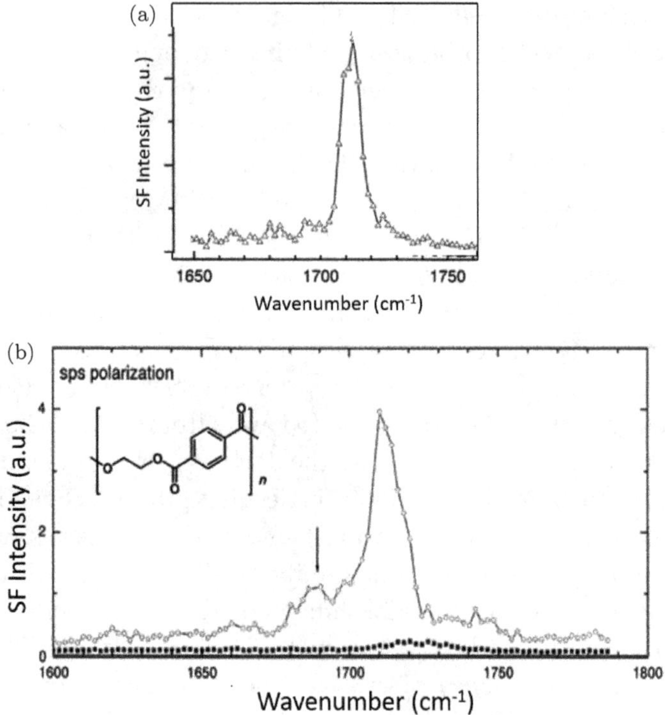

Fig. 9.12. (a) SPS SF spectrum of polyethylene terephthalate (PET) in air. The PET film was biaxially drawn, and the spectrum was taken with the incidence plane along the direction of aligned backbones. The C=O stretch mode at $\sim$1710 cm^{-1} is dominant in the spectrum. (b) SPS spectra of PET covered by a TiO$_2$ film taken with the incidence plan along the aligned backbones (open circles) and in the opposite direction (black dots). An additional mode at $\sim$1690 cm^{-1} appears in the former attributable to C=O bonded to Ti with a tilt from the surface. The molecular structure of PET is given in the inset (after Ref. [32]).

(polymer 2), (molecular formulae given in Fig. 9.13) as an example.[33] Individual polymer films were first deposited on silica. The observed SSP SF vibrational spectra for the two polymer/air interfaces are displayed in Fig. 9.13. They show the characteristic bending and stretching modes of NH$_2$ of the CO$_2$NH$_2$ side groups at 1635 and 3325 cm^{-1} from the polymer-1/air interface (Figs. 9.13(a) and 9.13(b)) and the C=O stretching band of the CHO side groups peaked at 1725 cm^{-1} from the polymer-2/air interface (Fig. 9.13(c)). When the two

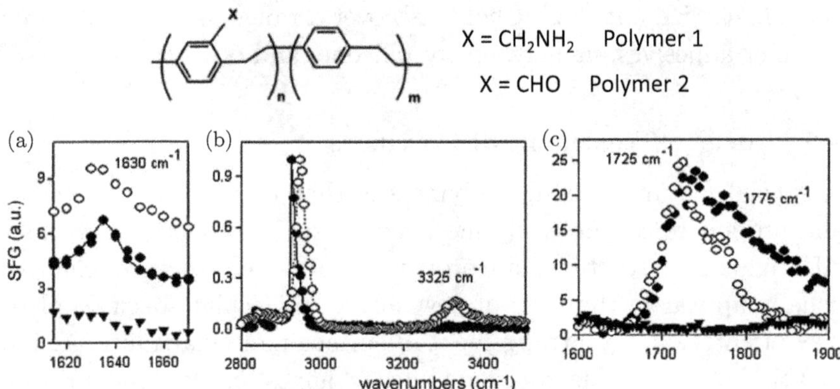

Fig. 9.13. SF vibrational spectra of polymers 1 and 2 before and after bonding; molecular structures of the polymers are described at the top. (a) Spectra of polymer 1 on silica exposed to air (empty circles), in contact with polymer 2 without bonding (filled circles), and bonded with polymer 2 (filled triangles), showing the NH_2 bending mode at 1635 cm^{-1} from the CH_2NH_2 side groups of polymer 1 before bonding and its disappearance after bonding. (b) Spectra of polymer 1 showing the NH_2 stretch mode at 3325 cm^{-1} before (empty circles) and after (filled circles) bonding with polymer 2. (c) Spectra of polymer 2 as deposited (empty diamonds), in contact with polymer 1 without bonding (filled diamonds), and with bonding (black triangles), showing the presence of the C=O stretch mode at 1725 cm^{-1} from the CHO side groups before bonding, and its disappearance after bonding (after Ref. [33]).

polymers were put in contact and heated to 140C for a few hours, these three modes all disappeared, indicating that chemical reaction between NH_2 and CHO had happened. The established bonding between the two polymers was found to be exceptionally strong. Clear separation of characteristic modes of opposite polymers in spectra is generally needed to glean information on polymer/polymer interfacial structure. For example, in the case of PBMA/PS, one has to resort to isotopic substitution to disentangle spectra of PBMA and PS.[34]

In general, adhesivity of polymers is important for polymer coating on materials. It would be helpful if we know how adhevisivity of a polymer on a selected material is related to its specific surface molecular units. In principle, this could be learned from systematic SFVS studies of polymer/solid interfaces. Research in this direction is being carried out in Z. Chen's lab over a range of commonly used polymer adhesives, namely, epoxy, silicone, and others.[10]

9.5. Surface Treatment of Polymers

A special feature of many polymers is that their surface structure and properties can be easily modified by surface treatment without affecting bulk structure and properties. Surface structural changes usually appear in the form of new surface molecular species and/or new orientation and arrangement of surface molecular units. Again, SFVS should be able to probe such changes *in situ* and provide molecular-level understanding of the changes. We describe here SFVS studies on polymers treated by several different means: wet etching, plasma etching, UV irradiation, and mechanical rubbing.

9.5.1. *Wet etching*

Surface chemical reaction alters surface polymer structure. It can happen on certain polymers immersed in acid or base solution, creating new surface molecular species that may or may not be dissolved away. Such an etching process could be monitored by SFVS. We take etching of polyimide (PI) in a base solution as an example. It is a scheme used to promote bonding of PI and metals in many industrial applications. As mentioned in Sec. 9.4, interaction of PI with some metals is weak. To improve PI/metal bonding, the PI surface needs to be modified. One frequently adopted method is to first dip PI in strong basic solution for some time and then have it joint with metal. It was believed that base solution would attack and cleave the imide rings and convert them to amides; the surface amides allow PI to bond well with metals.[35] The proposed imide–amide conversion process is illustrated in Fig. 9.14(a) for poly-n-alkyl-pyromellitic imide, $[-\mathrm{N(CO)_2C_6H_2(CO)_2N(CH_2)}_n-]$ with $n =$ 6 (labeled here as PI6); it was actually confirmed by SFVS.[36] The SF spectra for four spin-coated PI6 films on silica taken after they were immersed in a 2.5M NaOH solution for 0, 5, 10, and 30 seconds, respectively, are shown in Fig. 9.14(b). The two C=O symmetric and antisymmetric stretch modes at 1739 and 1775 cm^{-1}

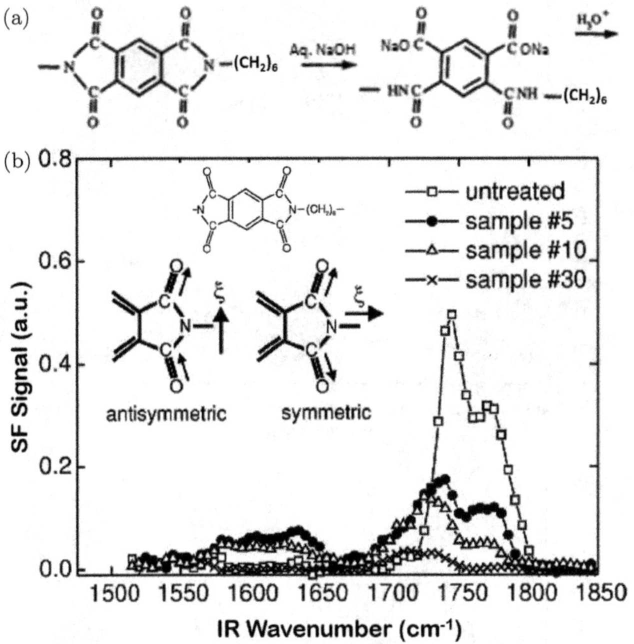

Fig. 9.14. (a) Sketch showing how NaOH in base solution can attack and open the imide ring of PI6 polyimide and convert it to amide. (b) SSP SF vibrational spectra of PI6/water interface in the C=O stretch range for a PI6 sample immersed in a 2.5M NaOH solution for various time durations: 5, 10, and 30 seconds (labeled as Sample #5, #10, and #30, respectively). The two C=O stretch modes of the imide ring at 1739 and 1775 cm^{-1} become weaker with increase of immersion time while a broad CO stretch band of amide around 1600 cm^{-1} emerges. The inset describes the molecular structure of PI6 and the antisymmetric and symmetric stretch modes of the C=O pair (after Ref. [36]).

are characteristic of the imide ring. They become weaker and more red-shifted with longer immersion time of PI in the NaOH solution. In the meantime, a broadband around 1600 cm^{-1} that can be assigned to the CO stretch of amides emerges. This amide band disappeared after $\sim$30 sec of immersion as amides were etched away by the solution.

9.5.2. *Plasma treatment*

Plasma etching is another commonly used method to modify polymer surfaces by surface reactions. It improves adhesion, friction,

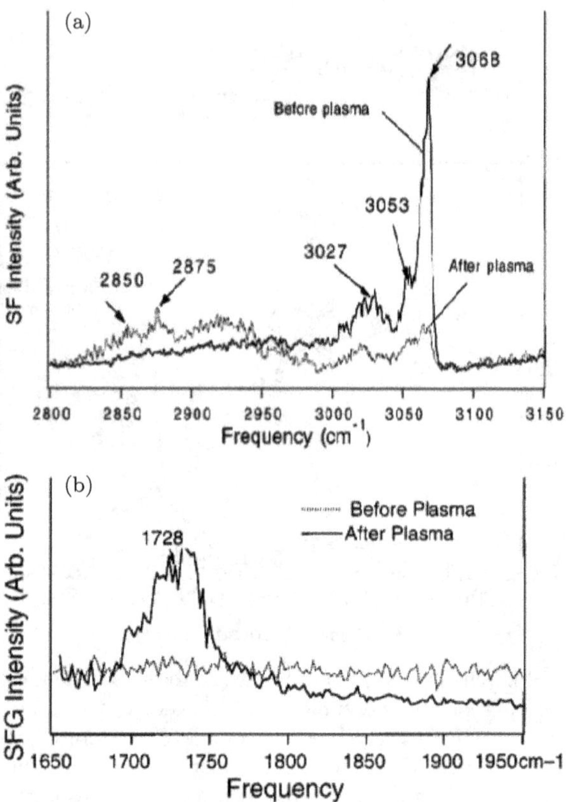

Fig. 9.15. PPP SF vibrational spectra of the polystyrene/air interface before and after plasma treatment. (a) Spectra show that plasma treatment suppresses the phenyl breathing modes, but brings out CH_2 and CH_3 stretch modes at $\sim$2850 and 2875 cm^{-1} due to fragmentation of the phenyl rings. (b) Spectra show the emergence of a C=O stretch mode after plasma treatment presumably resulting from ring opening and uptake of oxygen that lead to formation of aldehyde/carboxylate species (after Ref. [37]).

hydrophobicity, or other surface properties of polymers for specific applications. SFVS has been proven to be a viable means to investigate plasma etching of polymers. We take oxygen plasma treatment of PS as an example.[37] Figure 9.15(a) shows the PPP SF spectra of the PS/air interface in the CH stretch range before and after plasma treatment. The prominent aromatic stretch modes above 3000 cm^{-1} (Sec. 9.2) are drastically suppressed after the treatment and new CH_2 and CH_3 stretch modes at 2850 and 2875 cm^{-1} appear. In addition, a C=O stretch mode at 1675 cm^{-1} emerges (Fig. 9.15(b)). The spectral changes came from oxidation of PS that opened the phenyl rings and formed carbonyl/carboxyl species, some of which were volatile as prolonged plasma treatment could etch away PS. The new surface structure made PS hydrophilic. The result was consistent with the earlier finding of the presence of oxygen containing species at the plasma-treated PS surface layer by X-ray photoemission spectroscopy and the observation of change of the water contact angle on PS from 90° to ~10°.

Plasma treatment of other polymer surfaces has also been studied. In the case of polyimide, surface structural ordering, instead of surface reaction, caused by plasma treatment was observed.[38] It was found that at the treated PI/air interface, the PI backbones tilted more away from the surface, and after forming the PI/epoxy interface, had a somewhat larger twisted angle. Presumably, the plasma treatment also improved adhesion of PI to metals. This would require identification of plasma-induced active molecular units at the PI surface, which has not yet been done. How nitrogen plasma treatment of polypropylene surface can improve its adhesive properties has also been investigated.[39] Unfortunately, the SF spectra could only suggest that the surface was disordered by plasma attack, but no active functional groups at the surface could be identified.

9.5.3. *UV irradiation*

UV irradiation can also trigger oxidation of polymer surfaces and modify the surface structure. Again, we can use PS as an example.[37] After UV irradiation, the SF spectrum of the PS/air interface in the CH stretch range appeared essentially the same as the plasma-treated PS/air interface, i.e., suppression of the phenyl ring modes and emergence of CH_2 and CH_3 stretch modes (Fig. 9.15(a)). The result indicates that with both treatments, oxidation of the phenyl rings must have converted the rings to other species. However, unlike the plasma-treated PS, no CO stretch mode could be detected in the spectrum of UV-treated PS suggesting that no carbonyl species were formed. It was believed that UV irradiation was less invasive

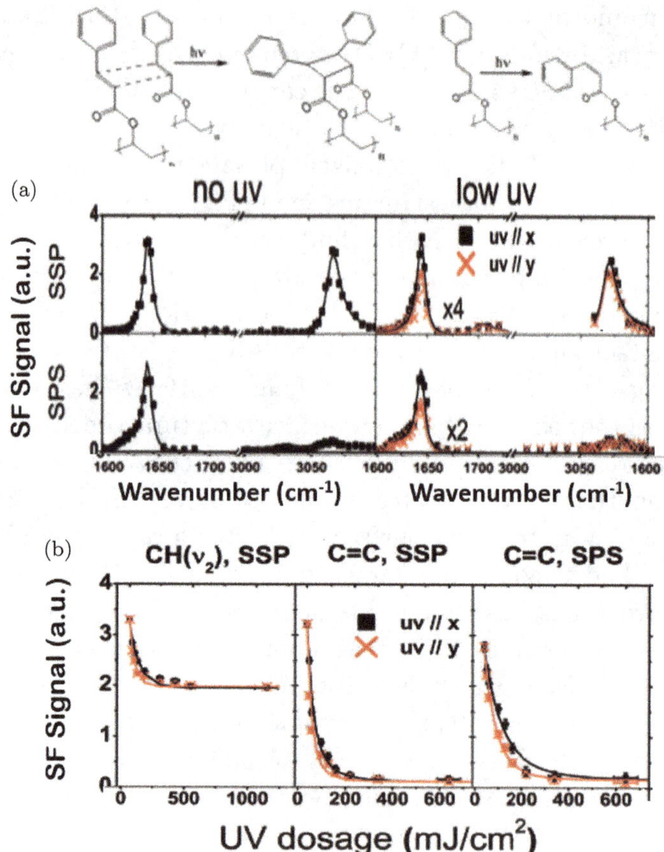

Fig. 9.16. (a) SSP and SPS SF vibrational spectra of CH(ν2) and C=C stretching modes of a polyvinyl cinnamate (PVCi) surface exposed to air with and without UV irradiation. The UV irradiation are linearly polarized along either x or y with x-z being the incidence plane. Difference in the spectra of samples irradiated by UV polarized along x and y is a manifestation of the induced surface structural anisotropy. (b) Decay of the CH(ν2) and C=C modes as functions of UV dosage. The proposed dimerization and isomerization processes of PVCi are illustrated above the spectra (after Ref. [40]).

than plasma and converted phenyl only to phenol-like species. The conclusion was supported by the observation that the water contact angle of PS was reduced to only $\sim$45° after UV treatment.

UV irradiation can also induce surface reconstruction and create or change anisotropy of certain polymer surfaces that can be used as templates to align, in a non-contact way, liquid crystal films or biological molecules adsorbed on them. Polyvinyl cinnamate ($[CH_2CH(O_2CCH=CHC_6H_5)]_n$, PVCi, Fig. 9.16) is such a photo-polymer. Two different mechanisms of surface reconstruction of PVCi leading to surface anisotropy were proposed: one is dimerization of neighboring cinnamoyl side chains that converts their C=C bonds to C-C bonds connecting the two chains, and the other is isomerization that converts the cinnamoyl chains from trans to cis, as sketched at the top of Fig. 9.16. Linearly polarized UV irradiation is expected to preferentially excite properly oriented cinnamoyl side groups and subsequently lead to directional reorientation of these side groups. An SFVS study of UV-irradiated PVCi was able to prove that the dimerization mechanism was correct.[40] Shown in Fig. 9.16(a) are the SSP and SPS SF spectra of the cinnamoyl side chains before and after a low UV dose. Two peaks stand out; one at 1640 cm^{-1} comes from the C=C stretch and the other at 3070 cm^{-1} from a CH stretch of the cinnamoyl ring. It is seen that both decrease in strength in all spectra upon UV irradiation, the change of C=C being more appreciable. The intensity of the two modes versus UV dose is described in Fig. 9.16(b); the C=C mode disappears almost completely when the UV dose is above 300 mJ/cm^2. The disappearance of the C=C mode clearly indicates that dimerization is the underlying mechanism for UV-induced surface reconstruction of PVCi. We note that SFVS studies of UV-induced surface reconstruction of polymers can be readily extended to buried polymer interfaces.

9.5.4. *Mechanical rubbing*

Rubbing of polymer surfaces can induce structural anisotropy on the surfaces through alignment of polymer chains along the rubbing direction. It is a general method used in the liquid crystal (LC) industry to align liquid crystal films for LC displays. It can also be used to align biomolecules on polymer-coated substrates. There was earlier controversy on the alignment mechanism: whether it was due to grooves created by rubbing on a polymer surface or due to directional interaction of adsorbed molecules with rubbing-aligned polymer chains. That macroscopic rubbing could lead to microscopic

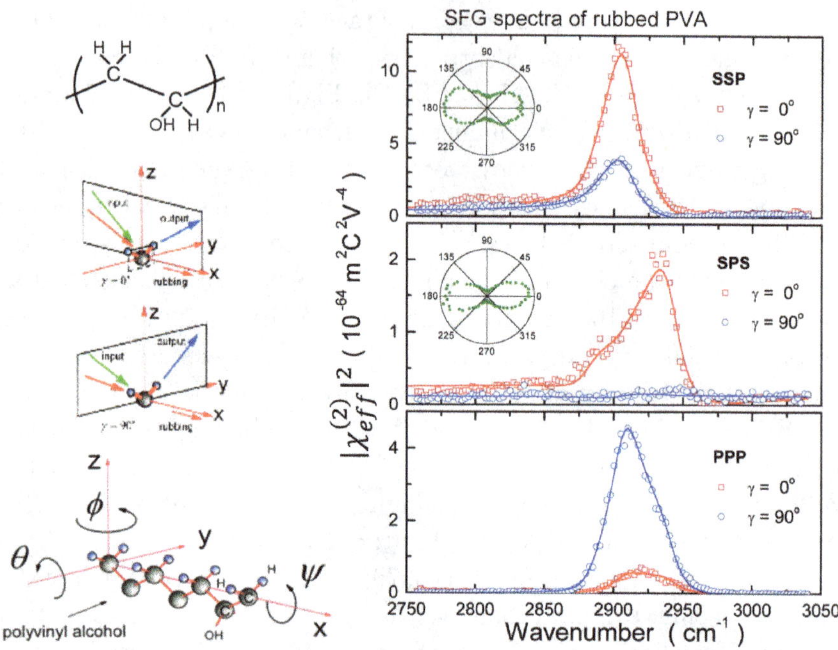

Fig. 9.17. SFG spectra of a rubbed polyvinyl alcohol (PVA) surface in the CH stretch range with three different polarizations, SSP, SPS, and PPP, and two beam geometries with the incidence plane parallel ($\gamma = 0°$) and perpendicular ($\gamma = 90°$) to the rubbing direction, sketched on the left. The chemical formula of PVA and cartoon of a PVA chain lying on the surface along the rubbing direction are also shown on the left. Solid lines over the data points in the figures are theoretical fits. The upper and lower insets in the figure describe the dependence of the resonance amplitudes of the CH_2 symmetric and antisymmetric stretch modes on γ (after Ref. [6]).

molecular rearrangement seemed difficult to conceive in the early days, but SFVS was able to prove that rubbing-induced surface alignment of polymer chains is generally possible.

The first SFVS experiment interrogating a rubbed polymer surface was on PVA($[CH_2CHOH]_n$).[6] As described in Fig. 9.1(c) and Sec. 9.2, the SSP SF spectrum for the PVA/air interface is dominated by the CH_2 symmetric stretch at 2907 cm^{-1}, indicating that the PVA surface has the CH_2 groups protruding into air closely along the surface normal and the backbones lying along the surface plane. The SSP, SPS, and PPP spectra of the rubbed PVA surface with the incident plane parallel and perpendicular to the rubbing direction are presented in Fig. 9.17. The beam-sample geometries are also sketched in Fig. 9.17 with z being the surface normal, x denoting the rubbing direction, and γ referring to the angle between the incident plane and the rubbing direction. We can learn directly from the SSP and SPS spectra the orientation of the CH_2 groups: The absence of the antisymmetric stretch at 2940 cm^{-1} in the SSP spectra at all γ shows that the CH_2 group must be oriented close to the surface normal; its absence in the SPS spectrum with the incident plane parallel to the rubbing direction ($\gamma = 0$) shows that the CH_2 plane must be oriented perpendicular to the rubbing direction. Accordingly, the backbones of PVA at the surface must be well aligned along the rubbing direction. Spectra of SSP and SPS polarizations at different γ were also measured. The resonant amplitudes of the symmetric and antisymmetric CH_2 stretch modes deduced from the spectra are plotted with respect to γ in the insets of Fig. 9.17. They further support the above picture. Quantitatively, we can follow the analytical procedure described in Sec. 6.4 to deduce the average orientation of the CH_2 groups and the backbones. It was found that the orientation of the CH_2 plane could be described by a Gaussian orientation distribution with $\theta = 2.5°$, $\phi = \psi = 0$, and Gaussian variances of $\sigma_\theta = 26°$, $\sigma_\phi = 27°$, and $\sigma_\psi = 35°$, respectively, where θ, ϕ, and ψ denote the tilt, twist, and rotation of the CH_2 plane from the surface normal, about the surface normal, and about the rubbing direction, respectively, as sketched in the lower left corner of Fig. 9.17. The backbones are along the rubbing direction but up-tilted by 2.5°.

Similar measurements on other rubbed polymer surfaces, namely, PS, PI, PVCi, teflon, nylon, and others, have been reported. In all cases, rubbing was seen to align the polymer backbones at the surface along the rubbing direction. The case of nylon (-[NH-CO-$(CH_2)_m]_n$-) with m + 1 carbons in each repeating unit is of special interest.[41] It is known that if m + 1 is an odd number, then such nylon can have a δ' phase that is ferroelectric:[42] As shown at the top of Fig. 9.18(a), two parallel nylon chains can have their successive NH and CO side groups along the chains link together and form pairs that

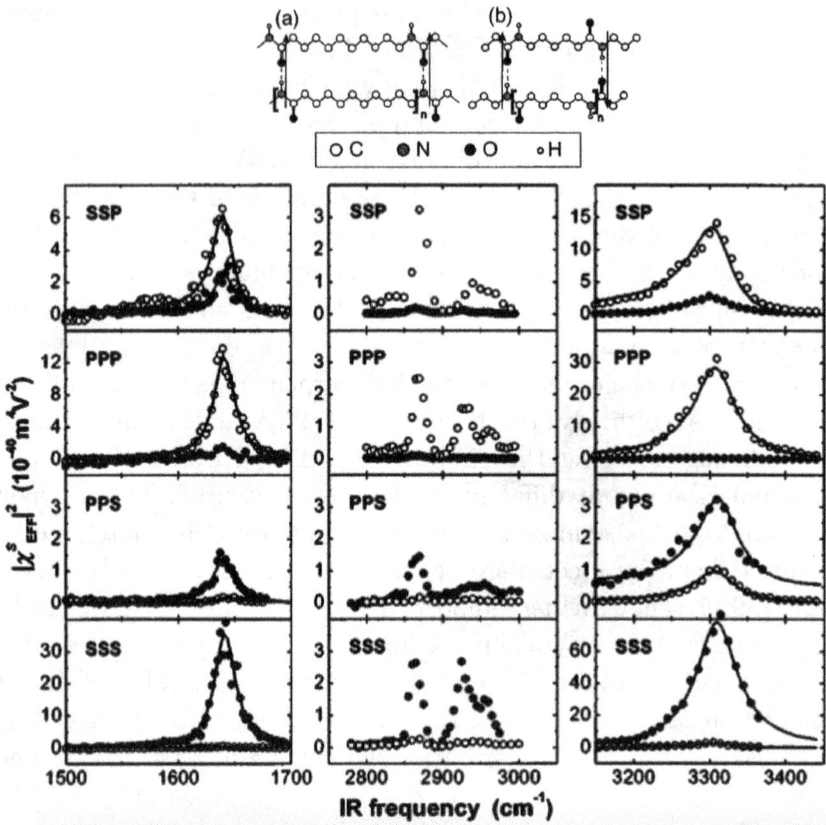

Fig. 9.18. SF spectra of rubbed nylon-11 with four different polarization combinations and two different beam geometries. Solid and open circles are data for the incidence plane parallel ($\parallel$) and perpendicular ($\perp$) to the rubbing direction, respectively. The solid lines are theoretical fit. Molecular structures of two side-by-side repeating units of (a) nylon-11 and (b) nylon-6 chains are described above the spectra showing that the former has NH-CO pairs pointing in the same direction, but the latter has neighboring pairs in opposite directions (after Ref. [41]).

have the same polar arrangement if m + 1 is odd, and alternatively opposite polar arrangements if m + 1 is even. The former is a ferroelectric structure having an electric polarization perpendicular to the chains. Rubbing of a nylon surface can align the nylon chains, and create a ferroelectric domain on the surface if m + 1 is odd. This was indeed observed on a nylon-11 (m + 1= 11) film. The SF spectra for different beam polarization combinations are displayed in Fig. 9.18(b) showing CO stretch at 1640 cm^{-1}, CH_2 symmetric stretch, antisymmetric stretch, and Fermi resonance at 2850, 2930, and 2960 cm^{-1}, respectively, and NH stretch at 3310 cm^{-1}. All modes are significantly weaker and many vanish if the incidence plane is perpendicular to the rubbing direction. This directly leads to the conclusion that CO, NH, and the CH_2 planes must be all oriented perpendicular to the rubbing direction and the nylon chains along the rubbing direction. The nonvanishing SSS spectra taken with the incident plane along the rubbing direction further indicate that CO and NH are polar oriented and lie close to the surface. Other relevant observations were the absence of CO and NH modes and orders-of-magnitude weaker CH_2 modes in the spectra of the unrubbed nylon surface. All these were clear evidences of the existence of rubbing-induced ferroelectricity (with broken inversion symmetry) in the surface layer of the nylon-11 film studied. In fact, the induced ferroelectricity was beyond the surface monolayers. This was seen from the observation that the transmitted SFVS from the film was two orders of magnitude higher than the reflected SFVS, a result possible only if the region without inversion symmetry had a depth larger than the coherent length for reflected SFVS. A rubbed nylon-6 film was also measured, and no induced ferroelectricity could be detected as expected.

9.6. Summary and Prospects

SFVS has been proven to be a unique and most viable tool for *in situ* probing of polymer interfaces, especially buried polymer interfaces, at the molecular level. It identifies surface molecular units, determines their average orientations, and learns about their structure and properties. The technique has promoted polymer surface science research in many areas:

- Provide experimental verification on design of polymer materials that requires specific molecular units appearing at polymer surfaces for intended applications.

- Monitor surface structural variation of polymers in response to change of environment.
- Acquire molecular structural information on how a polymer can be bonded to another material, weakly or strongly, with or without surface chemical reaction.
- Understand how surface treatment can change surface structure of polymers without affecting their bulk structure.

Further development and expansion of the scope of SFVS studies of polymers can be expected:

- Assignment of spectra features is always challenging, especially if they are tangled up. MD simulations that can provide clear interpretation of observed spectra are very much needed. Accumulation of spectra data on similar or closely related molecular structures in a data bank could be helpful.
- The spectral range ought to be extended in order to cover more possible structural variation, particularly in connection with polymer bonding with heavy materials such as metals and semiconductors.
- Phase-sensitive SFVS can provide more structural information, but has not yet been developed for buried polymer interfaces.
- *In situ* dynamic studies of surface reactions at polymer interfaces could be interesting and valuable; such studies have just started[43] and need to be further explored.
- SFVS can help resolve key issues related to polymer surfaces in practical applications, such as biocompatibility, antifouling, and plasma-metal bonding. Close collaborations with industry are likely to lead to many new important areas for SFVS to investigate.

References

1. Berkovic, G.; Rasing, T.; Shen, Y. R.: Study of Monolayer Polymerization Using Nonlinear Optics. *J. Chem. Phys.* 1986, 85, 7374–7376.
2. Zhuang, X.; Lackritz, H. S.; Shen, Y. R.: Photo-Isomerization of Polymer Monolayers and Multilayers on Water. *Chem. Phys. Lett.* 1995, 246, 279–284.

3. Zhang, D.; Shen, Y. R.; Somorjai, G. A.: Studies of Surface Structures and Compositions of Polyethylene and Polypropylene by IR Plus Visible Sum Frequency Vibrational spectroscopy. *Chem. Phys. Lett.* 1997, 281, 394–400.

4. Gautam, K. S.; Schwab, A. D.; Dhinojwala, A.; Zhang, D.; Dougal, S.M.; Yeganeh, M.S.: Molecular Structure of Polystyrene at Air/Polymer and Solid/Polymer Interfaces. *Phys. Rev. Lett.* 2000, 85, 3854–3857.

5. Wilson, P. T.; Briggman, K. A.; Wallace, W. E.; Stepheson, J. C.; Richter, L. J.: Selective Study of Polymer/Dielectric Interfaces with Vibrationally Resonant Sum-Frequency Generation via Thin-Film Interference. *Appl. Phys. Lett.* 2002, 80, 3084–3086.

6. Wei, X.; Hong, S. C.; Zhuang, X.; Goto, T.; Shen, Y. R.: Nonlinear Optical Studies of Liquid Crystal Alignment on a Rubbed Polyvinyl Alcohol Surface. *Phys. Rev. E.* 2000, 62, 5160–5172.

7. Hankett, J. M.; Liu, X.; Lu, Y.; Seeley, E.; Chen, A.: Interfacial Molecular Restructuring of Plasticized Polymers in Water. *Phys. Chem. Chem. Phys.* 2014, 16, 20097–20106.

8. Wang, J.; Chen, C.; Buck, S. M.; Chen, Z.: Molecular Chemical Structure on Poly(methal Methacrylate) (PMMA) Surface Studied by Sum Frequency Geneartion (SFG) Vibrational Spectroscopy. *J. Phys. Chem. C.* 2001, 105, 12118–12125.

9. Cowie, J. M. G.: *Polymers: Chemistry and Physics of Modern Materials.* Intext Educational Publishers: Now York, 1981, p. 265.

10. Lu, X.; Zhang, C.; Ulrich, N.; Xiao, M.; Ma, Y. H.; Chen, Z.: Studying Polymer Surfaces and Interfaces with Sum Frequency Generation Vibrational Spectroscopy. *Anal. Chem.* 2017, 89, 466–489.

11. Chen, Z.; Ward, R.; Tian, Y.; Baldelli, S.; Opdahl, A.; Shen, Y. R.; Somorjai, G. A.: Detection of Hydrophobic End Groups on Polymer Surfaces by Sum Frequency Generation Vibrational Spectroscopy. *J. Am. Chem. Soc.* 2000, 122, 10615–10620.

12. Hankett, J. M.; Zhang, C.; Chen, Z.: Sum Frequency Generation and Coherent Anti-Stokes Raman Spectroscopy Studies on Plasma-Treated Plasticized Polyvinyl Chloride Films. *Langmuir,* 2012, 28, 4654–4662.

13. Zhang, X.; Zhang, C.; Hankett, J/M/; Chen, Z.: Molecular Surface Structural Changes of Plasticized PVC Materials after Plasma Treatment. *Langmuir,* 2013, 29, 4008–4018.

14. Zhang, D.; Ward, R.; Shen, Y. R.; Somorjai, G. A.: Environment-Induced Surface Structural Changes of a Polymer: An In Sity IR + Visible Sum Frequency Spectroscopic Study. *J. Phys. Chem. B.* 1997, 101, 9060–9064.

15. Chen, Z.; Ward, R.; Tian, Y.; Eppler, A. S.; Shen, Y. R.; Somorjai, G. A.: Surface Composition of Biopolymer Blends Biospan-SP/Phenoxy and Biospan-F/Phenoxy Observed with SFG, XPS, and Contact Angle Goniometry. *J. Phys. Chem. B.* 1999, 103, 2935–2942.

16. Gracias, D. H.; Zhang, D.; Lianos, L.; Ibach, W.; Shen, Y. R.; Somorjai, G. A.: A Study of the Glass Transition of Polypropylene Surfaces by Sum-Frequency Vibrational Spectroscopy and Scanning Force Microscopy. *Chem. Phys.* 1999, 245, 277–284.

17. Zhang, C.; Hong, S. C.; Ji, N.; Wang, Y. P.; Wei, K. H.; Shen, Y. R.: Sum Frequency Vibrational Spectroscopic Study of Surface Glass Transition of Poly9vinyl alcohol. *Macromolecules,* 2003, 36, 3303–3306.

18. Schwab, A. D.; Dhinojawala, A.: Relaxation of a Rubbed Polystyrene Surface. *Phys. Rev. E.* 2003, 67, 021802.

19. Wang, J.; Woodcock, S. E.; Buck, S. M.; Chen, C.; Chen, Z.: Different Surface-Restructuring Behaviors of Poly(methacrylate)s Detected by SFG in Water. *J. Am. Chem. Soc.* 2001, 123, 9470–9471.

20. Li, G.; Ye, S.; Morita, S.; Nishida, T.; Osawa, M.: Hydrogen Bonding on the Surface of Poly(2-methoxyethyl Acrylate). *J. Am. Chem. Soc.* 2004, 126, 12198–12199.

21. Yang, C. S. C.; Wilson, P. T.; Richter, L. J.: Structure of Polystyrene at the Interfaace with Various Liquids. *Macromolecules,* 2004, 37, 7742–7746.

22. Chen, Q.; Zhang, D.; Somorjai, G. A.; Bertozzi, C. R.: Probing the Surface Rearrangement of Hydrogels by Sum-Frequency Generation Spectroscopy. *J. Am. Chem. Soc.* 1999, 121, 446–447.

23. Clarke, M. L.; Wang, J.; Chen, Z.: Sum Frequency Generation Studies on the Surface Structure of Plasticized and Unplasticized Polyurethane in Air and in Water. *Anal. Chem.* 2003, 75, 3275–3280.

24. Ye, S.; McClelland, A.; Majumdar, P.; Stafslien, S. J.; Daniels, J.; Chisholm, B.; Chen, Z.: Detection of Tethered Biocide Moiety Segregation to Silicone Surface Using Sum Frequency Generation Vibration Spectroscopy. *Langmuir,* 2008, 24, 9686–9694.

25. Liu, Y.; Leng, C.; Chisholm, B.; Stafslien, S.; Majumdar, P.; Chen, Z.: Surface Structures of PDMS Incorporated with Quaternary Ammonium Salts Designed for Antibiofouling and Fouling Release Applications. *Langmuir,* 2013, 29, 2897–2905.

26. Tsuruta, H.; Fujii, Y.; Kai, N.; Kataoka, H.; Ishizone, T.; Doi, M.; Morita, H.; Tanaka, K.: Local Conformation and Relaxation of Polystyrene at Substrate Interface. *Macromolcules,* 2012, 45, 4643–4649.

27. Lu, X.; Clarke, M. L.; Li, D.; Wang X.; Xue, G.; Chen, Z.: A Sum Frequency Generation Vibrational Study of the Interference Effect in

Poly(n-butyl methacrylate) Thin Films Sandwiched between Silica and Water. *J. Phys. Chem. C.* 2011, 115, 13759–13767.

28. Wang, J.; Paszti, Z.; Even, M. A.; Chen, Z.: Measuring Polymer Surface Ordering Differences in Air and in Water by Sum Frequency Vibrational Spectroscopy. *J. Am. Chem. Soc.* 2002, 124, 7016–7023.

29. Zhang, J. Y.; Shen, Y. R.; Soane, D. S.: Study of Dynamics of Diffusion and Cluster Formation of Copper Deposition on Polyimide by Optical Second Harmonic Generation. *Appl. Phys.* 1992, 71, 2655–2662.

30. Lu, X.; Shephard, N.; Han, J.; Xue, G.; Chen, Z.: Probing Molecular Structures of Polymer/Metal Interfaces by Sum Frequency Generation Vibrational Spectroscopy. *Macormolecules,* 2008, 41, 8770–8777.

31. Lu, X.; Li, D.; Kristalyn, C.; Han, J.; Shephard, N.; Rhodes, S.; Xue, G.; Chen, Z.: Directly Probing Molecular Ordering at the Buried Plymer/Metal Interface. *Macromolecules,* 2009, 42, 9052–9057.

32. Miyamae, T.; Yamada, Y.; Uyama, H.; Nozoye, H.: Molecular Orientation of Poly(ethylene terephthalate) and Buried Interface Characterization of TiO_2 films on Poly(ethylene terephthalate) by Using Infrared-Visible Sum-Frequency Generation. *Surf. Sci.* 2001, 493, 314–318.

33. Chen, H. Y.; McClelland, A. A.; Chen, Z.; Lahann, J.: Solventless Adhesive Bonding Using Reactive Polymer Coatings. *Anal. Chem.* 2008, 80, 4119–4124.

34. Chen, C. Y.; Wang, J.; Evan, M. A.; Chen, Z.: Sum Frequency Generation Vibrational Spectroscopy Studies on Buried Polymer/Polymer Interfaces. *Macromolecules,* 2001, 34, 8093–8097.

35. Lee, K. W.; Viehbeck, A.: Wet-Process Surface Modification of Dielectric Polymers: Adhesion Enhancement and Metallization. *IBM J. Res. Develop.* 1994, 38, 457–474.

36. Kim, D.; Shen, Y. R.: Study of Wet Treatment of Polyimide by Sum-Frequency Vibrational Spectroscopy. *Appl. Phys. Lett.* 1999, 74, 3314–3316.

37. Zhang, D.; Dougal, S. M.; Yeganeh, M. S.: Effects of UV Irradiation and Plasma Treatment on a Polystyrene Surface Studied by IR-Visible Sum-Frequency Generation Spectroscopy. *Langmuir,* 2000, 16, 4528–4532.

38. Myers, J. N.; Chen, Z.: Surface Plasma Treatment Effects on the Molecular Structure at Polyimide/Air and Buried Polyimide/Epoxy Interfaces. *Chin. Chem. Lett.* 2015, 26, 449–454.

39. Sato, T.; Akiyama, H.; Horiuchi, S.; Miyamae, T.: Characterization of the Polypropylene Surface After Atmospheric Pressure N_2 Plasma Irradiation. *Surf. Sci.* 2018, 677, 93–98.

40. Chen, C. Y.; Liu, W. T.; Pagliusi, P.; Shen, Y. R.: Sum-Frequency Vibrational Spectroscopy Study of Photoirradiated Polymer Surfaces. *Macromolecules,* 2009, 42, 2122–2126.
41. Hong, S. C.; Zhang, C.; Shen, Y. R.: Rubbing-Induced Polar Ordering in Nylon-11. *Appl. Phys. Lett.* 2003, 82, 3068–3070.
42. Lee, J. W.; Takase, Y.; Newman, B. A.; Scheinbeim, J. I.: Ferroelectric Polarization Switching in Nylon-11. *J. Poly. Sci. B.* 1991, 29, 273–277.
43. Li, B.; Andre, J. S.; Chen, X.; Walther, B.; Paradkar, P.; Feng, C.; Tucker, C.; Mohler, C.; Chen, Z.: Observing a Chemical Reaction at a Buried Solic/Solid Interface In Situ. *Anal. Chem.* 2020, 92, 14145–14152.

Review Articles

- Chen, Z.: Investigating Buried Polymer Interfaces Using Sum Frequency Generation Vibration Spectroscopy. *Prog. Polymer. Sci.* 2010, 35, 1376–1402.
- Hankett, J. M.; Liu, Y.; Zhang, Z.; Zhang, C.; Chen, Z.: Molecular Level Studies of Polymer Behaviors at the Water Interface using Sum Frequency Generation Vibrational Spectroscopy. *J. Poly. Sci. B: Pol. Phys.* 2013, 51, 311–328.
- Leng, C.; Sun, S.; Zhang, K.; Jiang, S.; Chen, Z.: Molecular Level Studies on Interfacial Hydration of Zwitterionic and Other Antibiofouling Polymers in Situ. *Acta Biomaterialia,* 2016, 40, 6–15.
- Lu, X.; Zhang, C.; Ulrich, N.; Xiao, M.; Ma, Y. H.; Chen, Z.: Studying Polymer Surfaces and Interfaces with Sum Frequency Generation Vibrational Spectroscopy. *Anal. Chem.* 2017, 89, 466–489.
- Zhang, C.: Sum Frequency Generation Vibrational Spectroscopy for Characterization of Buried Polymer Interfaces. *Appl. Spec.* 2017, 7, 1717–1749.

Chapter 10

Biological Interfaces

Optics has been instrumental in the rapid advances of bio/medical-science and technology in the past half century, ranging from single molecule detection/manipulation and super-resolution microscopy to sensing, tomography, therapy, and surgery. Novel spectroscopic techniques such as surface-enhanced Raman scattering, coherent anti-Stokes Raman scattering and multi-photon fluorescence spectroscopy have been developed for bioscience. They provide more mundane information, but greatly help in the general search for better understanding of biological systems. Surface-specific sum frequency spectroscopy is a new comer in the arsenal of tools. Many biological processes and functions take place at interfaces, and naturally, one would expect that SF spectroscopy could play a role in their studies. It actually does possess some capabilities that can yield unique information about bio-interfaces. In recent years, interest in SF spectroscopic studies of biological systems appears to be on the rise. However, like other spectroscopic techniques, it suffers from the fact that biomolecules are generally too complex. Spectral analysis is very challenging and tends to be qualitative or semi-quantitative. Most investigations have focused on simpler biological molecules and model systems; a few representative cases will be described in this chapter.

10.1. General Considerations

In studies of bio-interfaces, two types of problems are of general interest. One is of biomolecules adsorbed at interfaces. This subject matter is relevant to bio-sensing, bio-fouling/antifouling, bio-freezing/antifreezing, drug delivery, immunology, etc. One hopes to know structures of biomolecules at interfaces, including orientations and conformation of all their sub-molecular units preferably at the single molecule level and in real time, as well as their structural variation with surroundings that can provide information on their functional properties. Unfortunately, large biomolecules, such as proteins, are too complex for spectroscopy studies. We often have to select and focus on specific structural features of interfacial biomolecules. The other type of problems is of structures and functions of bio-membranes, intimately related to cell structures and functions. One hopes to learn how structures of bio-membranes may vary with environment or under external perturbation, especially when molecules wedge in and/or penetrate through membranes affecting cell functions and biological processes. Because of their orderly molecular structures, bio-membranes are more amiable to spectroscopy.

Although infrared and Raman spectroscopies have been widely used to study bio-interfaces, SF spectroscopy (including second harmonic generation, SHG) being surface-specific has some unique advantages, particularly in applications to bio-membranes because of its sensitivity to net polar orientation of molecules. As prototypes of bio-membranes, lipid bilayers have been investigated most extensively. In fact, lipid monolayers as Langmuir monolayers were first interrogated by SF vibrational spectroscopy (SFVS) as described in Sec. 6.6. An ideal lipid bilayer with perfect molecular ordering has inversion symmetry, and SFG from the bilayer, being forbidden under electric dipole approximation, is very weak. But SFG is sensitive to any perturbation that breaks the inversion symmetry of a bilayer, making it ideally suited to probe changes of bilayer structure. For example, molecular transportation across a bio-membrane can be monitored *in situ* by SFVS from spectral changes related to the bilayer structure and the behavior of the transported molecules.

SF study of biomolecules at interfaces is in principle the same as the study of molecular adsorption at interfaces described in Chapter 6. It is capable of providing information on the identity of surface molecular species and their orientation, conformation, and arrangement. However, biomolecules are usually orders-of-magnitude larger and structurally more complex than simple molecules discussed in Chapter 6. They are more like polymeric molecules with even higher complexity. In such cases, we generally have to be satisfied with limited information that can be extracted by SF spectroscopy. Focus of SFVS has been on sub-molecular units of biomolecules with characteristic vibrational modes such as CH, CO, NH, and OH stretches. Observed spectra can reveal which sub-molecular units of a biomolecule emerge at a specific interface with specific orientation and arrangement. From the information gathered on the sub-molecular units, we could hopefully learn about the whole structure and arrangement of the biomolecule. This is of course somewhat optimistic. It would require prior knowledge of the structure and conformation of the biomolecule like α-helix and β-sheet of proteins, both assumed to be rigid and fixed. Otherwise, drastic assumptions must be made, resulting in structural information of the molecule being less trustworthy. Even so, in a number of reported SFVS studies, useful information on biomolecules at interfaces has been acquired, as we shall describe in the following section.

There are many difficulties and challenges for SFVS studies on biomolecules caused by their structural complexity. The same molecular moiety can appear in different subunits of a biomolecule with slightly different vibrational frequencies. If these subunits could be separately identified in the vibrational spectrum, they would provide more detailed structural information about the biomolecule. Unfortunately, resolving them spectrally is difficult or nearly impossible.

Spectral analysis could also be a problem. At present, SFVS on bio-interfaces often measures only the intensity spectra, i.e., $|\chi^{(2)}_{S,eff}|^2$, where $\chi^{(2)}_{S,eff}$ is the effective surface nonlinear susceptibility that characterizes an interface and is a complex quantity near resonances. Extraction of $\chi^{(2)}_{S,eff}$ and resonance characteristics from $|\chi^{(2)}_{S,eff}|^2$ rely on spectral fitting with assumptions on the resonance profiles (Sec. 3.4). Such a practice may generate errors, especially in the presence of overlapping resonance modes. This problem can be solved with the use of phase-sensitive SF spectroscopy (Sec. 3.5), but the technique has not yet been adapted for buried bio-interfaces, which are more interesting from the practical point of view. Assignment of spectral features may also be a concern as currently it is based on prior knowledge attained from IR and Raman spectroscopy on biomolecules. Molecular dynamic simulations are needed to confirm the assignment and help relate spectra to structure of bio-interfaces. On the experimental side, we note that the low-frequency spectral range covered by SFVS needs to be extended beyond 600 cm^{-1}. The sensitivity of SFVS is usually limited by optical damage of bio-interfaces and may not permit detection of relatively weak resonance modes. Presumably because real bio-interfaces are even more complex, SFVS studies have so far been carried out mostly on model systems.

10.2. Biomolecules at Interfaces

As mentioned earlier, we can use SFVS, following the description in Secs. 6.2–6.5, to probe biomolecules adsorbed at interfaces. By focusing measurement on selected sub-molecular units of biomolecules and deducing their polar orientation and arrangement, we hope to learn about the structure of the whole molecules at interfaces. To illustrate such studies, we consider here a few representative cases. We start with the smallest and simplest biomolecules, amino acids; which as basic building blocks of proteins, their adsorption and orientation at various interfaces should certainly be of interest. We take adsorption of leucine from solution at the air/water interface as an example.[1] The SFVS study basically follows the procedure described in Sec. 6.5.

Leucine molecules dissolved in water may emerge at the air/water interface with a surface density and molecular orientation that depend on its bulk concentration in water and probably also pH and ionic concentration in water. As shown in Sec. 6.5, the information can be obtained by polarization-dependent SFVS. From the chemical formula of leucine ($[(CH_3)_2CHCH_2CH(NH_2)COOH]$, Fig. 10.1(a)), we expect that the molecules should have their hydrophilic end (COOH) in water and their hydrophobic end [$(CH_3)_2$] protruding out of water at the air/water interface, and SFVS should be able to detect CH stretch modes of the surface leucine molecules. Indeed, as seen in Fig. 10.1(b), the SF spectra of leucine at the air/water interface could be readily observed in three different polarization combinations, SSP, SPS, and PPP. The spectral intensity increases with bulk leucine concentration, and all spectra could be well fit by seven discrete modes: CH_3 symmetric stretch (r^+) and its Fermi resonance at 2873 and 2938 cm^{-1}, CH_3 asymmetric stretches parallel and perpendicular to CH_3 symmetric plane (r_a^- and r_b^-) at 2963 and 2953 cm^{-1}, relatively weak CH_2 symmetric and asymmetric stretches (d^+ and d^-) at 2889 and 2917 cm^{-1}, and CH stretch at 2902 cm^{-1}. Strong polarization dependence of the spectra suggests that the surface leucine molecule must be well ordered and oriented. The more prominent CH_3 symmetric and antisymmetric stretch modes in the spectra of different polarizations were used, following the description of Sec. 6.3, to deduce orientation and surface density of the $C(CH_3)_2$ groups of leucine, and hence orientation and surface density (N_s) of leucine. The $C(CH_3)_2$ groups were assumed to have an azimuthally isotropic distribution with their C_2 axis tilted by angle θ from the surface normal and the plane formed by the three carbons in $C(CH_3)_2$ twisted by angle ψ from the plane formed by the C_2 axis and the surface normal (Fig. 10.1(a)). By assuming δ-function distributions for θ and ψ, the strengths of the CH_3 stretch modes in Fig. 10.1(b) allowed deduction of N_s, θ, and ψ. Figure 10.1(c) depicts the deduced N_s versus leucine bulk concentration and shows that it can be fit by a simple Langmuir adsorption isotherm. The corresponding θ and ψ are plotted in the lower frame of Fig. 10.1(c), both decreasing with increase of surface density of leucine as one would expect from the steric geometry of monolayer assembly. Increase of pH should deprotonate more of COOH of leucine, but did not seem to affect the orientation much. From the above description, we may sense that SFVS study of amino acids is not different from other molecules discussed earlier in Sec. 6.3, and can be used to probe sub-molecular units of biomolecules if they can be approximated as independent units. SFVS studies have been extended to adsorption of amino acids at water/solid interfaces,[2,3] revealing some intriguing results: acidic amino acids form an ordered monolayer on TiO_2,

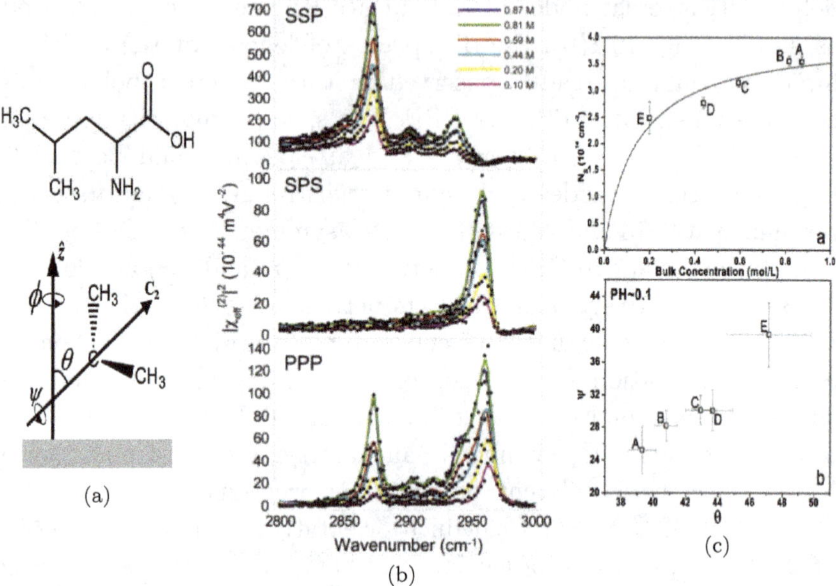

Fig. 10.1. (a) Chemical structure of leucine ($C_6H_{13}NO_2$) and the geometry that defines the orientation of the side group $C(CH_3)_2$ of leucine with z along the surface normal. (b) SF vibrational spectra of leucine molecules adsorbed at the air/water interface from acidic (pH 0.1) solutions with different bulk concentrations. SSP, SPS, and PPP denote the input/output polarization combinations. Lines are fits used to identify the discrete CH stretching modes. (c) Top frame: Surface density of leucine versus bulk concentration of leucine deduced from fitting of the spectra in (b). Bottom frame: Deduced orientation angles θ and ψ at several surface densities of leucine labeled A to E in the top frame (after Ref. [1]).

but non-acidic amino acids had little affinity toward TiO_2; all amino acids adsorbed on silica appeared to be disordered showing little SF signal, but they seemed to adsorb in an orderly way on hydrophobic interfaces. Amino acid molecules are chiral. Their SF chiral response in solution near electronic resonance transitions can be readily detected as described in Sec. 5.4, but their surface SF chiral response is much weaker than the achiral counterpart because of the much smaller nonlinear chiral molecular polarizability, as described in Sec. 5.7.

Peptides/proteins are formed through linkage of amino acids by peptide bonds (Fig. 10.2(a)). Their adsorption at various interfaces has been under intensive studies because of relevance to bio-functions and processes such as bio-sensing, bio-fouling, immune responses, bio-signal transmission, and many others. SFVS has been used to study peptides/proteins at interfaces by probing the spectral regions of amide-I (mostly composed of C=O stretch, 1600–1700 cm^{-1}) and amide-A (mostly N-H stretch, 3100–3300 cm^{-1}) associated with peptide bonds of backbones, and CH stretches (2700–3000 cm^{-1}) of side groups. The amide I and NH vibrational modes are sketched in Fig. 10.2(b). If the structure of a peptide is rigid and known, then its orientation can be determined from the orientations of the C=O and N-H stretches associated with the peptide bonds. Most SFVS

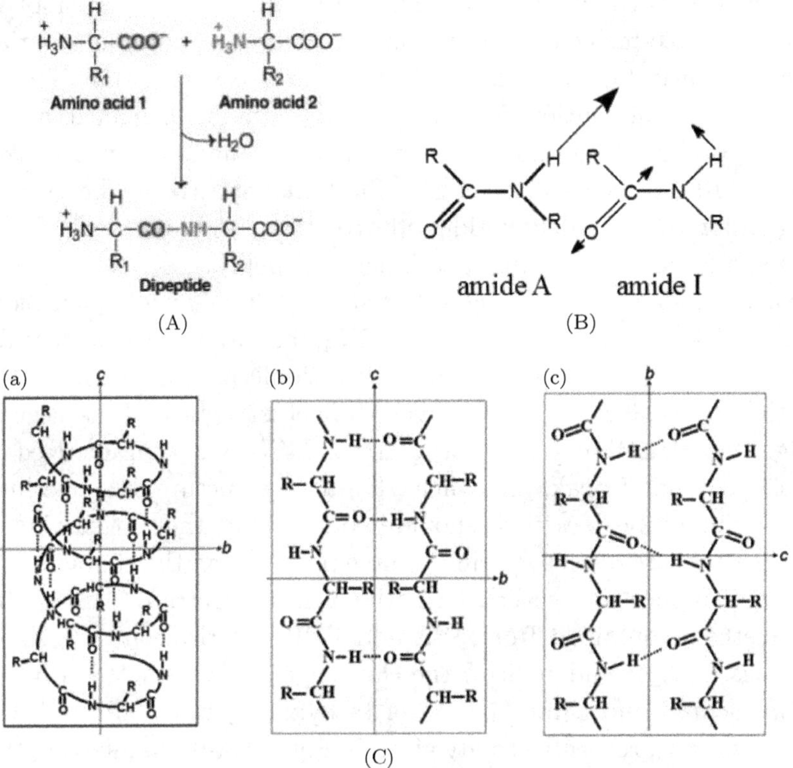

Fig. 10.2. Sketches illustrating (a) linking of amino acids by peptide bond, (b) stretch vibrations of amide-A and amide-I modes, and (c) structures of (A) α-helix, (B) parallel β-sheet, and (C) antiparallel β-sheet of peptides.

studies have been focused on model peptides with one of the four common secondary structures of peptides: α helix, parallel β sheet, anti-parallel β sheet (Fig. 10.2(c)), and 3_{10} helix (which is a helix with three amino residues per turn over a length of 0.2 nm along the helical axis and the NH group of an amino acid is hydrogen-bonded to CO of the amino acid three residues earlier). Disordered and random coil secondary structures are more difficult to study due to their irregular structures. Investigation of a protein at an interface usually aims at identifying its secondary structure and orientation at the interface *in situ* and possibly in real time.

Biomolecules are essentially all chiral and like to have chiral molecular arrangement. Following the description in Sec. 5.7, we expect both chiral and achiral SF vibrational spectra of an adsorbed bio-molecular layer can be measured with appropriate beam polarization combinations; the chiral SF spectrum is dominated by contribution from chiral molecular arrangement as the chiral response from individual sub-molecular units is much weaker. Figure 10.3 presents a set of SF vibrational spectra in the amide I region for three model peptides adsorbed on a lipid bilayer:[4] MSI-78 being a 22-redisue synthetic antimicrobial peptide with an α-helical structure, alamethicin being a 21-residue channel-forming antibiotic peptide with separate α-helical and 3_{10}-helical domain structures, and tachyplesin I being a 17-residue cationic antimicrobial peptide with a typical cyclic antiparallel β-sheet structure. It is seen in Fig. 10.3 that the α-helix structure of MSI-78 is characterized by a single amide I peak, the α-helix/3_{10}-helix structure of alamethicin by a similar peak with a shoulder bump, and the antiparallel β-sheet structure of tachyplesin by three peaks (form the characteristic amide I modes of β-sheet). Only the achiral spectra of MSI-78 and alamethicin obtained from SSP and PPP polarizations are shown in Figs. 10.3(a) and 10.3(b); the chiral spectra were not detectable. Both achiral and chiral spectra of tachyplesin, from SSP and SPP polarizations, respectively, are given in Fig. 10.3(c), manifesting that the antiparallel β-sheet structure has a chiral molecular arrangement. The strong polarization dependence of the spectra indicates that the peptides are fairly well oriented. In the NH stretch region,

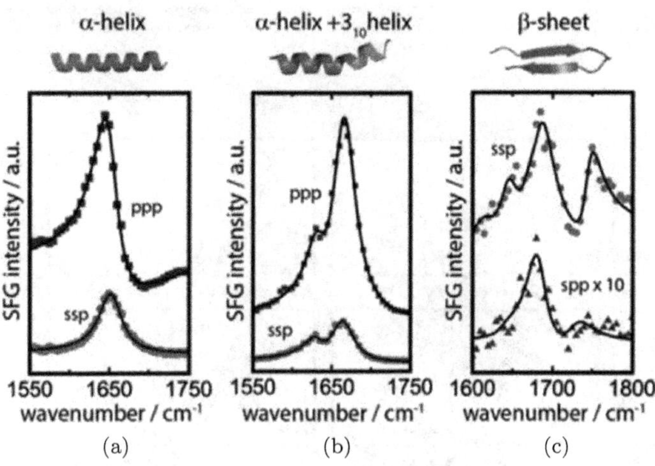

Fig. 10.3. Sum frequency vibrational spectra of three peptides adsorbed on a lipid bilayer: (a) MSI-78 with an α-helical structure on DPPG/d-DPPG, (b) alamethicin with α-helical and 3_{10} helical segments on d-DMPC/DMPC, and (c) tachyplesin with antiparallel β-sheet structure on DPPG/d-DPPG. The labels PPP, SSP, and SPP denote input/output polarization combinations (after Ref. [4]).

the achiral NH spectrum of a peptide generally suffers from mode overlapping with OH stretches of water if water is present, and is difficult to detect. However, the chiral spectrum of peptides adsorbed at, for example, a lipid/water interface, has no chiral background from either lipid or water and can be readily observed. Figure 10.4 shows the chiral NH spectra of six different model peptides at the air/water interface:[5] LK_n-α and LK_n-β are sequential leucine-lysine peptides with n residues and α-helix and β-sheet secondary structures, respectively, and hIAPP (rIAPP) denotes human (rat) islet amyloid polypeptide, which is a 37-residue peptide hormone. In all cases except for hIAPP and rIAPP, the NH stretch mode appears prominently. It is expected that chirality is lost in the disordered structure of rIAPP, but ought to be present in the parallel β-sheet structure of hIAPP. Further investigation found that it was the orientation of hIAPP at the air/water interface that prevented the chiral NH mode from observation; the mode could be detected when hIAPP was adsorbed on glass.[5]

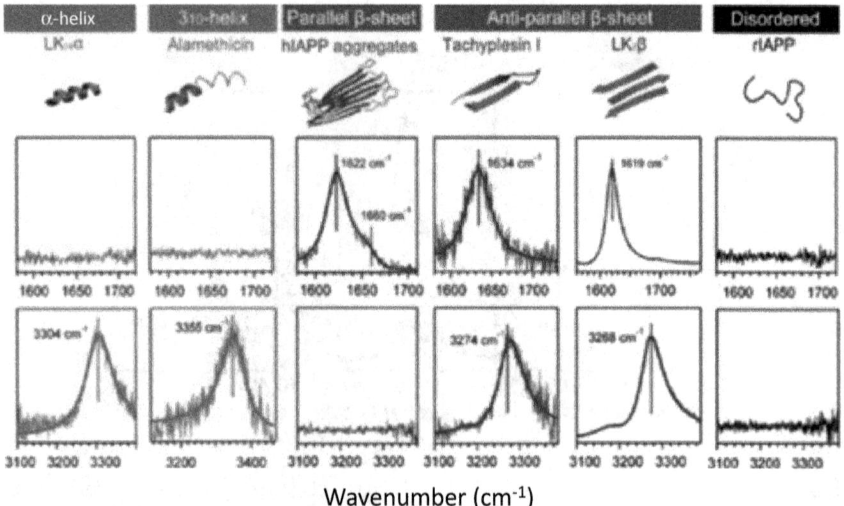

Fig. 10.4. Chiral SF vibrational spectra in the amide-I and NH stretch ranges of six peptides adsorbed at a lipid/water interface. Names of the peptides and their molecular structures are sketched on the top of the spectra. The vanishing NH spectrum of hIAAP is not intrinsic, but is due to the preferred orientation of hIAAP at the lipid/water interface (after Ref. [5]).

The spectra of different polarizations of a peptide at an interface allow determination of its orientation. Generally, as described in Sec. 6.3, the nonlinear susceptibility $\overleftrightarrow{\chi}^{(2)}$ of an adsorbed molecular monolayer is related to the nonlinear polarizability $\overleftrightarrow{\alpha}^{(2)}$ of the molecules by a coordinate transformation and an orientation average. Explicit expressions relating $\overleftrightarrow{\chi}^{(2)}$ and $\overleftrightarrow{\alpha}^{(2)}$ have been worked out by Moad and Simpson in terms of the Euler angles in transforming the molecular coordinates to the lab coordinates.[6] Measurements of n independent $\overleftrightarrow{\chi}^{(2)}$ elements at a discrete resonance (corrected by removal of Fresnel factors that connect the measured $\overleftrightarrow{\chi}^{(2)}$ with the intrinsic $\overleftrightarrow{\chi}^{(2)}$) allow determination of n parameters that can be used to describe the orientation distribution in terms of the Euler angles if the molecular density of the monolayer and $\overleftrightarrow{\alpha}^{(2)}$ are known. For peptides, specifically, the molecular coordinates are often chosen to have, for the helical structure, the molecular c axis along the helical axis, and for the β-sheet structure, the c axis along the sheet

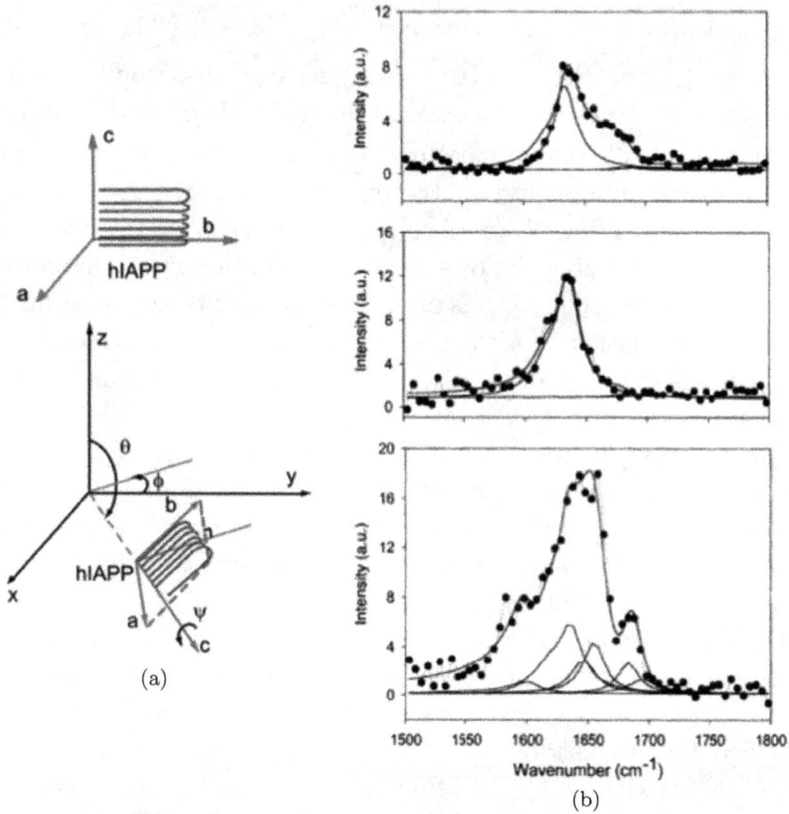

Fig. 10.5. (a) Sketches describing the β-sheet structure of hIAAP protein (top) and its molecular coordinates rotated from the lab coordinates in terms of the Euler angles θ, ϕ, and ψ. (b) SF vibrational spectra in the amide-I region of tachyplesin I adsorbed at the polystyrene/water interface with input/output polarization combinations, PSP (top), SPP (middle), and SSP (lower) (after Ref. [10]).

buildup direction and the b axis perpendicular to c along the chains as sketched in Fig. 10.5(a).

Deduction of the orientation of adsorbed peptides from polarization-dependent SFVS measurement is similar to the procedure discussed in Sec. 6.3 and usually proceeds as follows. From measurements with n independent polarization combinations, one can deduce n independent $\chi^{(2)}_{ijk}$ elements, or more specifically, the amplitudes of a selected resonance, A_{ijk}, in $\chi_{ijk}^{(2)}$. The expression relating $\chi_{ijk}^{(2)}$ and the set of nonvanishing $\alpha_{\xi\eta\zeta}^{(2)}$ also applies to A_{ijk} and the set of $a_{\xi\eta\zeta}$ that denotes the amplitude of the selected resonance in $\alpha_{\xi\eta\zeta}^{(2)}$. The n measured A_{ijk} values allow determination of n unknown parameters in the expression relating A_{ijk} and the set of $a_{\xi\eta\zeta}$, including N_s (surface molecular density), $a_{\xi\eta\zeta}$ and some parameters describing the orientation distribution of the molecules. In the case of peptides, $a_{\xi\eta\zeta}$ is often separately obtained by either an ab initio calculation or a calculation from known Raman scattering and IR absorption cross-sections on the selected resonance. Since A_{ijk} is proportional to N_s, we can eliminate N_s by taking ratios of A_{ijk}. Thus, the n-1 independent values of $A_{ijk}/A_{i'j'k'}$ can all be used to specify the orientation distribution. For peptides with a known secondary structure containing a number of molecular moieties with the same resonance, each moiety is assumed to be independent and contribute to $\alpha_{\xi\eta\zeta}^{(2)}$ in a geometric sum; the phase difference in the optical responses of different moieties is neglected. In the simple case of an isotropic monolayer of peptides adsorbed at an interface, the orientation of peptides with helix structure is specified by the distribution of the tilt angle θ of the helical axis from the surface normal,[7] and that of peptides with β-sheet structure by the distributions of two Euler angles, θ and ψ[8] (Fig. 10.5(a)). If we assume a δ-function distribution for each angle, then we only need one $\chi_{ijk}^{(2)}/\chi_{i'j'k'}^{(2)}$ ratio to determine the orientation of helical peptides but two to determine the orientation of β-sheet peptides. For example, as demonstrated by $Z.$ Chen and coworkers, the amide-I PPP and SSP spectra of α-helical MSI-78 peptide on a bilayer in Fig. 10.3(a) can be used to find the orientation of the peptide,[9] but for β-sheet tachyplesin (Fig. 10.3(c)), an additional spectrum of a different polarization combination is needed, as demonstrated in Fig. 10.5(b) for the case of tachyplesin at a polystyrene/water interface.[10] More generally, with chiral and achiral SFVS, one can measure more than three independent $\chi_{ijk}^{(2)}/\chi_{i'j'k'}^{(2)}$ ratios at each vibrational resonance. They allow deduction of more parameters to specify the orientation distribution. Furthermore, measurement of different vibrational resonances of the same peptide, for example, amide I and NH stretches, also provides additional information on the orientation distribution of the peptide.[11] In more refined experiments, isotopic labeling can be used to determine the structure and orientation of selected sections of a peptide at an interface.[12]

It is possible to use SFVS to monitor kinetics of structural changes of peptides at interfaces. Evolution of hIAPP adsorbed at an air/water interface and self-assembled into a β sheet in the presence of DPPG in water that catalyzes the assembly process is an example.[13] As seen from Fig. 10.6, both achiral and chiral spectra of the amide-I modes of hIAPP (Figs. 10.6(a) and 10.6(b)) increase in strength with time and saturate after $\sim$5 hours, but the chiral

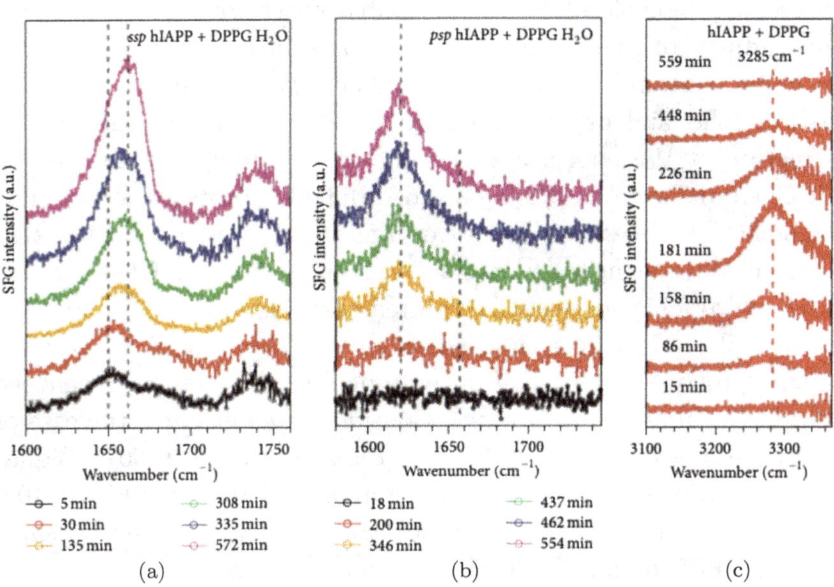

Fig. 10.6.   Kinetics of hIAPP aggregation and assembly at the air/H_2O interface, in the presence of DPPG as catalyst, probed by SF vibrational spectra in the amide I region using (a) SSP (achiral) and (b) PSP (chiral) polarizations and (c) in the N-H stretching region using the PSP (chiral) polarization. Both achiral and chiral spectra of the amide-I modes increase with time and saturate after $\sim$5 hours, but the chiral spectrum of NH first increases to a maximum at $\sim$3 hours and then decreases to zero after $\sim$10 hours (after Ref. [13]).

spectrum of NH (Fig. 10.6(c)) first increases to a maximum around 3 hours and then decreases to zero after $\sim$10 hours. The result was understood as follows. Initially, hIAPP adsorbed at the lipid/water interface as random coils that exhibited no chiral spectrum, then assembled into an intermediate more ordered α-helical structure showing increasingly strong achiral and chiral spectra, and finally, transformed into the stable antiparallel β-sheet structure that had a very weak chiral NH spectrum (Fig. 10.4). Ultrafast dynamics of peptide structural changes could in principle also be studied by pump/probe SFVS (Sec. 11.2).

Proteins with many amino acid residues and many possible conformations (or other large bio-macromolecules) are obviously too complex for optical spectroscopy, but we can still use SFVS to identify molecular units exposed at an interface. Like polymeric molecules, biomolecules have a flexible structure that can change and adjust to the environment. At a solid/water interface, for example, certain parts of a protein may like to interact and bond to the solid and other parts may like to face water; the protein structure at the interface must conform accordingly. Using SFVS, Z. Chen and coworkers have studied different proteins adsorbed at different interfaces and observed changes of protein structures as environments changed.[14] Most experiments focused on CH stretches. The hydrophilic groups of a protein are expected to bond to a hydrophilic substrate such as silica and the hydrophobic CH groups should protrude out into air or hydrophobic solvent such as benzene and CCl_4, but hide from water. Consider as an example the case of bovine serum albumin (BSA), which is a protein with 607 residues and has hydrophobic side chains rich in methyl groups.[14] Figure 10.7 shows the SSP SF CH stretch spectra of BSA adsorbed on silica and exposed to air, hydrophobic fluorinated solvent FC-75, and water, respectively. It is seen that the CH stretch peaks of BSA are most prominent at the air/silica interface, less at the silica/FC-75 interface, and invisible at the water/silica interface. The result can be readily understood if we know that air is more hydrophobic than FC-75 and water is as hydrophilic as silica. The CH groups of BSA tend to hide from the hydrophilic interface and protrude out more at

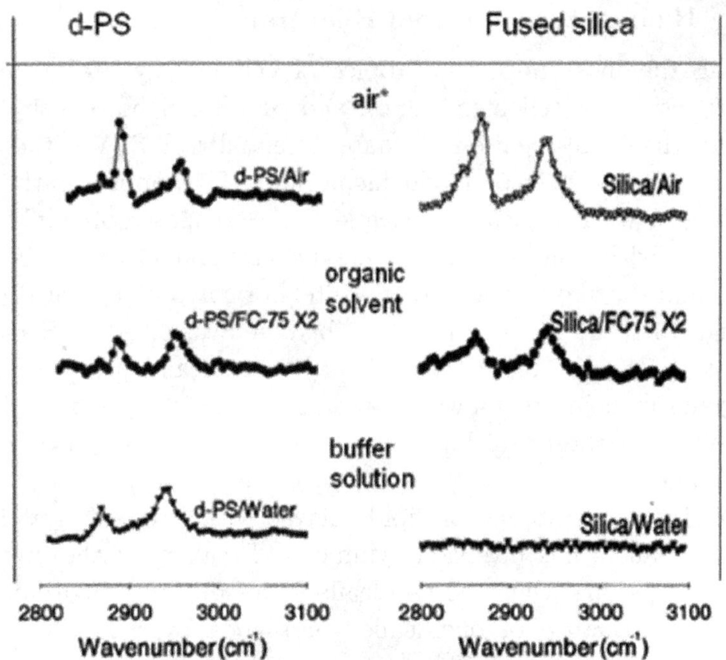

Fig. 10.7. SF CH stretch spectra of adsorbed BSA on deuterated polystyrene (PS) (left column) and fused silica (right column) exposed to media of different hydrophobicity: air (top row), hydrophobic fluorinated solvent FC-75 (middle row), and water (bottom row). Differences in the spectra can all be understood by the preferred orientation of hydrophobic CH chains of BSA toward more hydrophobic media. The very weak spectrum for BSA at the silica/water interface results from both silica and water being hydrophilic (after Ref. [20]).

the more hydrophobic interface. It was their asymmetric orientations at the silica/air and silica/FC-75 interfaces that led to the strong and less strong SF CH stretch spectra from the two interfaces, respectively, and their symmetric orientation at the silica/water interface that resulted in a null spectrum.

There are very few reported SFVS studies on nucleotides and nuclei acids presumably because their structures and spectra are much more complicated. Spectral changes of adsorbed nuclei acids induced by environmental changes could be observed, as demonstrated by single-strand DNA on gold,[15] but quantitative spectral analysis is very challenging.

10.3. Biomembranes (Lipid Bilayers)

Among the most important topics of cell biology are structures and functions of cell membranes, and SFVS can be effective as a tool for their interrogation. To date, essentially all SFVS studies on membranes have been on model membranes. They appear in the form of a bilayer with two leaflets; each leaflet is composed of amphiphilic molecules with long hydrophobic alkyl chains and a hydrophilic head group and the two leaflets are oppositely oriented and aligned (inset of Fig. 10.8(a)). As already described in Sec. 6.6, SFVS can be used to probe the orientation of alkyl chains and head groups of Langmuir monolayers (on water) and Langmuir–Blodgett monolayers (on solid substrate), which can be regarded as single leaflets of model membranes.

A simple membrane or lipid bilayer with two identical head-to-head leaflets has inversion symmetry. Breaking of the inversion symmetry occurs when the two leaflets face different environments, have the structure of one leaflet perturbed by molecule or ion adsorption, have one leaflet isotope-labeled, or have two leaflets made of different molecules. SFVS can be used to monitor the broken symmetry and the structural changes. We describe here a few representative cases.

Conboy and coworkers have conducted extensive SFVS investigation on lipid bilayers, especially on the flip-flop process of molecular exchanges between two leaflets.[16] In their first experiment, they prepared initially an asymmetric phospholipid bilayer of 1,2distearoyl-sn-glycero-3phosphocholine (DSPC, $C_{44}H_{88}NO_8P$) in the gel phase, with one leaflet perdeuterated, sandwiched between water and silica[17] (inset of Fig. 10.8(a)). Such a bilayer would flip-flop and transform into a symmetric one, but before this happened, they still had time to probe the structure of each leaflet. The SSP SF spectrum in the CH stretch range they obtained from the undeuterated leaflet was shown in Fig. 10.8(a). It is very similar to that of the DOAC monolayer on water given in Fig. 6.10(c) with the CH_3 symmetric stretch and its Fermi resonance appearing at 2875 and 2940 cm^{-1}, respectively, and the symmetric CH_2 stretch (from gauche defects of

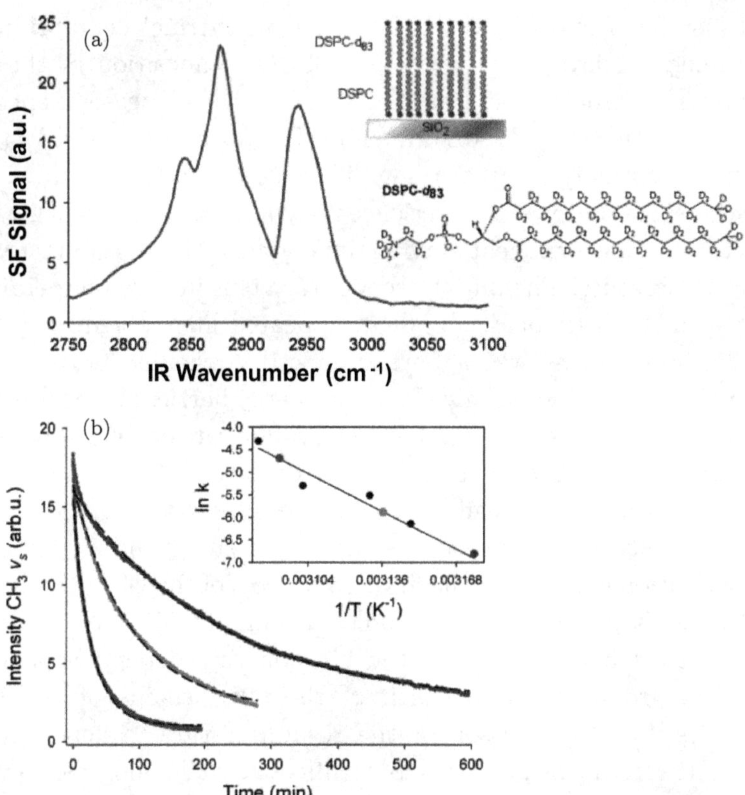

Fig. 10.8. (a) SSP SF spectrum in the CH stretching range of a DSPC/DSPC-d$_{83}$ bilayer at 24°C. All resonant peaks come from the fatty acid chains of DSPC. The insets show a schematic representation of the bilayer structure with one leaflet perdeuterated and the molecular structure of the perdeuterated DSPC-d$_{83}$. (b) Decay of the intensity of the symmetric CH$_3$ stretch of the bilayer with time at various temperatures: 41.7 °C (upper curve), 45.7 °C (middle curve), and 50.3 °C (lower curve). The dashed lines are exponential fits. The inset shows that the data on the initial decay rate as a function of temperature can be fit by a straight line, indicating that flip/flop of the bilayer is an Arrhenius process (after Ref. [17]).

chains) at 2840 cm^{-1}. Polarization dependence of the spectrum indicated that the alkyl chains were nearly perpendicular to the surface. They also made measurements on the deuterated leaflet and found that it had nearly the same conformation and orientation (in opposite direction) as expected from a lipid bilayer. Flip-flop switched the deuterated and undeuterated DSPC molecules in the two leaflets back and forth and eventually appeared in thermal equilibrium as a symmetric bilayer with the same DSPC composition in the two leaflets. This process could be monitored by the decrease of spectral intensity with time, as described in Fig. 10.8(b); the SF signal did not eventually approach zero because the bilayer sandwiched between silica and water experienced asymmetric environment. The signal decay was exponential with time and the decay rate increased with temperature. The initial decay rate versus inverse temperature, plotted in the inset of Fig. 10.8(b), appeared linear, signifying that the flip-flop process was a barrier-limited Arrhenius process. The linear slope of the plot then yielded the energy barrier for the flip-flop. Conboy and coworkers found that the flip-flop rate depended not only on temperature, but also on the surface density of the leaflets (or the surface pressure of the leaflets). From the results, they could deduce from a transition state theory the activation energy, activation area, and activation entropy for the flip-flop process of the DSPC bilayer.[17]

Conboy's group has also studied a number of other phospholipid bilayers and found that the flip-flop rate depended on both the head group and the length of the alkyl chains of the lipid molecules.[18, 19] Comparison of the results of three different lipids having the same head group but different chain lengths, DMPC (1,2-dimyristoyl-*sn*-glycero-3-phosphocholine, $C_{36}H_{72}NO_8P$), DPPC (1,2-dipalmitoyl-*sn*-glycero-3-phosphocholine, $C_{40}H_{80}NO_8P$), and DSPE (1,2-distearoyl-*sn*-glycero-3-phosphoethanolamine, $C_{41}H_{82}NO_8P$), showed that the observed flip-flop rate increased with decrease of the chain length, but the energy barrier appeared nearly the same. The chain length only affected the Arrhenius pre-exponential factor, and it must be the head group that controlled the activation energy barrier. It was indeed found in comparing DSPC and DSPE bilayers, which have different head groups but the same chain length, that the former had a flip-flop rate two orders of magnitude faster than the latter. The difference presumably came from the packing density of the lipid bilayer that affected the energy barrier through its dependence on the size and hydrophobicity of the head groups. Wedging peptide/protein molecules into leaflets could change the packing structure of a leaflet and facilitate flip-flop of bilayers. Anglin *et al.* observed that incorporation of 1 mol % of a model hydrophobic peptide, WALP23, into a DSPC bilayer significantly reduced the flip-flop energy barrier, while incorporation of the same amount of an amphiphilic model peptide, melittin, only had a moderate effect.

Lipid bilayers become more mobile with increase of temperature and may undergo a phase transition from the more ordered gel phase to the fluidic, more disordered liquid or liquid crystal phase. Conboy and coworkers devised a method of using SFVS to monitor the phase transition.[20,21] They realized that there should be a temperature range in which the gel and liquid phases of a bilayer coexist. As temperature increases, the gel domains shrink and the liquid domains grow, and the area of the domains of different phases of the two leaflets facing each other should first increase and then decrease (Illustrated in Fig. 10.9(a)). Since the SF signal is generated only from the area where different phases of the two leaflets face each other, it must also first increase and then decrease with increase of temperature. The experimental demonstration was on a DSPC bilayer sandwiched between silica and water. Figure10.9(b) displays the SSP SF spectra in the CH stretch region at three different temperatures. At 49°C, the bilayer was still in the gel phase and supposedly would have a symmetric structure, but was actually asymmetric because the two leaflets face silica and water, respectively. The nonvanishing spectrum was somewhat different from the one in Fig. 10.8(a) taken at 27°C because of more temperature-induced disordering; an additional CH stretch peak appearing above 2950 cm^{-1} was from the choline head groups. The CH$_3$ stretches of the alkyl chains became stronger at 58°C, but significantly weaker at 66°C. The peak intensity of the CH$_3$ symmetric stretch versus temperature was measured and plotted as the bell curve on the far right in Fig. 10.9(c), and the temperature where the maximum appeared was set as the phase transition temperature (in the middle of the phase coexistence range). Similar results were obtained from other phospholipid bilayers; those from DPPC (1,2-dipalmitoyl-*sn*-glycero-3-phosphocholine, $C_{40}H_{80}NO_8P$) and DHPC (1,2-diheptadecanoyl-*sn*-glycero-3- phosphocholine, $C_{22}H_{44}NO_8P$) are plotted in Fig. 10.9(c)

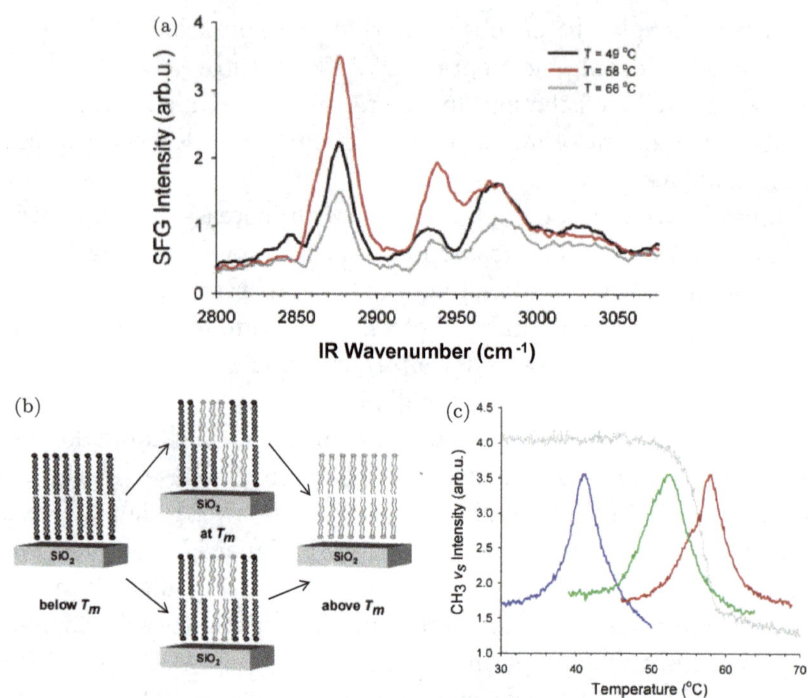

Fig. 10.9. (a) SSP SF spectra in the CH stretch range of a DSPC bilayer sandwiched between silica and water in the gel phase (49 °C), the gel/liquid crystal coexistent phase (58 °C), and the liquid crystal phase (66 °C). The peaks in the spectra all come from the fatty acid chains of DSPC except the one at 2974 cm^{-1}, which is partly from the CH$_3$ symmetric stretch of the choline head group. (b) Schematic representation describing the gel to liquid crystalline phase transition, illustrating domain dislocation (upper middle) and domain disparity (bottom middle) between the two leaflets at the phase-coexistence stage. (c) Intensity of the CH$_3$ symmetric stretch as a function of temperature obtained from SFVS for DPPC (left curve), DHPC (middle curve), and DSPC (right curve) bilayers. The result for a DSPC monolayer (peak-less top curve) is also shown for comparison (after Ref. [20]).

as the left and middle bell curves. The same plot for a single leaflet of DSPC is also given in Fig. 10.9(c) for comparison, showing very different temperature dependence with no trace of any maximum. Incorporation of large molecules in the leaflets of bilayers would perturb the lipid structure and lead to a lower phase transition temperature.[22]

Since phospholipid head groups are charged, they tend to attract ions to adsorb on lipid bilayers that can perturb their structure. This was demonstrated by Brown and Conboy.[23] They prepared a mixed deuterated DSPC and undeuterated DPPS (1,2-dipalmitoyl-sn-glycero-3-phospho-L-serine) symmetric bilayer sandwiched between silica and water as sketched in the inset of Fig. 10.10. DPPS has a more negatively charged head group than DSPC. At equilibrium, the bilayer was structurally symmetric and the CH symmetric stretch and its Fermi resonance mode from DPPS were very weak in the SSP SF spectrum, but when a 1 μM positively charged poly-L-lysine solution was injected into water at 50°C (to facilitate flip-flop), the peaks grew with time and eventually saturated (Fig. 10.10). Adsorption of positively charged poly-L-lysine on the leaflet that faced water must have induced flip-flop of the more negatively charged DPPS to be more populated in that leaflet. This is an example showing how ion adsorption can control the structure of a lipid bilayer.

Molecular interactions with membranes and molecular transport through membranes are most important for cell biology. We already discussed briefly in Sec. 10.2 how SFVS can be employed to probe peptides adsorbed on membranes. It is also possible to use SFVS

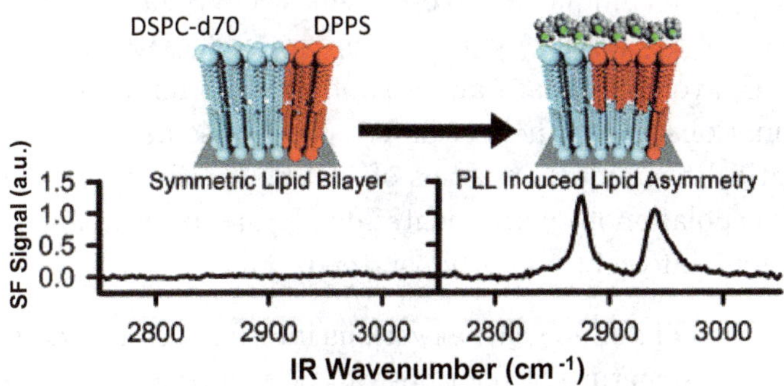

Fig. 10.10. Top: Schematic representation of a mixed gel-phase bilayer composed of 65 mol % DSPC-d70 (lighter part) and 35 mol % DPPS sandwiched between silica and water in the absence (left) and presence (right) of poly-L-lysine (PLL) adsorbed on the top leaflet. The adsorbed PLL induces an asymmetric bilayer structure. Bottom: the corresponding SF spectra for the symmetric (left) and asymmetric (right) bilayer; the former is very weak (after Ref. [23]).

to monitor molecular transport across a membrane. Biomolecules interacting with lipid bilayers generally are polar/amphiphilic; they adsorb on a leaflet with a preferred orientation and can be detected by SFVS. If they pass through a lipid bilayer, they would adsorb on the other leaflet with opposite orientation and can be monitored by a decrease of the SF signal. Actually, in this case, if identification of molecules is not needed, second harmonic generation (SHG) can be equally effective as a probe, as first demonstrated by Srivastava and Eisenthal.[24] To be closer to studies of real cell biology, they used liposomes in their experiment instead of a planar model membrane. The SHG signal from such small spherical particles was not highly directional, but appeared with a scattering pattern. We will postpone the detailed discussion of SHG studies of molecular transport through liposomes to Sec. 11.1, where SHG/SFG scattering from colloidal particles will be described.

10.4. Summary and Prospects

Applications of SFVS to bio-interfaces are still in the rudimentary stage with focus on two areas, biomolecules at interfaces and bio-membranes. Studies are mostly on model systems, and basically follow the discussions in Chapter 6 (Secs. 6.2, 6.3, and 6.6). As a spectroscopic technique for bio-systems, SFVS faces the common difficulties of spectroscopy in having to deal with very complex entities. Unavoidably, data analysis often relies on assumptions and approximations. Nevertheless, SFVS still has some advantages over other optical techniques in terms of surface specificity and ability to determine polar molecular orientation. It has been able to provide some unique information on bio-systems.

- Orientation of amino acids at various interfaces can be determined.
- Secondary structure of peptides can be identified.
- Orientation of peptides of known secondary structure at various interfaces can be obtained if the peptide structure is rigid.
- Chiral SF spectroscopy is surface-specific for peptides because of their chiral molecular arrangement.

- SF spectra on proteins are difficult to analyze, but sub-molecular units that preferentially appear at selected interfaces can be identified.
- Structure of leaflets of lipid bilayers and changes induced by perturbation can be learned.
- Flip-flop of molecules between two leaflets of lipid bilayers and the kinetics can be monitored.
- Phase transition from the gel phase to the liquid crystal phase of membranes can be observed.
- Interaction of adsorbed molecules with lipid leaflets, their influence on the structure and flip-flop process of bilayers, and their transportation across bilayers can be studied.

Sum frequency spectroscopy is yet to be developed into a powerful tool for studies of bio-interfaces. Asides from the problem that bio-systems are generally too complex for spectroscopy, the complexity of current experimental setups is also an obstacle, but hopefully can be improved with advance of laser technology. Development of label-less chiral/achiral SF microscopy with sufficient sensitivity and phase discrimination for bioscience would be very welcome. Finally, it is most important to have a few successful demonstrations to show that SF spectroscopy can help solve impactful problems in biology.

References

1. Ji, N.; Shen, Y. R.: Sum Frequency Vibrational Spectroscopy of Leucine Molecules Adsorbed at Air-Water Interface. *J. Chem. Phys.* 2004, 120, 7107–7112.
2. Paszti, Z.; Keszthelyi, T.; Hakkel, O.; Guczi, L.: Adsorption of Amino Acids on Hydrophilic Surfaces. *J. Phys. Cond. Matter* 2008, 20, 224014.
3. Holinga, F. J.; York, R. L.; Onorato, R. M.; Thompson, C. M.; Webb, N. E.; Yoon, A. P.; Somorjai, G. A.: An SFG Study of Interfacial Amino Acids at the Hydrophilic SiO_2 and Hydrophobic Deuterated Polystyrene Surfaces. *J. Am. Chem. Soc.* 2011, 133, 6243–6253.
4. Ding, B.; Jasensky, J.; Li, Y.; Chen, Z.: Engineering and Characterization of Peptides and Proteins at Surfaces and Interfaces: A Case Study in Surface-Sensitive Vibrational Spectroscopy. *Acc. Chem. Res.* 2016, 49, 1149–1157.

5. Yan, E. C. Y.; Wang, Z.; Fu, L.: Proteins at Interfaces Probed by Chiral Vibrational Sum Frequency Generation Spectroscopy. *J. Phys. Chem. B* 2015, 119, 2769–2785.

6. Moad, A. J.; Simpson, G. J.: A Unified Treatment of Selection Rules and Symetry Relations for Sum-Frequency and Second harmonic Spectroscopies. *J. Phys. Chem. B* 2004, 108, 3548–3562.

7. Nguyen, K. T.; Le Clair, S. V.; Ye, S.; Chen, Z.: Orientation Determination of Protein Helical Secondary Structure Using Linear and Nonlinear Vibrational Spectroscopy. *J. Phys. Chem. B* 2009, 113, 12169–12180.

8. Nguyen, K. T.; Kung, J. T.; Chen, Z.: Orientation Determination of Interfacial Beta-Sheet Structures *In Situ*. *J. Phys. Chem. B* 2010, 114, 8291–8300.

9. Yang, P.; Ramamoorthy, A.; Chen, Z.: Orientation of MSI-78 Measured by Sum Frequency Generation Vibrational Spectroscopy. *Langmuir* 2011, 27, 7760–7767.

10. Wang, J.; Chen, X.; Clarke, M. L.; Chen, Z.: Detection of Chiral Sum Frequency Generation Vibrational Spectra of Proteins and Peptides at Interfaces *In Situ*. *Proc. Nat. Acad. Sci.* 2005, 102, 4978–4983.

11. Yan, E. C. Y.; Fu, L.; Wang, Z.; Liu, W.: Biological Macromolecules at Interfaces Probed by Chiral Vibrational Sum Frequency Generation Spectroscopy. *Chem. Rev.* 2014, 114, 8471–8498.

12. Ding, B.; Panahi, A.; Ho, J. J.; Lasser, J. A.; Brooks, C. L.; Zanni, M. T.; Chen, Z.: Probing Site-Specific Structural Information of Peptides at Model membrane Interface *In Situ*. *J. Am. Chem. Soc.* 2015, 137, 10190–10198.

13. Fu, L.; Liu, J.; Yan, E. C. Y.: Chiral Sum Frequency Generation Spectroscopy for Characterizing Protein Secondary Structure at Interfaces. *J. Am. Chem. Soc.* 2011, 133, 8094–8097.

14. Chen, X. Y.; Clarke, M. L.; Wang, J.; Chen, Z.: Sum Frequency Generation Vibrational Spectroscopy Studies on Molecular Conformation and Orientation of Biological Molecules at Interfaces. *Int. J. Mod. Phys. B* 2005, 19, 691–713.

15. Howell, C.; Schmidt, R.; Kurz, V.; Koelsch, P.: Sum-Frequency-Generation Spectroscopy of DNA Films in Air and Aqueous Environments. *Biointerfaces* 2008, 3, FC47-51.

16. Alhusen, J. S.; Conboy, J. C.: The Ins and Outs of Lipid Flip-Flop. *Acc. Chem. Res.* 2017, 50, 58–65.

17. Liu, J.; Conboy, J. C.: Direct Measurement of the Transbilayer Movement of Phospholipids by Sum-frequency Vibrational Spectroscopy. *J. Am. Chem. Soc.* 2004, 126, 8376–8377.

18. Liu, J.; Conboy, J. C.: 1,2-diacyl-phosphatidylcholine Flip-Flop Measured Directly by Sum-Frequency Vibrational Spectroscopy. *Biophys. J.* 2005, 89, 2522–2532.

19. Anglin, T. C.; Conboy, J. C.: Kinetics and Thermodynamics of Flip-Flop in Binary Phospholipid Membranes Measured by Sum-Frequency Vibrational Spectroscopy. *Biochemistry-US* 2009, 48, 10220–10234.
20. Liu, J.; Conboy, J. C.: Phase Transition of a Single Lipid Bilayer Measured by Sum-Frequency Vibrational Spectroscopy. *J. Am. Chem. Soc.* 2004, 126, 8894–8895.
21. Liu, J.; Conboy, J. C.: Asymmetric Distribution of Lipids in a Phase Segregated Phospholipid Bilayer Observed by Sum-Frequency Vibrational Spectroscopy. *J. Phys. Chem. C* 2007, 111, 8988–8999.
22. Liu, J.; Conboy, J. C.: Phase Behavior of Planar Supported Lipid Membranes Composed of Cholesterol and 1,2-distearoyl-sn-glycerol-3-phosphocholine Examined by Sum-Frequency Vibrational Spectroscopy. *Vib Spectros* 2009, 50, 106–115.
23. Brown, K. L.; Conboy, J. C.: Electrostatic Induction of Lipid Asymmetry. *J. Am. Chem. Soc.* 2011, 133, 8794–8797.
24. Srivastava, A.; Eisenthal, K. B.: Kinetics of Molecular Transport across A Liposome Bilayer. *Chem. Phys. Lett.* 1998, 292, 345–351.

Review Articles

- Ye, S.; Nguyen, K. T.; Le Clair, S. V.; Chen, Z.: In Situ Molecular Level Studies on Membrane Related Peptides and Proteins in Real Time Using Sum Frequency Generation Vibrational Spectroscopy. *J. Struct. Biol.* 2009, 168, 61–77.
- Fu, L.; Wang, Z.; Yan, E. C. Y.: Chiral Vibrational Structures of Proteins at Interfaces Probed by Sum-Frequency Generation Spectroscopy. *Int. J. Mod. Sci.* 2011, 17, 9404–9425.
- Zhang, X. X.; Han, X. F.; Wu, F. G.; Jasensky, J.; Chen, Z.: Nano-Bio Interfaces Probed by Advanced Optical Spectroscopy: From Model System Studies to Optical Biosensors. *Chin. Sci. Bull.* 2013, 58, 2537–2556.
- Ye, S.; Wei, F.; Li, H.; Tian, K.; Luo, Y.: Structural and Orientation of Interfacial Proteins Determined by Sum Frequency Generation Vibrational Spectroscopy: Method and Application. *Adv. Protein Chem. Struct. Biol.* 2013, 93, 213–255.
- Yan, E. C. Y.; Fu, L.; Wang, Z.; Liu, W.: Biological Macro-molecules at Interfaces Probed by Chiral Vibrational Sum Frequency Generation Spectroscopy. *Chem. Rev.* 2014, 114, 8471–8498.

- Ding, B.; Jasensky, J.; Li, Y.; Chen, Z.: Engineering and Characterization of Peptides and Proteins at Surfaces and Interfaces: A Case Study in Surface-Sensitive Vibrational Spectroscopy. *Acc. Chem. Res.* 2016, 49, 1149–1157.
- Fu, L.; Wang, Z.; Batista, V. S.; Yan, E. C. Y.: New Insights from Sum Frequency Generation Vibrational Spectroscopy into the Interactions of Islet Amyloid Polypeptides with Lipid Membranes. *J. Biabetes Res.* 2016, 2016, 7293063.
- Tran, R. J.; Sly, K. L.; Conboy, J. C.: Applications of Surface Second Harmonic Generation in Biological Sensing. *Ann. Rev. Anal. Chem.* 2017, 10, 387–414.
- Hosseinpour, S.; Roeters, S. J.; Bonn, M.; Peukert, W.; Woutersen, S.; Weidner, T.: Structure and Dynamics of Interfacial Peptides and Proteins from Vibrational Sum-Frequency Generation Spectroscopy. *Chem. Rev.* 2020, 120, 3420–3465.
- Guo, W.; Lu, T.; Gandhi, Z.; Chen, Z.: Probing Orientatins and Conformations of Peptides and Proteins at Buried Interfaces. *J. Chem. Phys. Lett.* 2021, 12, 10144–10155.

Chapter 11

Miscellaneous

11.1. Interfaces of Colloidal Particles

Colloidal particles are ubiquitous. They appear everywhere in nature and play essential roles in atmospheric and environmental sciences, such as pollution, and in geoscience, such as nutrient transportation. They are also most important in many areas of bio- and medical sciences and in industry because their large surface-to-volume ratio can effectively enhance efficiency of reactive processes. Yet for lack of available tools for *in situ* probing, surface structures of colloidal particles in real environments are practically unknown. One may wonder if surface-specific SHG/SFG can be a viable technique to probe colloids. It has indeed been successfully demonstrated in the past decades for colloidal size ranging from ~ 10 nm to a few μm. We present below a sketch of the underlying theory and a few representative cases of possible applications.

Our discussion on SHG/SFG so far has been limited to plane surfaces although extension to large curved surfaces is fairly straightforward. Extension to surfaces of colloidal particles is in principle similar, but the mathematics involved is more complicated. At first glance, one might suspect that SHG/SFG from colloidal particles with inversion symmetry would be very weak. This is however not true. If the particle size is comparable to input/output wavelength, the fields radiated from the polarizations, $\vec{P}_{eff}(\vec{r})$ and $\vec{P}_{eff}(-\vec{r})$, induced in/on a particle do not cancel each other because although $|\vec{P}_{eff}(\vec{r})| = |\vec{P}_{eff}(-\vec{r})|$, their phase difference is generally not π or odd multiples of π. Interference of radiation fields from all $\vec{P}_{eff}(\vec{r})'s$

actually results in a far-field radiation pattern that depends on the particle size and shape and can be calculated by integrating the dipole radiation fields from all $\vec{P}_{eff}(\vec{r})'s$. The total radiation field from a collection of independent, identical colloidal particles is the incoherent sum of radiation from all particles and has approximately the same radiation pattern as that from a single particle if the radiation is detected at a distance much larger than the excitation spot size. Such an induced radiation process is known as Mie scattering. Linear and nonlinear Mie scattering arise, respectively, from $\vec{P}_{eff}(\vec{r})$ induced linearly and nonlinearly by the input electric fields, i.e., $\vec{P}_{eff}^{(1)}(\vec{r}) = N\overset{\leftrightarrow}{\chi}^{(1)} \cdot \vec{E}(\vec{r})$ for linear scattering and $\vec{P}_{eff}^{(2)}(\vec{r}) = N\overset{\leftrightarrow}{\chi}^{(2)} : \vec{E}(\vec{r})\vec{E}(\vec{r})$ for second-order nonlinear scattering (see Sec. 2.1). Specifically, SH/SF Mie scattering refers to radiation from $\vec{P}_{eff}^{(2)}(\vec{r})$ induced at the SH/SF frequency. Comparison of experiment with theory could allow us to deduce $\overset{\leftrightarrow}{\chi}^{(2)}$ and hence the structure of the colloidal particles.

In dealing with the theory of Mie scattering, we generally assume all colloidal particles are identical and their shapes are known (often assumed to be spherical for convenience). The complexity of the theory lies in connecting fields inside and outside of each particle through boundary conditions. In the rigorous Mie theory, this is done by expanding the fields into series of partial waves and having each partial wave component satisfy the boundary conditions. If the particles are small, approximation can be made by taking only a few leading terms of the partial wave series. For particle size much smaller than the reduced wavelengths and the refractive index of the particles close to that of the surrounding, the so-called Rayleigh–Gann–Debye (RGD) approximation that neglects the boundary effect and retains only the leading two terms of the series can be adopted. With connection between fields inside and outside a particle known and $\vec{P}_{eff}(\vec{r})$ given, the expression of the far-field radiation pattern from the particle can be found. In the case of SH/SF Mie scattering, the expression is a function of $\overset{\leftrightarrow}{\chi}^{(2)}$, and matching of the theoretical and experimental radiation patterns allows determination of $\overset{\leftrightarrow}{\chi}^{(2)}$. We consider here only cases of molecular adsorption on surfaces/interfaces of colloidal particles that dominate the surface nonlinear susceptibility, $\overset{\leftrightarrow}{\chi}_S^{(2)}$, of colloidal particles when resonantly excited. Deduction of the nonvanishing elements of $\overset{\leftrightarrow}{\chi}_S^{(2)}$ provides information on the interfacial structure of adsorbates on colloidal particles. We will not go into more details on the theory of SH/SF Mie scattering here, but refer the readers to Appendix of Chapter 12 in Ref. [12.1] for an outline of the theory and Refs. [12.2–12.5] for mathematical formalism and derivations.

The beam geometry of Mie scattering is sketched in Fig. 11.1(a). To record the scattered radiation pattern, a detection arm rotatable around the center of the sample cell is generally used. We consider a solution with dispersed colloidal particles and molecular solutes, from which SH Mie scattering (SHMS) arises mainly from two-photon resonant $\overset{\leftrightarrow}{\chi}_S^{(2)}$ of molecules adsorbed on the colloidal particles. There is however also signal from incoherent hyper (SH) Rayleigh scattering of solvent molecules in the solution that is simultaneously collected by the detector. This background signal can be separately measured from a solution without colloidal particles and subtracted off from the detected signal. One may wonder if linear Mie scattering with resonant excitation of the molecules could probe the adsorbed molecules on colloidal particles equally well. Unfortunately, linear absorption of incident light and resonant Rayleigh scattering by solvent molecules in the solution posts serious problems that strongly deteriorate the signal-to-background ratio. The two-photon-resonant, surface-specific SHMS has the advantage of not being much affected by absorption or Rayleigh scattering.

SH Mie scattering (SHMS) was first demonstrated by Eisenthal and coworkers on polystyrene (PS) microspheres ($\sim$1 μm in diameter) in water solution with malachite green (MG, structural formula in Fig. 11.1(b)) dissolved in it.[6,7] They measured SHMS from the solution using input pulses at 854 nm from a Ti:sapphire laser and observed increase of the SH signal with increase of MG concentration in the solution. Obviously, the SH signal was from adsorbed MG on PS spheres (Fig. 11.1(b)) since $\overset{\leftrightarrow}{\chi}_S^{(2)}$ of PS microspheres is enhanced by two-photon electronic resonance of MG on PS. From the result

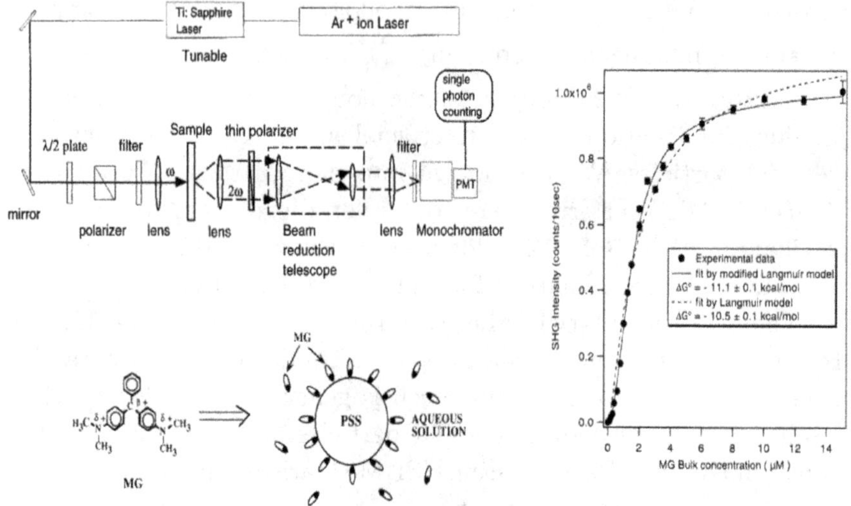

Fig. 11.1. (a) Schematic representation of beam arrangement for SH Mie scattering (SHMS). (b) Sketch describing adsorption on malachite green (MG) adsorbed on polystyrene spheres (PSS). (c) Adsorption isotherm of MG on PSS in water solution. The dotted and solid lines are theoretical fits of the experimental data from a simple Langmuir and a modified Langmuir model, respectively (after Ref. [7]).

of the SH field strength (which is proportional to the square root of the SH intensity and the surface density of MG) versus MG concentration, they found that the data could be fit by a Langmuir adsorption isotherm (Fig. 11.1(c)), from which the adsorption energy and the maximum surface density ($N_{S,max}$) of MG could be deduced. Yang *et al.* extended the work to PS microspheres of different diameters; their observed adsorption isotherms are displayed in Fig. 11.2(a), showing that $N_{S,max}$ is higher for larger PS microspheres as expected.[8] They also recorded the SHMS radiation patterns with P-in/P-out polarization (Fig. 11.2(b)) and found that the patterns could be fit fairly well by the theory of SHMS with RGD approximation and the assumption that $\overset{\leftrightarrow}{\chi}_S^{(2)}$ of MG was dominated by $\alpha_{\xi\xi\xi}^{(2)}$ (ξ being the C2 molecular axis) with ξ lying along the surface normal. Gonella and Dai did a more careful investigation on the same system with four different sizes of PS microspheres and two

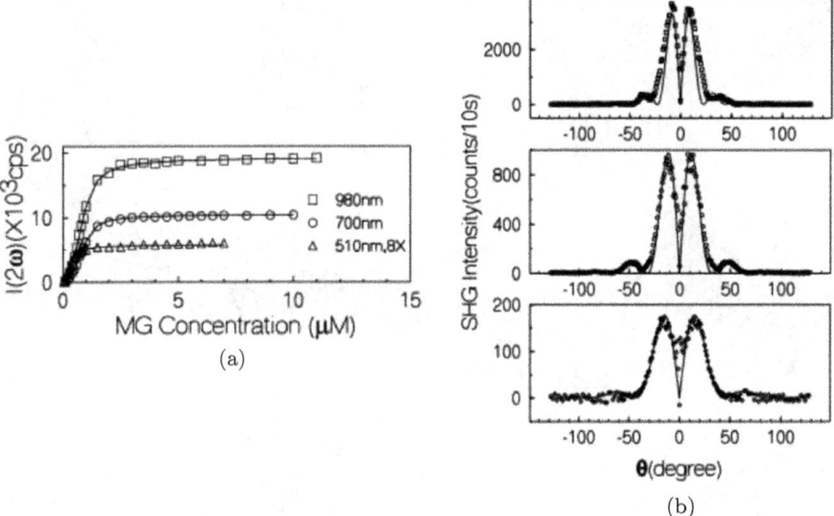

Fig. 11.2. (a) Adsorption isotherms obtained by SHMS for MG adsorbed from solution on PS spheres of three different sizes, 980, 700, and 510 nm in diameter. The solid lines are fits to a modified Langmuir model. (b) SHMS signal versus scattering angle θ away from the forward scattering direction obtained with P-in/P-out polarization combination. The three frames from top to bottom are for particles with diameters of 980, 700, and 510 nm, respectively. The lines are fits from the Mie scattering theory with RGD approximation (after Ref. [8]).

different polarization combinations (P-in/P-out and S-in/P-out).[9] They used the more rigorous theory of SHMS with four nonvanishing elements of $\overset{\leftrightarrow}{\chi}_S^{(2)}$ of MG, $\chi_{S,\perp\perp\perp}^{(2)}, \chi_{S,\parallel\parallel\perp}^{(2)}, \chi_{S,\parallel\perp\parallel}^{(2)} \simeq \chi_{S,\perp\parallel\parallel}^{(2)}$ ($\perp$ and $\parallel$ denoting directions perpendicular and parallel to the surface) to fit the observed angular patterns of SHMS output. As seen in Fig. 11.3, the theory fits the experimental data very well. The fit allowed deduction of the nonvanishing $\overset{\leftrightarrow}{\chi}_S^{(2)}$ elements, and with the relation between $\overset{\leftrightarrow}{\chi}_S^{(2)}$ and $\alpha_{\xi\xi\xi}^{(2)}$ (the dominating element of $\overset{\leftrightarrow}{\alpha}^{(2)}$) known from coordinate transformation, the orientation of MG at the PS surface can be extracted (Sec. 6.3). They found that the C2 axis of MG was along the surface normal with an orientation spread of $\sim$40°.

Perhaps the most interesting application of SHMS is on kinetics of molecular adsorption on and transportation through membranes of liposomes or model cells. This was also first demonstrated by

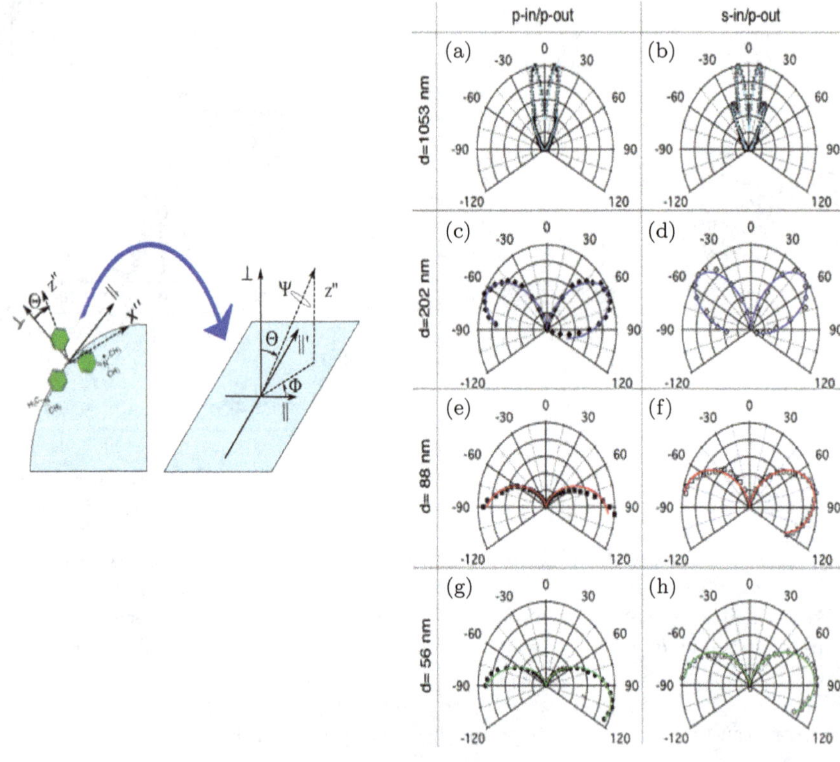

Fig. 11.3. (a) MG structure and relation between the local coordinate system ($\parallel$, $\parallel'$, $\perp$) on a curved surface and the coordinate system (x'', y'', z'') on a plane surface. (b) Angular radiation patterns of SHMS with P-in/P-out (left) and S-in/P-out (right) polarizations from MG adsorbed on PS spherical particles of four different diameters d in water. Lines are fits obtained from a more rigorous SH Mie scattering theory (after Ref. [9]).

Eisenthal's group using MG and liposomes composed of DOPG (dioleoylphosphatidylglycerol, chemical structure in Fig. 11.4(a)) bilayers in water solution.[10] DOPG has a negatively charged head group, while MG in neutral water is partially ionized and positively charged; the Coulomb interaction attracts MG to DOPG. Upon rapid mixing of the solution after injection of MG, the initial condition was set up to have a maximum amount of MG^+ molecules adsorbed on liposomes, exhibiting a maximum SHMS signal. Afterwards, the SH signal decreased exponentially with time to a constant (Fig. 11.4(b)).

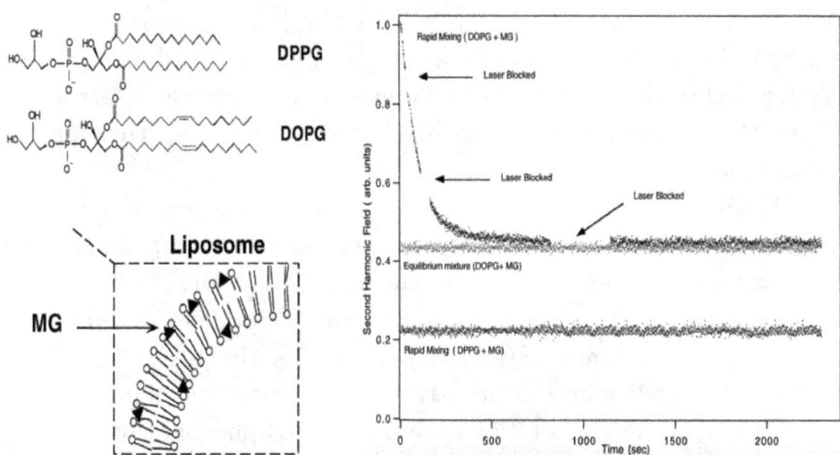

Fig. 11.4. Transport kinetics of MG molecules across liposome membranes. (a) Structural formula for DOPG and DPPG constituting the liposome bilayers. (b) Cartoon describing adsorption of MG on the outer and inner walls of a liposome. (c) Experimental results on amplitude of SHMS output versus time from three sets of measurements labeled in the figure. The decay curve describes MG transport through the DOPG bilayer. No transport was observed if DPPG replaced DOPG (after Ref. [10]).

This could be understood from the kinetics of MG molecules traversing the liposome membranes. Initially, MG molecules were adsorbed only on the outer wall of liposomes. After passing through the membranes, they had to flip-flop and adsorb on the inner wall of liposomes with opposite orientations (Fig. 11.4(c)). Because the oppositely oriented MG molecules on the two sides of membranes contributed destructively to SHG, the SH signal decreased as more MG adsorbed on the inner wall until it approached a constant towards equilibrium; the residual signal was not zero because the inner wall had a smaller area and hence less adsorbed MG^+ molecules on it than the outer wall. From the measured SH field as a function of time, the transport rate of MG across the DOPG membrane could be extracted. It was interesting to note that if DOPG was replaced by DPPG (dipalmitoylphosphatidylglycerol, chemical structure in Fig. 11.4(a)), the SH signal from the solution remained unchanged with time, indicating no MG transport across the membranes. This was due to the structural difference of DOPG and DPPG membranes.

It is known that at room temperature, the DOPG membrane is in a fluidic liquid crystal phase, while the DPPG membrane having tighter packing is in a solid-like gel phase. The more ordered structure of the DPPG membrane would block transport of MG through the membrane.

SHMS can be used to study how to control transport of molecules across membranes of liposomes/cells. Eisenthal's group found that cholesterol molecules wedged into the leaflets of a DOPG membrane of liposomes could partially block transport channels of molecules; they could slow down MG transport across the membrane by 6 times if the cholesterol concentration in solution changes from 0 to 50 mol%.[11] When POPC (palmitoyloleoyl-phosphatidylcholine) was mixed into the DOPG membrane, the MG transport across the membrane also slowed down, more so with increase of POPC proportion.[12] This is because POPC is zwitterionic with a positive charge in the head group. Its presence in the membrane electrostatically impeded transport of the positively charged MG through the membrane. The transport rate of molecules through membranes also depends on structural compatibility of molecules with membranes. Pons *et al.* showed that the flip-flop transport rate of stilbazolium molecules across a DOPC (dioleoylphosphatidylcholine) membrane increased by 1000 times when they were photo-isomerized from trans to cis conformation.[13]

Dai and coworkers extended the SHMS technique to studies of molecular transport across live cells, bacterium E coli, that had a rod shape $\sim$2 μm long and $\sim$0.5 μm in diameter. The E coli cells comprise two membranes.[14] The experiment was to monitor transport of MG across the two membranes by SHMS. The result for a set of different concentrations of MG is presented in Fig. 11.5. It is seen that after rapid mixing of MG and E coli in a water solution for MG^+ to adsorb on the outer membrane of E coli, the SHMS signal versus time first rises to a maximum and then drops rapidly, but later turns around and increases to another maximum before decreasing again more gradually to a steady state. Like the case of molecules traversing a single membrane discussed earlier, the initial rapid drop was due to MG^+ passing through the outer membrane and adsorbing on the inner wall of the outer membrane. The later signal increase was the result of having part of the penetrated MG^+ adsorbing on the outer wall of the inner membrane and the subsequent signal decay came from MG^+ adsorbing on the inner wall of the inner membrane (sketch on the left of Fig. 11.5). The faster decay of the first stage was because the outer membrane of E coli is more loosely structured. As shown in Fig. 11.5, the experimental results could be well fit by a Langmuir adsorption/kinetic transport model; from the fit, the transport rates of MG across the two

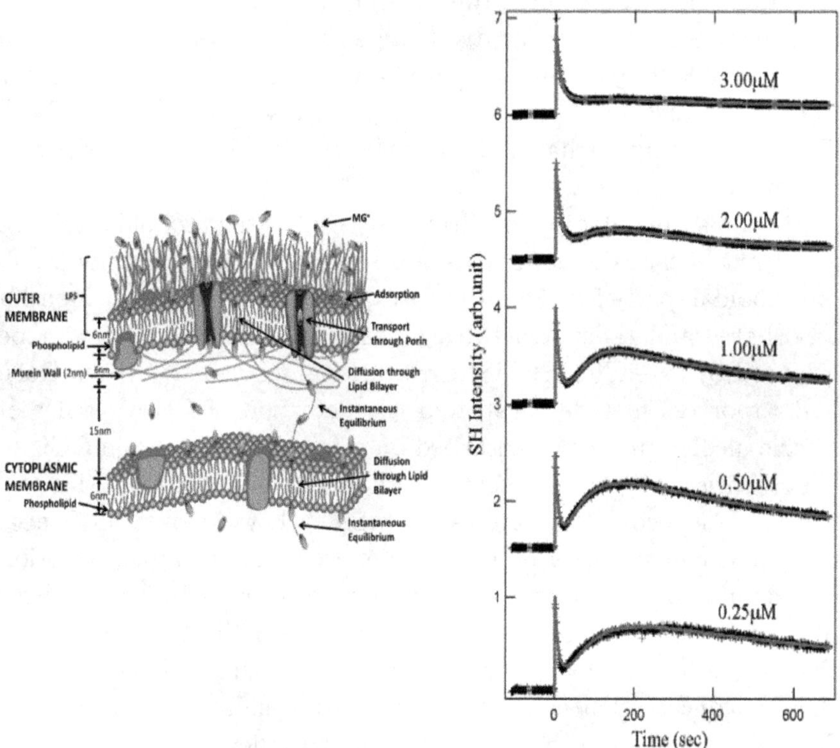

Fig. 11.5. Output intensity of SHMS versus time from adsorption of MS on E coli cells in solution with five different MG concentrations. The solid lines are theoretical fits from a modified Langmuir adsorption model together with kinetic equations for transport across the two membranes of E coli. Toward equilibrium, adsorbed MS molecules appear on both outer and inner walls of E coli's dual membranes as sketched on the left (after Ref. [14]).

membranes of E coli could be deduced. Dai's group also showed that antibiotic molecules could induce change to E coli membranes and improve their permeability.[15] They found that if E coli cells were in contact with azithromycin (AZM, an antibiotic compound) of sufficient concentration in water for a sufficiently long time, the transport rate of MG across the membranes increased significantly. Presumably, AZM adsorbed into leaflets of membranes distorted the membrane structure and facilitated transport of molecules including themselves; AZM needs to get to the cytoplasma of a cell to exert the antibiotic effect. In another case, they found with SHMS that the famous Gram staining protocol to characterize bacteria actually has the crystal violet (CV) dye trapped in the peptidoglycan mesh (PM), unable to penetrate the cytoplastic membrane (CM), contrary to the old belief that CV would readily traverse both PM and CM.[16] These are examples that SHMS can be employed to study practical cell biology problems.

It is also possible to use SHMS to study other colloidal science problems such as competitive adsorption, surface reaction, etc. on colloidal particles. In such cases, one would need to identify adsorbates and their structures at colloidal surfaces. This can be achieved by extending SHMS to vibrationally resonant SFMS. Roke and coworkers first demonstrated the detection of CH stretches of stearic alcohol molecules adsorbed on silica nanoparticles in CCl_4.[17] The theory for SFMS is essentially the same as that for SHMS.[4]

There is also strong interest in probing how colloidal interfaces may change in water (or liquid) in response to, for example, variation of pH and salt concentration in water, problems similar to those discussed in Sec. 8.9 on planar interfaces. SHMS/SFMS is likely to be the only technique that can be used to study such problems although there is concern that colloidal surfaces are usually not well defined. So far, the technique has not yet been fully developed and widely adopted to interrogate interfaces of colloidal particles in general. Complexity of data analysis is a drawback, but with advance of algorithm and computing power, it could be alleviated. Sensitivity of the technique is an issue. It is currently limited to applications to colloidal particles in liquid with a particle density higher than

$\sim 10^9/\mathrm{cm}^3$. Extension to the gas phase, for example, for studies of aerosols, is difficult and would be at a disadvantage when compared with linear Mie scattering.

11.2. Ultrafast Surface Dynamics

The primary step of events in nature or man-made systems often starts locally in ultrashort time, but it dictates the subsequent processes and determines the end result. Understanding ultrafast dynamics of a relevant process can be of key importance in pursuit of controlling an event. Laser spectroscopy with ultrashort pulses has become the effective means to probe ultrafast dynamics. Naturally, we should consider adoption of SHG/SFG spectroscopy for studies of ultrafast surface dynamics. It is possible to use linear optical spectroscopy in some cases, but as we repeatedly mentioned earlier, surface specificity of SHG/SFG provides many advantages.

Surface dynamics describes how a surface or interface, after being perturbed, spontaneously returns to its initial equilibrium state or to a new equilibrium state. Surface reactions and interfacial electron transfers are examples. They usually occur on ps and fs time scales, and require ps/fs laser pulses to monitor the time-varying dynamic processes. The pump/probe scheme is generally used with the pump pulse perturbing the surface under study and the time-delayed probe pulses monitoring the surface changes in time. The schematic representation of experimental arrangement is given in Fig. 11.6: The pump pulse excites an interface at time zero, and the probe pulse (for SHG) or pulse pair (for SFG) interrogates the interface at time t; the probe signal versus t provides information on the dynamics of the surface structural change.

There are different scenarios of pump/probe dynamical processes at interfaces. We can simplify the picture by compartmentalizing an interfacial system into the interfacial layer (A) and the two bulk media (B and B$'$) connecting with the interfacial layer (Fig. 11.6). We assume that with SFG/SFG spectroscopy, we are able to probe only the interfacial layer A and monitor its relaxation dynamics. The pump excitation is however not surface-specific, and can excite both

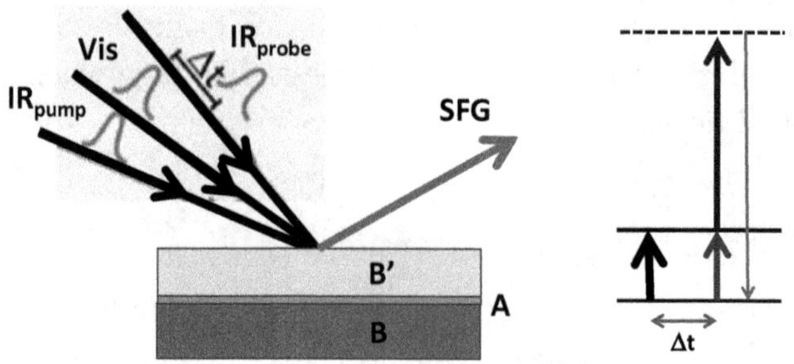

Fig. 11.6.   Experimental arrangement of pump/probe TR-SFVS on an interfacial system with the interface layer A sandwiched between two media, B and B'. Transitions of the pump/probe process are sketched on the right.

A and B/B$'$ if the latter also absorbs. We then have roughly the following cases.

I. Only A is excited by pump and selected resonant transitions of A are probed to find ensuing changes of states: If A is weakly coupled to B/B$'$, its excitation relaxes essentially within A before thermalization steps in. If A is strongly coupled to B or B$'$, then excitation of A is expected to relax mainly into B or B$'$. There is of course also the intermediate case with A excitation relaxing partly within A and partly into B or B$'$.

II. Only B is excited by pump and selected resonant transitions of A are probed: Relaxation of the excitation of a bulk medium (B) is expected to be essentially the same with or without the interfacial layer (A) because B is much larger than A, but A can be excited by excitation transfer from B with a time variation intimately related to relaxation of excitation in B.

III. Both A and B are excited by pump and selected resonant transitions of A are probed:

(a) If the coupling between A and B is weak due to little overlap between the resonant transition bands of the two so that the Forster resonant energy transfer (FRET) is much slower than relaxation of excitations in A and B, then relaxation of A has little effect from relaxation of B.

(b) If the coupling between A and B is strong due to strong overlap of their resonant transition bands so that FRET is faster than relaxations in A and B, then relaxation of A basically follows relaxation of B.

(c) If FRET is comparable to relaxation of A, then the latter is expected to be influenced by relaxation of B.

We discuss in the following a number of representative cases of ultrafast surface dynamics studied by TR-SHG/SFG spectroscopy.

SHG was first employed to probe ultrafast surface melting dynamics. Shank *et al.* used time-resolved (TR) SHG to monitor the beginning stage of fs-laser-induced surface melting of Si(111).[18] As electrons in the penetration depth of the laser light were excited, their excess energy distribution led to lattice disordering and melting. The reflected SHG signal from the surface dropped in time and became azimuthally isotropic. Since SHG probed the surface layer and the excitation was on both surface and bulk, this could be considered as Case IIIb.

Eisenthal and coworkers used TR-SHG to monitor photo-induced isomerization of DODCI (3,3'-diethyloxadicarbocyanine iodide) at the air/water interface and found that it was significantly faster than in bulk water presumably because of less spatial hindrance at the surface.[19] They also studied orientation dynamics of molecules at the air/water interface with an experimental scheme that has been followed by others.[20] The idea behind the scheme is that one can pump-excite a selected group of molecules with specific orientations and probe how excited and unexcited molecules reach equilibrium orientation distribution by orientation diffusion. It was assumed that the lifetime of excited molecules was much longer than the orientation relaxation time and the orientation dynamics of ground-state and excited-state molecules were the same. Consider the case of coumarin 314 (C314)) adsorbed at the air/water interface studied by Eisenthal and coworkers.[21] The structural formula of C314 on the right of Fig. 11.7(a) suggested that its nonlinear polarizability $\overset{\leftrightarrow}{\alpha}{}^{(2)}$ is dominated by $\alpha^{(2)}_{\xi\xi\xi}$ along the molecular axis ($\vec{\mu}$), and the surface nonlinear susceptibility of the unexcited C314 layer can be expressed as $\chi^{(2)}_{S,ijk} = N_S \langle A(\theta,\phi)\rangle (\alpha^{(2)}_{\xi\xi\xi})_g$. Here, $A(\theta,\phi)$ is an angular function relating $\chi^{(2)}_{S,ijk}$ and $\alpha^{(2)}_{\xi\xi\xi}$ though the coordinate transformation with

θ and ϕ being the polar and azimuthal angles of $\hat{\xi}$ in the lab coordinates, N_S is the surface density of molecules, and the angular brackets denote an average over the orientation distribution of the molecules. For example, with $\hat{z}$ along the surface normal, we have $\chi^{(2)}_{S,zxx} = \chi^{(2)}_{S,xzx} = N_S < \cos\theta \sin^2\theta \cos^2\phi > \alpha^{(2)}_{\xi\xi\xi}$. If molecules are excited and a hole appears in the orientation distribution so that a change of $\langle \cos\theta \sin^2\theta \cos^2\phi \rangle$ is created at t = 0, the corresponding change of $\chi^{(2)}_{S,xzx}$ is $\Delta\chi^{(2)}_{S,xzx} = N_S\Delta[\langle \cos\theta \sin^2\theta \cos^2\phi \rangle](\alpha^{(2)}_{\xi\xi\xi,e} - \alpha^{(2)}_{\xi\xi\xi,g})$ with the subindices g and e referring to ground and excited states. For t>0, $\Delta[\langle \cos\theta \sin^2\theta \cos^2\phi \rangle]$, changes with time as the orientation distribution return to equilibrium. If we assume the orientation dynamics in θ and ϕ are uncorrelated, then we can write $\Delta[\langle \cos\theta \sin^2\theta \cos^2\phi \rangle_t] = c(\theta,t)c(\phi,t)$ and $\Delta\chi^{(2)}_{S,xzx}(t) = N_S c(\theta,t)c(\phi,t)(\alpha^{(2)}_{\xi\xi\xi,e} - \alpha^{(2)}_{\xi\xi\xi,g})$.

Eisenthal and coworkers used SHG with S-in/P-out polarization to measure $\Delta\chi^{(2)}_{S,xzx}(t)$, where $\hat{z}$ is along the surface normal and $\hat{z}-\hat{y}$ is the incidence plane.[21] With circularly polarized pump beam incident along the surface normal, molecules were uniformly excited in the azimuthal plane, but in the polar direction, those with $\vec{\mu}$ more inclined toward the surface were more preferentially excited. As a result, only $c(\theta,t)$ was time-dependent, reflecting the out-of-plane orientation dynamics of molecules back to equilibrium, and SHG measurement of $\Delta\chi^{(2)}_{S,xzx}(t) \propto c(\theta,t)$ recorded the dynamics. Their result is shown in Fig. 11.7(a), from which the out-of-plane orientation relaxation time was found to be $\sim$350 ps. This relaxation time is much shorter than the excitation lifetime of 4.7 ns, justifying the assumption that relaxation of excitation is negligible. Linearly polarized pumps were used to probe in-plane orientation dynamics. Pump excitations along x and y induced time-dependent $c_x(\phi,t)$ and $c_y(\phi,t)$, respectively, in the azimuthal plane, but the same $c(\theta,t)$ out of the plane. Results of TR-SHG measurements of $[\Delta\chi^{(2)}_{S,xzx}(t)]_x \propto c(\theta,t)c_x(\phi,t)$ and $[\Delta\chi^{(2)}_{S,xzx}(t)]_y \propto c(\theta,t)c_y(\phi,t)$ after the pump excitation are displayed in Fig. 11.7(b) exhibiting two similar relaxation curves. The normalized difference of the two,

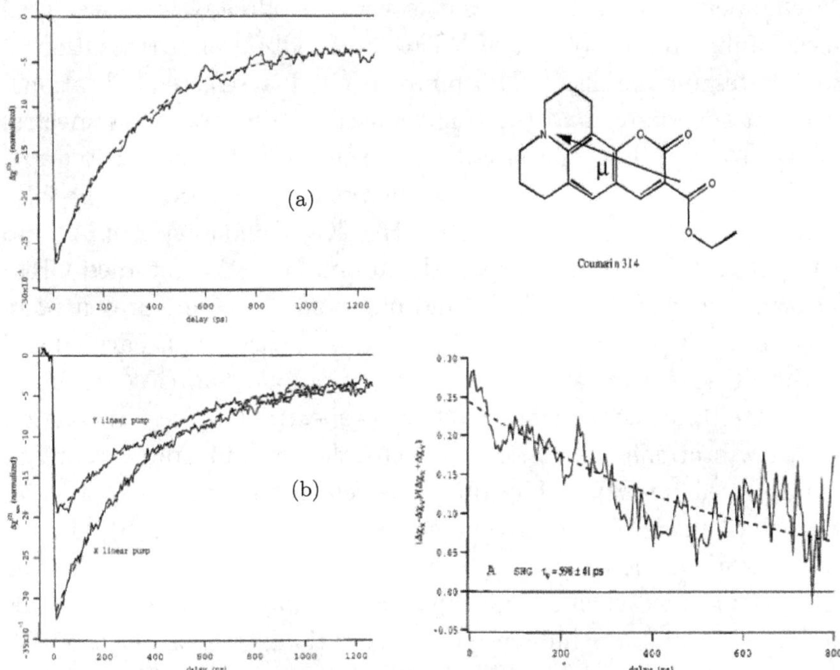

Fig. 11.7. Normalized $\Delta\chi_S^{(2)}$ versus pump/probe delay measured by SHG from adsorbed coumarin 314 at the water/air interface excited by normally incident pump of (a) circular polarization, and (b) linear polarizations along x and y. The normalized difference of the two curves in (b) is shown on the right of (b). The pump was at 420 nm and the probe at 840 nm. The dashed lines are fits to the data. The structural formula of coumarin 314 is shown on the right of (a) with the arrow $\vec{\mu}$ denoting the molecular axis (after Ref. [21]).

expressed by $\dfrac{[\Delta\chi_{S,xzx}^{(2)}(t)]_x - [\Delta\chi_{S,xzx}^{(2)}(t)]_y}{[\Delta\chi_{S,xzx}^{(2)}(t)]_x + [\Delta\chi_{S,xzx}^{(2)}(t)]_y} = \dfrac{c_x(\phi,t) - c_y(\phi,t)}{c_x(\phi,t) + c_y(\phi,t)}$ is plotted in Fig. 11.7(c). It depends only on (ϕ, t) and therefore describes the in-plane orientation relaxation dynamics. The above cases belong to the Case I category.

Ultrafast vibrational relaxation at interfaces can be investigated by time-resolved SF vibrational spectroscopy (TR-SFVS). An ultrashort IR pulse excites an interfacial vibrational mode, and the ultrafast relaxation of the excitation is monitored by surface-specific TR-SFVS (Fig. 11.6). Even at the very beginning of SFVS development stage, Harris and coworkers already demonstrated successfully the use of TR-SFVS to study vibrational relaxation of adsorbates on metals.[22] The pump excited a selected vibrational mode of adsorbates and the time-delayed SFVS probed its spectral change to learn how the vibrational excitation returned to thermal equilibrium. They observed multi-component relaxation of the CH_3 symmetric stretch of a stearate ($C_{17}H_{35}CO_2H$) monolayer on Ag and attributed it to intramolecular relaxation. They then studied CH_3S adsorbed on Ag(111) in UHV and obtained the result presented in Fig. 11.8.[23] The CH_3 symmetric stretch appeared as a narrowband in the SF spectrum with proper sample preparation (Fig. 11.8(a)); pump excitation of the mode reduced its spectral intensity; relaxation of the vibrational excitation back to the ground state recovered the spectral intensity. The observed relaxation was biexponential with the longer relaxation time around 63 ps (Fig. 11.8(b)). Harris and coworkers believed it was also mainly due to intramolecular relaxation; pump excitation of Ag decayed away in less than a few ps and would not have influenced the vibrational relaxation of CH_3S (Case IIIa described above). However, they found CO relaxation on Cu(111) in UHV was different;[24] the vibrational relaxation time

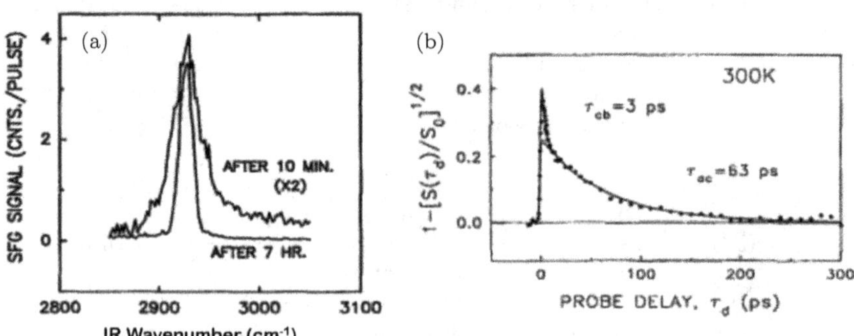

Fig. 11.8. (a) SF spectra of the symmetric stretch of the methyl group of CH_3S on Ag(111) in UHV at 300K taken at two different times after dosing. The sharp peak is an indication of the good sample quality. (b) Normalized signal change, $1-[S(\tau_d)/S_0]^{1/2}$, of SFG from the symmetric stretch as a function of pump–probe delay, τ_d, for the good sample in (a). $S(\tau_d)$ and S_0 are the signals at delay time τ_d and 0, respectively. The data can be fit by a bi-exponential curve (solid line) (after Ref. [23]).

of the CO stretch was only a few ps, indicating that its relaxation must have gone to Ag through vibration(CO)–electron(Ag) coupling (Case IIIc). Presumably, the link established between O and Cu was much more effective for energy transfer than that between S and Ag. Schmidt and Guyot-Sionnest found that the vibrational lifetime of CO on Pt(111) in anhydrous acetonitrile could be varied by applying potential on Pt;[25] shift of the Fermi level of Pt with respect to the CO stretch must have changed the electron–vibration coupling and affected CO relaxation. Excitation transfer from adsorbates to semiconductors is expected to be much less effective because of the lower quasi-free electron density in semiconductors. The lifetime of H-Si stretch vibration on Si was observed to be 1.4 ns at 95K and 550 ps at 460K, and the dominant relaxation mechanism was not electron–vibration coupling, but phonon (Si)–vibration (H-Si) coupling.[26]

Rapid excitation transfer between adsorbates can happen. Morin *et al.* found by TR-SFVS that the lifetime of H-Si on terraces of a vicinal Si(111) surface was significantly shortened by the presence of dihydride species at steps.[27] It was understood that Si-H and Si-H_2 stretches were strongly coupled via dipole–dipole interactions, but Si-H_2 had a much shorter lifetime, effectively draining the vibrational excitation from Si-H and making it appear to relax faster. This could be considered an example of Case IIc with Si-H and SiH_2 being A and B, respectively, although in this case, A and B refer to two surface parts of comparable size.

In studies of adsorbate dynamics, the pump light exciting the adsorbates may unavoidably also excite the substrate. In the case of metals, the excited electrons rapidly relax to heat and increase the local temperature in much less than a ps; they have little effect on relaxation of the adsorbates although eventually the adsorbates would relax toward a higher equilibrium temperature. This is Case IIIa described earlier. For adsorbates on insulators and semiconductors, pump excitation of adsorbates may also excite the substrate, and the two may relax on the same time scale. How much the adsorbate relaxation is influenced by the substrate relaxation depends on coupling of the two. This is Case IIIc. Molecular

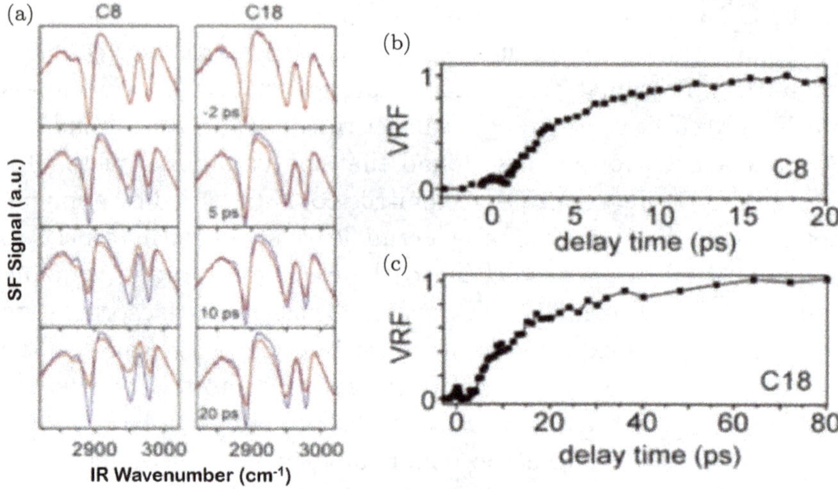

Fig. 11.9. (a) SF spectra of the methyl group of two self-assembled alkane thiol monolayers on an Au film with chain lengths C8 ($n = 7$) and C18 ($n = 17$) taken at different time delays after flash heating of Au to 800°C (red curves). Spectra taken without flash heating (blue curves) are shown for comparison. (b)VRF for C8 and (c)VRF for C18 as a function of delay time, with VRF denoting the fractional peak intensity change of the methyl symmetric stretch with reference to the intensity at final equilibrium (after Ref. [28]).

adsorbates at water interfaces are examples. How strongly adsorbate vibrational relaxation is affected by water relaxation depends on how strongly the excited vibrational band of adsorbates overlaps with the vibrational band of interfacial water.

For Case II described earlier, we consider pump excitation of only a substrate and monitor subsequent excitation and decay of adsorbates upon transfer of excitation from the substrate to the adsorbates. Dlott and coworkers studied two alkane thiol monolayers of different chain lengths self-assembled on Au steps.[28] They irradiated Au by 500 fs pump pulses, practically flash heating Au instantaneously to $\sim$800°C because the excited electrons in Au relaxed and converted the excitation energy to heat in much less than a ps. They measured the time-delayed CH stretch spectrum of the adsorbed alkane thiol and observed the characteristic CH_3 stretch modes from the terminal methyl group (Sec. 6.6). All modes were found to decrease in intensity with time, as displayed in Fig. 11.9(a)

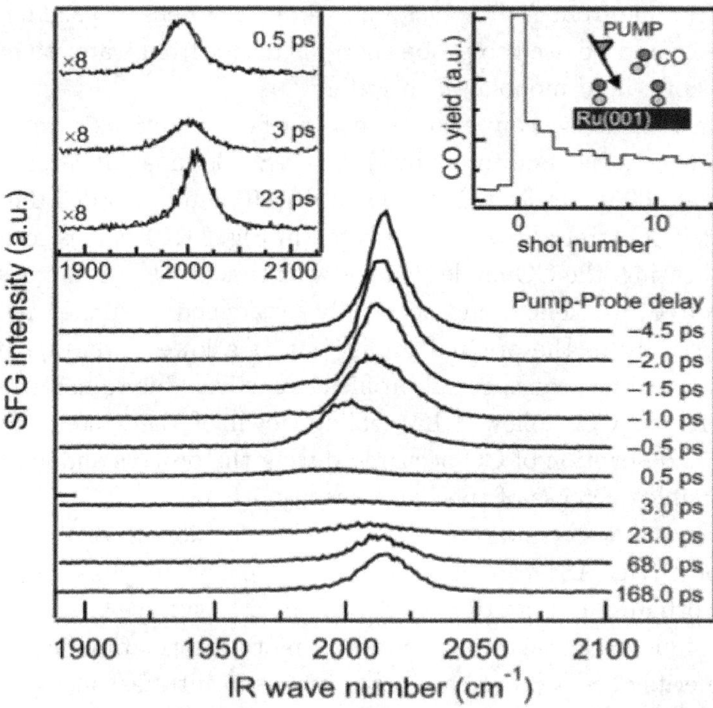

Fig. 11.10. SF vibrational spectra of the stretching mode of CO on Ru(001) taken at different delay times after pumping of the system by a 110 fs pulse at 800 nm, showing that the spectral intensity first decreases and then recovers to roughly half of the initial magnitude. Inset at the top right corner describes the CO yield from desorption after a given number of pump pulses (after Ref. [30]).

for alkane thiols of C8 and C18 chain lengths. (Because of interference with the broad background from Au, the CH_3 stretches appear as dips in the spectra.) This was due to heat transfer from Au to thiols followed by heat propagation along the alkyl chains. As the heat pulse reached the end of the chains, the methyl groups became orientation-disordered and the stretching mode intensity dropped. The fractional change of the peak intensity of the CH_3 symmetric stretch (at 2880 cm^{-1}) with time is plotted in Fig. 11.9(b). The two curves for C8 and C18 alkane thiols in Fig. 11.9(b) are fairly similar, suggesting that heat transfer from Au to the methyl groups was limited more by Au-S contact than by propagation along the chains. The heat transfer time from Au to thios was around few tens of ps. The same scheme could be used to probe energy transfer from water to methyl groups of a lipid monolayer on water.[29]

Flash heating can also trigger surface reactions, which can subsequently be monitored by TR-SFVS. Bonn *et al.* used 110-fs pulses at 800 nm to flash-heat CO on Ru(001) and recorded the time-resolved CO stretch spectra presented in Fig. 11.10.[30] It is seen that upon heating, the CO mode changes with time; first, it is increasingly red-shifted, weakened, and spectrally broadened, and then recovers progressively to the original profile but at a lower intensity. It was understood that the spectral profile change was due to heat transfer from Ru to CO, followed by cooling down of the system. Partial thermal desorption of CO occurred during the process and led to the irreversible decrease of spectral intensity.

We should remark that infrared reflection-absorption spectroscopy (IRRAS) can also be used to probe ultrafast dynamics of absorbates, but as discussed in Sec. 6.1, surface-specific SFVS has unique advantages and provides more information in general. The merits of SFVS become more obvious in ultrafast measurement because the TR signal is much weaker than in the static case.

Ultrafast surface dynamics of neat materials can be probed by TR-SFVS, but interpretation of results is more complex. Surface optical transitions of a neat material usually overlap with bulk transitions, and ultrashort pump pulses with an intrinsic broad bandwidth unavoidably excite both. Yet, it is not necessarily clear how strongly the two are coupled, although relaxation measurement of the surface excitation may give some clue. Such systems belong to Case IIIc described earlier; water interfaces are examples. Ultrafast surface vibrational dynamics at water interfaces has been fairly extensively studied because of its relevance. We provide here a summary discussion.

The free OH stretch at $\sim$3700 cm^{-1} at the air/water interface is a distinct surface mode (Sec. 7.2) and is the simplest to study. The stretch mode can be resonantly excited by $\sim$100 fs pulses while there are only very weak bonded OH stretch modes of the bulk in the pump bandwidth (Case I). Thus, we can expect vibration relaxation of the

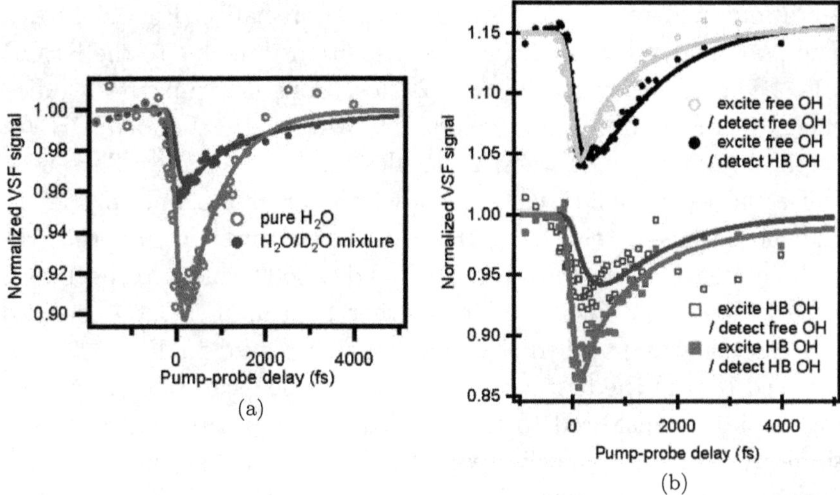

Fig. 11.11. (a) Normalized SSP SF probe signal varying with time from free OH stretch at the air/water interfaces after P-polarized pump excitation of the free OH stretch for pure H_2O (open circles) and 1:1 isotopic mixture of H_2O and D_2O (filled circles). The solid lines are theoretical fits. (b) Normalized SSP SF probe data varying with time from free OH and hydrogen-bonded (HB) OHs around 3500 cm^{-1} after they are separately pumped as labeled. The solid curves are theoretical fits (after Ref. [32]).

free OH mode not much affected by bulk excitation (Case I). There are two possible relaxation pathways for the vibrational excitation of free OH.[31] One is through intramolecular vibrational relaxation (IVR) into lower excited states within the water molecule itself. The other is by transferring excitation to the hydrogen-bonded OH of the same molecule to be further relaxed away. The latter process occurs in two possible ways, an intramolecular energy transfer (IET) and a rotational jump of the excited free OH to form an excited bonded OH (labeled as REOR). Bonn and coworkers used SFVS to probe the excitation decay of the free OH at the air/water interface and observed a time-dependent excitation/relaxation curve depicted in Fig. 11.11(a), from which a relaxation time of $\sim$840 fs was deduced. When H_2O was replaced by 1:1 H_2O/D_2O isotopic mixture, the pathways of REOR and IET for relaxation of excited OH of HDO molecules were effectively shut off by energy mismatch of OH and OD vibrations and relaxation slowed down appreciably (Fig. 11.11(a)). The data analysis showed that the relaxation time due to the REOR relaxation pathway alone was ~2.6 ps. Bonn's group also studied energy transfer from the excited free OH to the bonded OH (of neighboring interfacial water molecules) and vice versa.[32] They used fs IR pulses with 300 cm^{-1} bandwidth for pump and probe. In one measurement, the free OH stretch was pumped and the free and bonded OH stretches at 3700 and ~3500 cm^{-1}, respectively, were probed by TR-SFVS; in another measurement, the bonded OH stretch was pumped. Their results displayed in Fig. 11.11(b) revealed that in the first case nearly half of the excitation of free OH was rapidly transferred to the bonded OHs and then both relaxed similarly and more slowly away. In the second case, only a small fraction of the excitation of bonded OHs was transferred to the free OH and later the two relaxed away together. These results indicate that energy transfer from free OH to bonded OHs is fast but the reverse is not, and IVR of free OH is not important.

The problem of pump excitation of the bonded OH stretches is actually much more complex since molecules at the interface and in the bulk are both necessarily excited, although TR-SFVS still only probes relaxation dynamics of OH stretches at the interface. Figure 11.12 shows the results of such an experiment on a neat water/silica interface.[33] The SFVS spectrum of the interface before pump excitation and the spectra of the 100-fs IR pump pulses at 3200 and 3400 cm^{-1} are described in the top frame of Fig. 11.12(a). Almost instantaneously after the pump excitation, a spectral hole significantly broader than the pump pulse spectrum appeared as seen in the lower two frames of Fig. 11.12(a). Later, the profile of the spectral hole changed more gradually into a new quasi-steady-state spectral shape. Variations of spectral intensities with time recorded at different IR frequencies are presented in Figs. 11.12(b) and 11.12(c). They behave quite differently, but interestingly, they can all be fit by a simple relaxation equation with the same relaxation times:

$$S(t) = 1 - (1 - S_0)e^{-(t-100)/T_v} + \Delta S[1 - e^{-(t-100)/T_t}] \qquad (11.1)$$

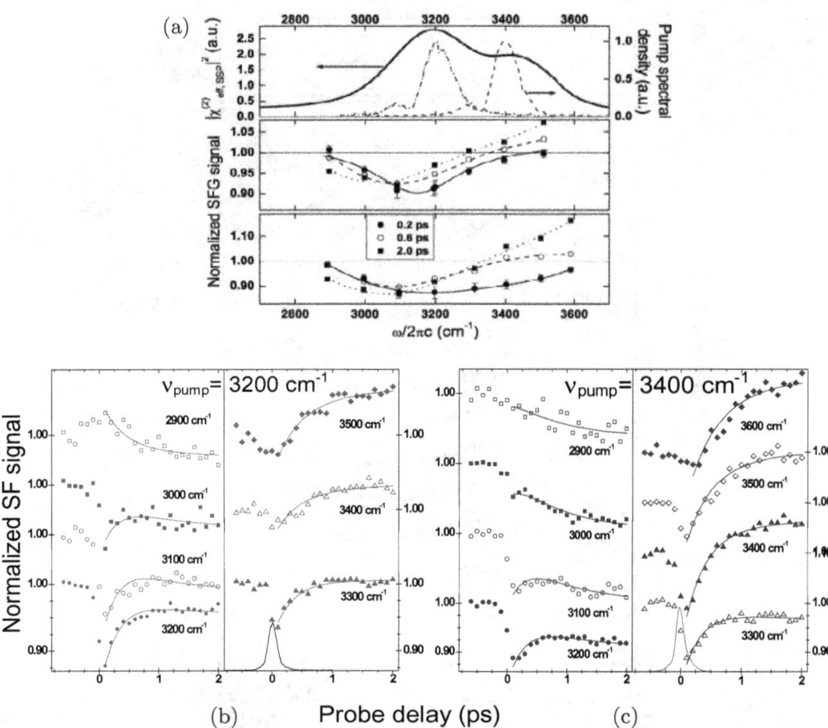

Fig. 11.12. (a) Top frame: SF vibrational spectrum of the water/silica interface at room temperature (solid curve) together with the spectral profiles of the pump pulses at 3200 cm^{-1} (dash dotted curve) and 3400 cm^{-1} (dashed curve). Middle and bottom frames: Spectral holes of the water/silica interface induced by 100-fs pump pulses at 3200 cm^{-1} and 3400 cm^{-1}, respectively, at three pump/probe delay times. (b) and (c) SF probe signals versus pump/probe delay time obtained at different probe frequencies for pump frequencies at 3200 cm^{-1} and 3400 cm^{-1}, respectively. The solid curves over the data points are theoretical fits using Eq. (11.1). The traces for different probe frequencies are displaced vertically for clarity. The solid bell curves at the bottom of (b) and (c) describe the third-order cross-correlation traces of the IR pump and the SF probe pulses (after Ref. [33]).

where $S(t)$ is the SF signal at a given IR probe frequency, S_0 is the signal at t = 100 fs with t = 0 referring to the time when the peak of the pump pulse hit the water/silica interface, ΔS is the final signal change as the system approaches quasi-steady state, $T_\nu = 700$ fs, and $T_t = 300$ fs. The good fitting of the data by Eq. (11.1) prompted a simple interpretation of the relaxation dynamics: Relaxation of

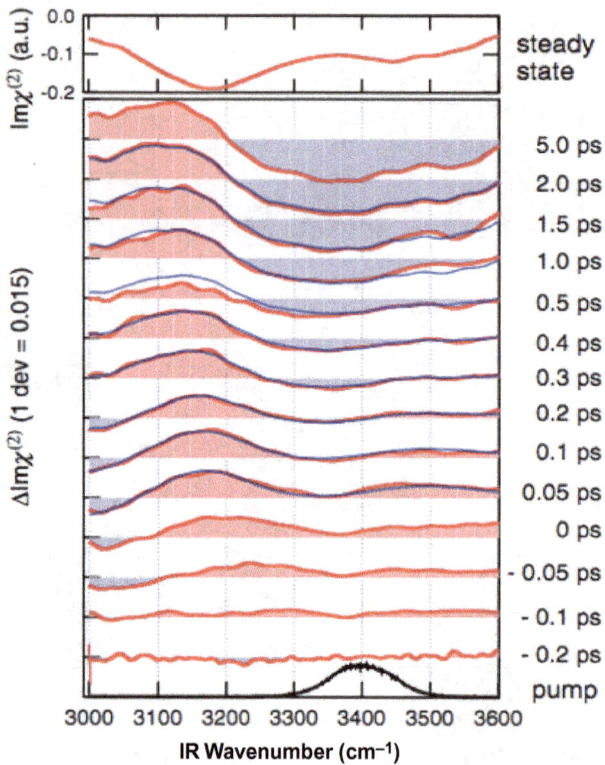

Fig. 11.13. Pump-induced changes of the SF spectrum, $\Delta\mathrm{Im}\chi_S^{(2)}(\omega_{IR})$, of the CTAB ($C_{16}H_{33}N(CH_3)_3Br$, cetyltrimethylamonium bromide)-monolayer-covered/water interface in the OH stretching region. Spectra of Im $\chi_S^{(2)}(\omega_{IR})$ were obtained by phase-sensitive SFVS; the one without pump is presented in the top panel for comparison, and the spectral profile of the pump is shown in the bottom panel (after Ref. [39]).

the bonded OH excitation at the interface appears to proceed in three steps: (1) A spectral hole is created in $\sim$100 fs by the 100-fs pump pulse due to rapid near-resonant energy transfer from excited OHs to their unexcited neighbors, a mechanism known as spectral diffusion or FRET. Enhanced absorption from v = 1 to v = 2 of the excited vibrational states also contributes to the spectral change. (2) Population of v = 1 of all excited OH species at the interface is drained through intra- and inter-molecular vibrational coupling to low-lying excitations with a time constant T_ν. (3) Excitation then goes through thermal relaxation and approaches a thermal equilibrium at a higher temperature reflected by the spectral change ΔS. This vibrational relaxation process is very similar to what was proposed to explain vibrational relaxation of bulk water molecules,[34] and seems to be characteristic of molecules in a hydrogen-bonding network.

Bonn and coworkers extended the measurements to the air/water interface with more different IR pump frequencies over the bonded OH stretch range, and provided a 2D map on SF signal as a function of pump and probe IR frequencies that gave a clearer picture on correlation between the pumped and probed modes.[35] Borguet and coworkers carried out a series of ultrafast dynamics measurements on OH stretches at silica/water and Al_2O_3/water interfaces with different pHs and salt concentrations in water, and ion adsorptions at interfaces.[36−38] Tahara and coworkers applied phase-sensitive TR-SFVS to probe vibrational relaxation of OH stretches at a water interface underneath a charged surfactant monolayer.[39] The observed change of the Im $\chi_S^{(2)}$ spectrum with time, displayed in Fig. 11.13, provided more direct information on the vibrational relaxation. In all cases, the results could be explained by the relaxation process described above although the values of T_ν and T_t changed. The differences were believed to be due to different interfacial water structures.

The interpretations of the above-mentioned results in the literature however had some difficulties. Ultrafast vibrational relaxation dynamics of water interfaces clearly belong to case III b or c described earlier. It was usually assumed that only the interfacial bonded OHs were excited and the excitation energy was transferred and relaxed away, but excitation of the bulk within the pump absorption length (~ 1 μm) was not considered. Because the OH stretch bands of water surface and bulk strongly overlap, vibrational excitations of the two are expected to be strongly coupled by FRET. Energy transfer from bulk OH to surface OH and vice versa ought to have significant effect on relaxation of the excited surface OH. So far, this has been ignored in the interpretation and also in the MD simulations attempting to explain experimental results. It would be interesting to know how different interfacial water structures have different relaxation times despite their excitation being strongly coupled with bulk excitation. Another problem is that the structure of the interfacial water layer probed by SFVS was often not clearly specified. For example, at a charged water interface, the SFVS probe signal comes from both the bonded interface layer (BIL) and the electric-double layer (EDL); the observed relaxation dynamics has contributions from both layers and the structure of EDL is bulk-like. Indeed, Borguet's group found that the vibrational relaxation of bonded OH at a neutral water/silica interface (pH 2) was two times slower than that of the negatively charged interface at pH 6;[36] EDL is absent in the former case but present in the latter. It would be nice if the TR-SFVS measurement could separately monitor the relaxation dynamics of BL and EDL. Ultrafast dynamics of water interfaces under various circumstances is obviously interesting, but both theory and experiment clearly need further development. This comment applies in general to cases where excitations of both surface and bulk are unavoidable. Finally, to have a thorough understanding of vibrational relaxation, we need to be able to follow the relaxation pathway, i.e., to know how the excitation relaxes to lower states in time. In principle, this can be achieved by probing time-varying populations in lower states, but the experiment is very challenging for lack of low-frequency coherent light sources.

In addition to the lifetime T_ν, vibrational relaxation is also characterized by the coherent dephasing time T_2, the inverse of which is the homogeneous linewidth of the excited vibrational mode. For surface vibrational excitation, T_2 can be measured by IR photon echoes probed by surface-specific TR-SFVS, as first demonstrated by Guyot-Sionnest on the H-Si stretch of H/Si(111).[40] Measurement was also performed on bonded OH stretches at a water/silica interface; the observed $T_2 < 100$ fs was consistent with the spectral hole bandwidth detection.[33]

11.3. SHG/SFG Microscopy

Optical microscopy has always been an essential tool for material and biological studies. Advances of nonlinear optics have led to the invention of a host of different microscopic techniques; each has its own unique features that can provide useful information in microscopic studies. Two-photon-fluorescence microscopy (TPFM) in particular has radically transformed bioscience in recent years. Stimulated emission depletion microscopy and coherent stimulated antiStokes Raman microscopy have also become popular for molecular biology. SHG/SFG, being the earliest discovered nonlinear optical process, naturally was first developed as a nonlinear optical microscopic tool. Hellwarth and Christensen used SH microscopy (SHM) to map out grain boundaries and defects in polycrystalline thin films of III–V and II–VI semiconductors in early 1975.[41]

SHG microscopy is based on the expectation that different parts of a heterogeneous material have different characteristic SH responses. As discussed in Chapter 2, SHG is electric-dipole allowed in media without inversion symmetry and is surface-specific in media with inversion symmetry. As a coherent process, it is highly directional and its phase and polarization dependence provides information on absolute orientation of molecules or microstructures in a medium. Structural labeling is not required. Moreover, SHG has monolayer sensitivity. Following Eq. (2.20) for estimation of SHG signal strength in Sec. 2.5, we find that for a 1-W, 100-fs, 100-MHz pulsed input at 800 nm focused to a spot of 100 μm in diameter on a surface monolayer with $\chi^{(2)} \sim 10^{-15}$ esu, the SHG output from the monolayer is 3×10^6 photons/sec, or 300 photons/sec/pixel if the 100-μm spot is partitioned into 10^4 pixels. Microscopic images with such signal strength should be easily detectable by a CCD camera. SHG from thin films or bulks of non-centrosymmetric media is expected to be even orders of magnitude stronger. As a microscopic technique, SHM has the advantages of being non-invasive, label-free, and capable of providing information on polar orientation of molecules and microstructures; it also has low noise background and high signal strength, showing great potential for surface microscopic studies.

The experimental arrangement of SHM is not much more complicated than linear optical microscopy, both requiring only a single input beam, although in SHM two separate input beams at the same frequency can also be used (Fig. 11.14). In the simple one input beam case, the conventional optical microscope can be

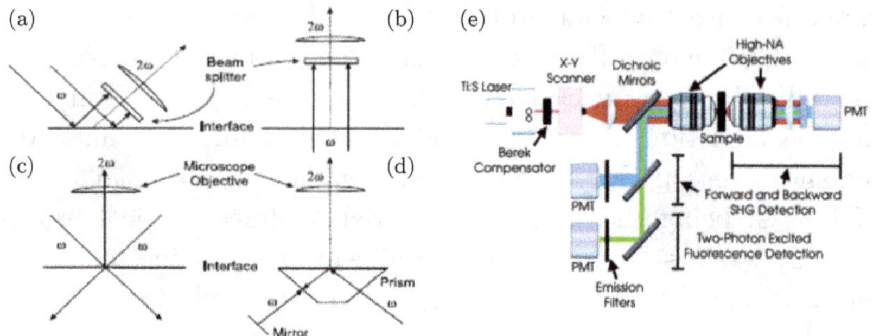

Fig. 11.14. (a)–(d) Schematic representations of four different beam geometries that can be adopted for SH microscopy (after Ref. [42]). (e) A practical SH microscope setup that allows transmission and reflection SH microscopy as well as two-photon fluorescence microscopy (after Ref. [43]).

easily adapted for SHM. Shown in Fig. 11.14(a)–11.14(d) are four possible beam geometries for transmission/reflection SHM described by Florsheimer.[42] The transmission geometry generally has much higher signal strength if the sample has no inversion symmetry and its thickness is comparable to the coherent length (inverse of phase mismatch) for transmitted SHG. Two schemes can be used to record SHM images, a wide-field no-scan scheme with a 2D CCD camera for detection and a focused point-by-point scanning scheme. A practical setup of the latter scheme that allows simultaneous recording of SHM and TPFM images of a biological sample constructed by Dombeck is depicted in Fig. 11.14(e).[39] Spectral filters can be used to effectively suppress the background. SHM and TPFM are complementary to each other. Both have the merits of low scattering background from microstructures, high optical damage threshold, and large penetration depth on a sample that renders 3D imaging of samples possible, but SHM relies on structure without inversion symmetry and TPFM does not. SHM has the additional advantages of requiring no exogenous labeling and being able to provide orientation information of microstructures.

SHM has been applied to studies of domain structures of ferroelectric thin films, orientations of micro-crystals, and others.[44] It is however in bio-imaging that SHM has found most applications, targeting particularly non-centrosymmetric biological materials like

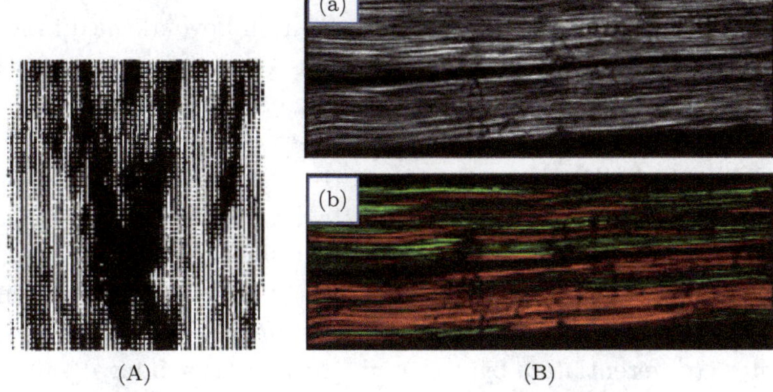

Fig. 11.15. (A) SH microscopic images of a three-month-old rat-tail tendon taken with the wide-field scheme using beam geometry of Fig. 11.14(b). Collagen fibrils appear as long aligned structures along the tendon axis. The image is negative, i.e., dark areas are regions of strong SH generation, and the pixel size is 50 μm^2 (after Ref. [45]). (B) Regular (top frame, (a) and phase-sensitive (bottom frame), (b) SH microscopic images ($500 \times 150\mu m^2$) of an 8-week mouse tail tendon taken with the point-by-point scanning scheme using the transmitted beam geometry. Red and green colors indicate a nearly 180° phase difference between fibrils oppositely polar-oriented (after Ref. [46]).

collagens, myofibrils, myosin filaments, and microtubules. Freund and Deutsch first reported SHM images of a rat-tail tendon[45] (Fig. 11.15(A)). The well-aligned collagen fibrils along the tendon axis are clearly seen in the image taken with P-in and P-out polarization combination along the axis. Since then, SHM has been widely adopted to study structures of all kinds of tissues, image cells in vitro and in vivo, still and live identify and monitor tumors/cancers, control quality in tissue engineering, probe cellulose structures of plant cells, and so on. With advances in optics and computer technology, the technique has become much more viable over the years. On the one hand, the time required for recording an image is greatly reduced. On the other hand, image quality is greatly improved through computer-aided data analysis. Two technical developments directly related to optics of SHM should be mentioned. One is the development of phase-sensitive SHM[46] (Sec. 3.5). It can provide phase-contrasted images that not only have better spatial resolution, but also allow determination of absolute orientation of microstructures in images. The other is the development of near-field SHM that improves the spatial resolution beyond the diffraction limit.[47] Details of advances on SHM can be found in a number of recent review articles listed at the end of the chapter. Here, we describe a few selected representative cases on applications of SHM.

Figure 11.15(B) shows the comparison between two images of a mouse tail tendon taken by ordinary SHM (top frame) and phase-sensitive SHM (bottom frame). From the latter, it is seen that the collagen fibrils are not all oriented in the same way; similarly oriented fibrils bunch together, but neighboring bunches can be oppositely oriented (differentiated by red and green in the image), and the orientation also varies along single fibrils as evidenced from the SH intensity variation.[46]

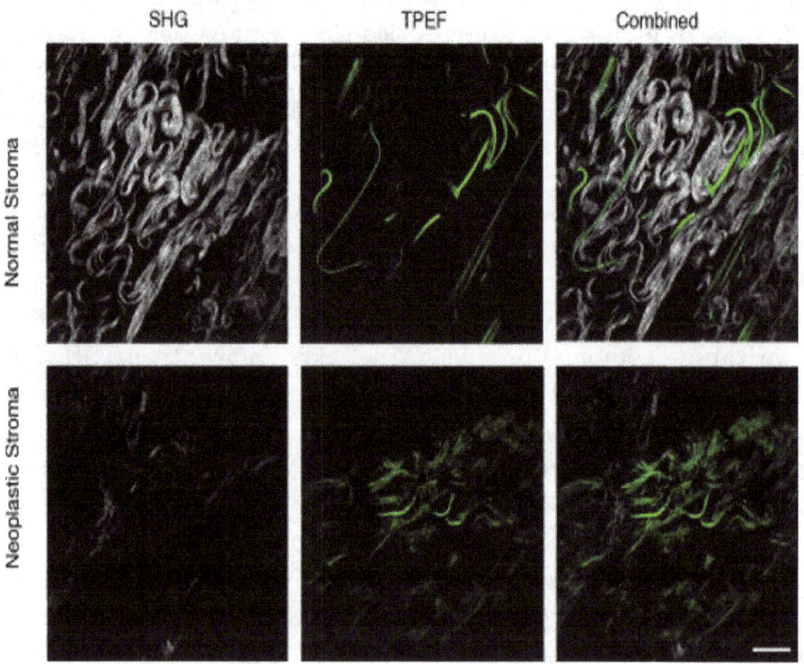

Fig. 11.16. Microscopic images of human esophageal stroma. From left to right, the images are from SHM, TPFM, and combined SHM and TPFM, respectively, taken with excitation wavelength at 850 nm. The scale bar on the lower right image corresponds to 20 μm (after Ref. [48]).

Tumors and their development stages can be identified by their collagen structures and densities, and SHM is effective in providing histology of biopsies. An example is described in Fig. 11.16, where the microscopic images of a normal and a neoplastic human esophageal stroma are presented: from left to right, the images were from SHM, TPFM, and combined SHM and TPFM, respectively.[48] As complementary techniques, SHM images are sensitive to collagens and TPFM images to elastins. It is seen that the normal stroma has a denser and more organized collagen matrix, longer and more distinct elastin fibers, and a lower proportion of elastin versus collagen fibers, whereas the neoplastic stroma loses fine organization structure, has shorter and more fragmental fibers, and a higher proportion of elastin versus collagen fibers.

Another example of SHM for cancer diagnosis is shown in Fig. 11.17 where the SHM images from normal ovary and cancerous ovary biopsies are compared.[49] Again, the normal and malignant ovaries have clear differences in their extracellular matrix structures. In this case, the malignant one appears to have more highly packed and bunched together collagen fibrils, resulting in a brighter SHM image.

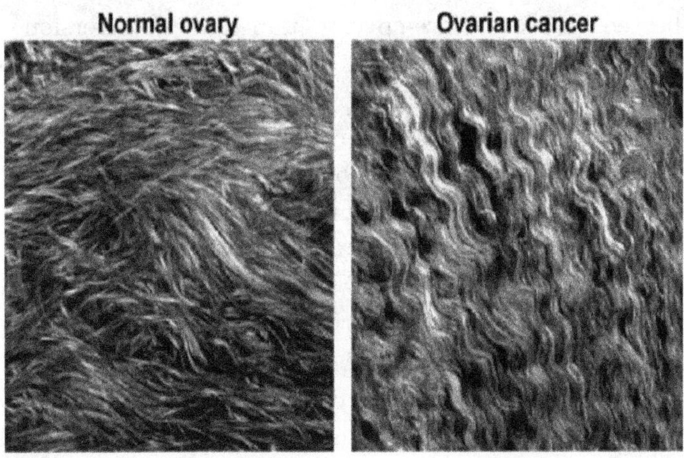

Fig. 11.17. SH microscopic images of sections (field size 170 μm) of collagen fibers in normal human ovary (left) and malignant ovary (right) taken with the point-by-point scanning scheme. Structural differences of the two are obvious (after Ref. [49]).

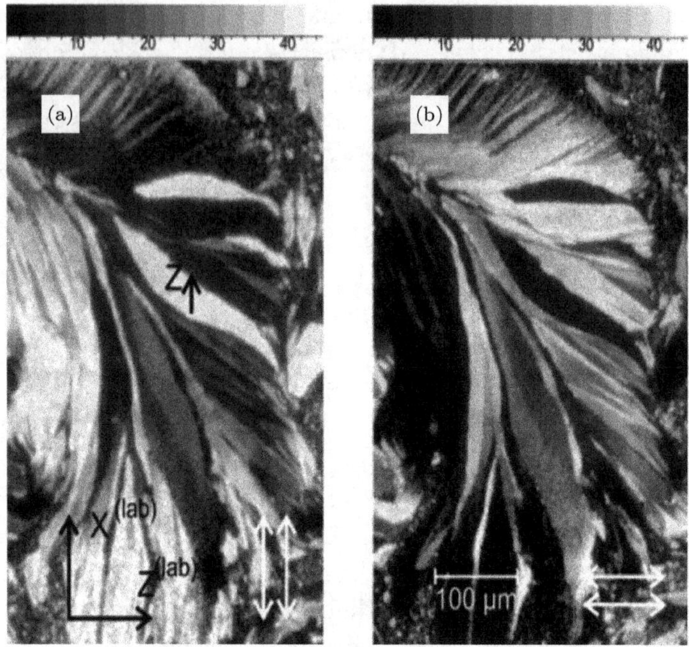

Fig. 11.18.   SH microscopic images of a DCANP Langmuir film on water taken with the beam geometry described in Fig. 11.14(b). Beam polarizations used are indicated by white arrows in the two images (after Ref. [42]).

Reflected SHG is surface-specific in media with inversion symmetry and has monolayer sensitivity. Thus, in reflection geometry, SHM can be used for surface microscopy. This was first demonstrated on a monolayer of rhodamine-6G dye molecules on silica.[50] Florsheimer later showed that surface SHM could be used to image a Langmuir monolayer of 2-docosylamino-5-nitropyridine (DCANP) on water remotely with the beam geometry of Fig. 11.13(b), as displayed in Fig. 11.18.[42] (The SH nonlinearity of DCANP originates mainly from the nitropyridine part.) This experiment set the stage for applications of SHM to probe bio-membranes, including molecular adsorption on and into membranes. In materials science, SHM can be used to probe surface microstructure and changes. Recently, its monolayer sensitivity has prompted its use to visualize domain structures of 2D materials.[51,52] Time-resolved SHM has also been developed to study ultrafast dynamics in 2D materials.[53]

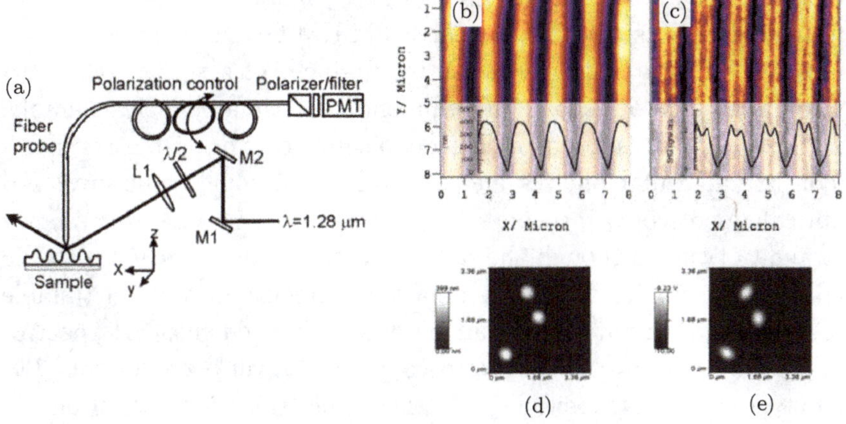

Fig. 11.19.   (a) Schematic setup of a near-field scanning SH microscope (after Ref. [50]). (b) Topographical and (c) SH images of a grating with a periodicity of 1.37 μm and a modulation depth of 400 nm inscribed on an azobenzene polymer thin film by holography through photoisomerization (after Ref. [54]). (d) Topographical and (e) SH images of NPP nanocrystals taken with the beam geometry of Fig. 11.14(d), but the SH output was collected by a fiber tip (after Ref. [55]).

Near-field SHM was first employed by Smolyaninov *et al.* to probe domain structures of a ferroelectric thin film and a poled $BiTiO_3$ crystal with a spatial resolution of 80 nm.[47] A typical setup for near-field SHM is sketched in Fig. 11.19(a). The input beam is obliquely incident on a sample and the SH signal is collected by a tapered fiber tip that can also be used as the scanning tip for atomic force microscopy (AFM). With proper arrangement, scanning of the fiber tip generates not only an SHM image but also a topography image of a surface. Two examples are presented in Figs. 11.19(b) and 11.19(c): Left and right frames of Fig. 11.19(b) describe the topographical image and the SHM image, respectively, of a grating inscribed on an azobenzene polymer film by photo-isomerization.[54] Both images exhibit hills and valleys, but the SHM image taken with the beam geometry of Fig. 11.13(a) shows an additional dip at the tops of the hills. Figure 11.19(c) describes the AFM image and the SHM image (taken with the beam geometry of Fig. 11.13(d)) of a set of N-(4-nitrophenyl)-L-prolinol (NPP) nanocrystals; the two images look identical.[55] Study of nanostructure dynamics by time-dependent near-field SFM is possible.[56] Near-field SHM for bio-imaging has not received much attention yet; the requirement of a tip to collect SH signal may be too much an impediment to image a biological sample.

Spectral information on microscopic images is desirable as it helps identify molecular species and unravel microscopic structures. For optical microscopy, incorporation of this feature seems to be fairly straightforward although the data collection time is longer with the addition of optical frequency as a new dimension. With a tunable or broadband input, SHM can be easily changed into SH spectro-microscopy. However, SH spectroscopy has not yet been developed to cover vibrational transitions, and electronic transitions are often not capable of distinguishing different species. Sum frequency microscopy (SFM) with emphasis on vibrational resonances is clearly the alternative.

In principle, SFM is just an extension of SHG with two input beams at frequencies at ω_1 and ω_2. Any of the four beam geometries in Fig. 11.14 can be used in SFM with either the wide-field or the focused point-by-point scanning scheme. Transmission SFM is preferred if the sample has no inversion symmetry and the thickness is comparable to the coherent length. The wide-field scheme for SFM actually has complications due to tilting of the SF output wavefront along the incidence plane if the inputs are obliquely incident on the sample surface, but the tilt can be compensated by inserting a proper reflection grating in the SF beam path, as sketched in Fig. 11.20(a):[57] If the exit angle of SFG is $\sim 60°$ and the grating is blazed at $30°$, then after a 1:1 imaging of the surface plane to the grating, the reflected SF field from the grating has the corrected wavefront perpendicular to the propagating direction and the image can be properly recorded by a CCD camera. To acquire spectral information, the IR input needs to be scanned over vibrational resonances of interest.

The confocal point-by-point scanning scheme is essentially the same for SHM and SFM. A typical setup with collinear input beams is depicted in Fig. 11.20(b),[58] which is actually a system that allows phase-sensitive SFM with the presence of a quartz plate and a wedged ZnSe window in the beam path before the sample, quartz as a

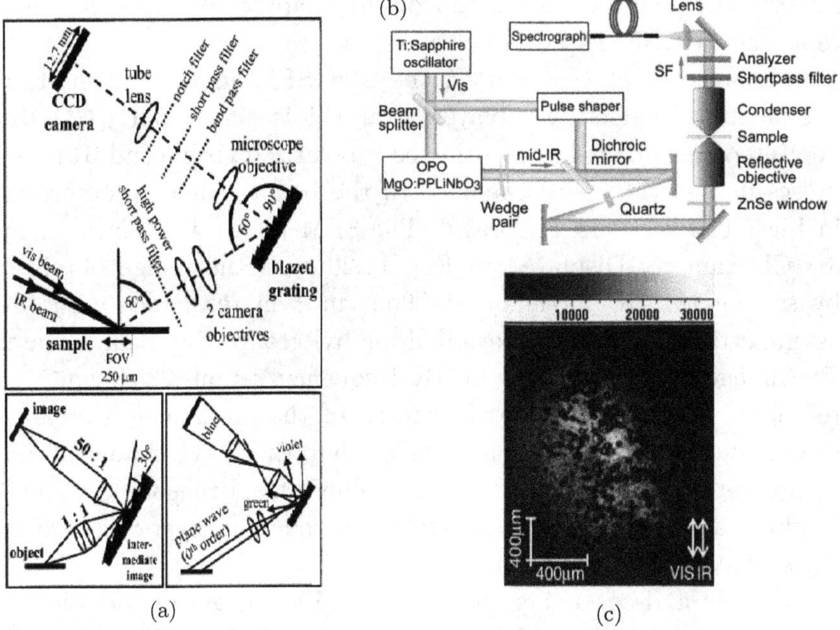

Fig. 11.20. (a) Sketch of beam geometry for a wide-field SF microscope showing that a reflection grating is needed to correct the SF wavefront tilt from the sample surface. The bottom frames show more details on the wavefront correction scheme (after Ref. [57]). (b) Schematic of a phase-sensitive SFVS microscopy setup. Quartz and ZnSe serve as the local oscillator and time delay element for phase interference measurement, respectively; without them, the setup is a common SF intensity microscope. (c) SF microscopic image of a Langmuir–Blodgett monolayer of arachidic acid deposited on the prism surface taken with the beam geometry of Fig. 11.14(d) using the wild-field scheme and having the IR frequency set at 2962 cm^{-1} (after Ref. [59]).

local oscillator for interference and ZnSe for variation of time delay between signal and reference (from the local oscillator) pulses (see Sec. 3.5). In the point-by-point scanning scheme, frequency scanning is not needed if IR input pulses have a bandwidth covering the vibrational resonances of interest. It is possible to use a combined scheme with wide-field irradiation of the sample by broadband IR input pulses and focused point-by-point scanning of narrowband visible input pulses on the IR-irradiated area of the sample. Details of experimental setups of various schemes can be found in the review articles listed at the end of the chapter. We present a few representative SFM studies below.

Florsheimer *et al.* reported the first SFM images taken from a Langmuir–Blodgett monolayer of arachidic acid ($C_{19}H_{39}COOH$) deposited on a prism surface using ps P-polarized visible and IR input pulses in the wide-field scheme with the beam geometry described in Figs. 11.14(d) and 11.20(a).[59] The SFM image was captured by a CCD camera. Displayed in Fig. 11.20(c) is the image obtained by setting the IR frequency at 2962 cm^{-1} in resonance with the asymmetric stretch of the terminal methyl group of the alkyl chain. The image obtained with the IR frequency set at 2850 cm^{-1} in resonance with the symmetric stretch of the methylene groups of the chain was very weak, indicating that the alkyl chains of the monolayer had little gauche defects. The inhomogeneous dark and bright image in Fig. 11.20(b) reveals that the monolayer coverage on the surface was not uniform.

Baldelli and coworkers recorded the SFM images of corrosion of an Au film on a silicon wafer by CN$^-$ in a 0.5M NaCN solution.[60] They took essentially the same experimental setup of the wide-field scheme adopted by Florsheimer *et al.*, but used the reflection beam geometry of Fig. 11.14(a). Two sets of observed images of the Au/water interface with the IR frequency set at (i) 2105 and (ii) 2225 cm^{-1} are given in Fig. 11.21 with (a) for Au at the early stage of reaction with CN- and (b) for Au having immersed in the solution for 8 hours. The vibrational spectra taken at regions marked by A and D in Figs. 11.21(a) and 11.21(b) are displayed in Figs. 11.21(c) and 11.21(d), respectively. Four peaks at 2105, 2140, 2170, and 2225 cm^{-1} were used to fit the spectra; the 2105 cm^{-1} peak was attributed to the stretch mode of CN attached to an Au atom and the other three peaks to Au-CN complexes produced in the reaction. The images are brighter for the second set presumably because roughening of the Au surface led to stronger overall SF output by local field enhancement.

Ge and coworkers first developed phase-sensitive SFM and applied it to imaging of a rat-tail tendon.[58] They used the point-by-point scanning scheme with a ps cw mode-locked laser and associated optical parametric oscillator system in an experimental arrangement

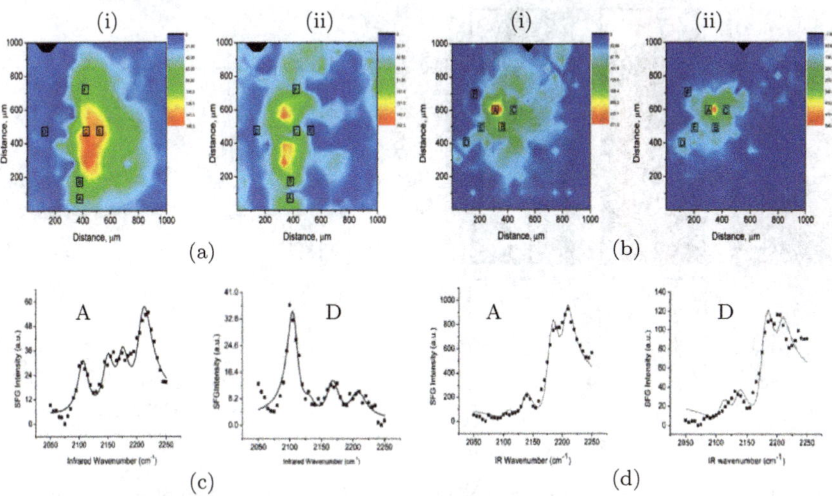

Fig. 11.21.   SM microscopic images taken from an interface of Au/0.5M NaCN solution using the wide-field scheme and the beam geometry described in Fig. 11.14(a): (a) images taken at the early stage of Au-CN reaction, and (b) images taken after Au was immersed in the solution for 8 hours. In both (a) and (b), the images were taken with the IR frequency set at 2105 cm^{-1} in (i) and at 2225 cm^{-1} in (ii). SF spectra obtained in regions marked by A and D on the images of (a) are shown in (c) and those marked by A and D on the images of (b) are shown in (d). The solid lines are fits to the spectra assuming four discrete peaks at 2105, 2140, 2170, and 2225 cm^{-1} (after Ref. [60]).

similar to the one described in Fig. 11.20(b). The scheme yields not only SFM intensity images, but also SFM amplitude and phase images, as exemplified by the images of a rat-tail tendon on a z-cut quartz plate in Figs. 11.22(Aa)–11.22(Ac) taken by setting the IR input at 2950 cm^{-1} in resonance with a stretch mode of the CH$_2$ groups on the collagen fibrils. The observed SF vibrational spectra of the tendon in the CH stretch range with the incident plane along the tendon are given in Fig. 11.22(B). The SFM images were clear when all beams were P-polarized along the tendon, but weak if perpendicular to the tendon, indicating that the CH$_2$ groups of collagen fibers had a net polar orientation along the fibers. The amplitude and phase images in Figs. 11.22(Ab) and 11.22(Ac) are clearly better resolved; the latter particularly shows that neighboring strands can have a 180° phase difference revealing that they were

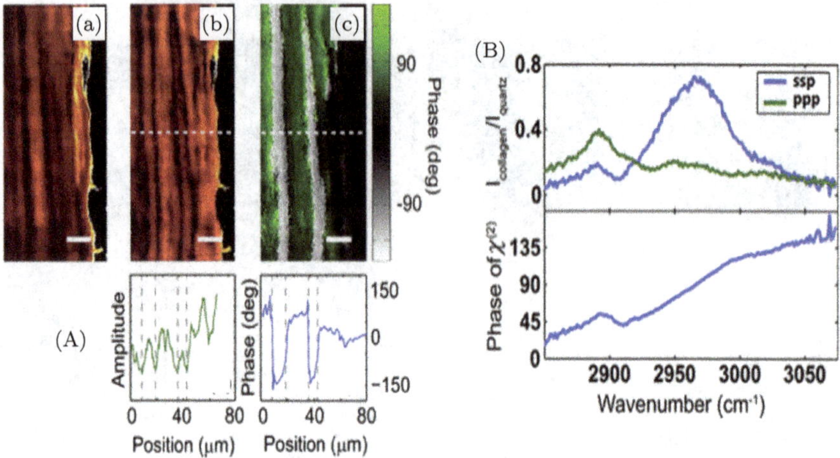

Fig. 11.22.   (A) Microscopic images of a rat-tail tendon on top of a z-cut quartz: (a) SF intensity image, (b) SF amplitude image, and (c) SF phase image taken with the point-by-point scanning scheme and the IR frequency set at 2950 cm^{-1}. The scale bar is 20 μm. The amplitude and phase variations with position along the horizontal dashed lines in (b) and (c) are shown underneath (b) and (c). (B) Normalized SSP and PPP SF intensity spectra and the SSP phase spectrum of the tendon in the CH stretch range taken with the incident plane along the tendon. (After Ref. [58]).

oppositely polar oriented, but along each strand, the phase change is small. Generally, a polar orientation map of molecular groups identified in an SFM image can be deduced from the polarization dependence of the image (Sec. 6.3).

Huang *et al.* studied the microstructure of cellulose microfibril assembly in plant cell walls.[61] They used the point-by- point scanning scheme with a laser and optical parametric system that generated fs broadband IR pulses and ps narrowband visible pulses as inputs for SFM (Sec. 3.3). They showed that SFM could resolve different tissue regions, namely, interfascicular fiber, xylem, phloem, cortex, and epidermis, of the 6- and 8-weeks old influorescence stems from their spectra in the CH and OH stretch range, as described in Fig. 11.23. The results for the two samples are obviously different.

Since SFM requires scanning of either frequency or position, time needed to take an image is always a concern. Following the information retrieval idea, Baldelli's group recently devised a

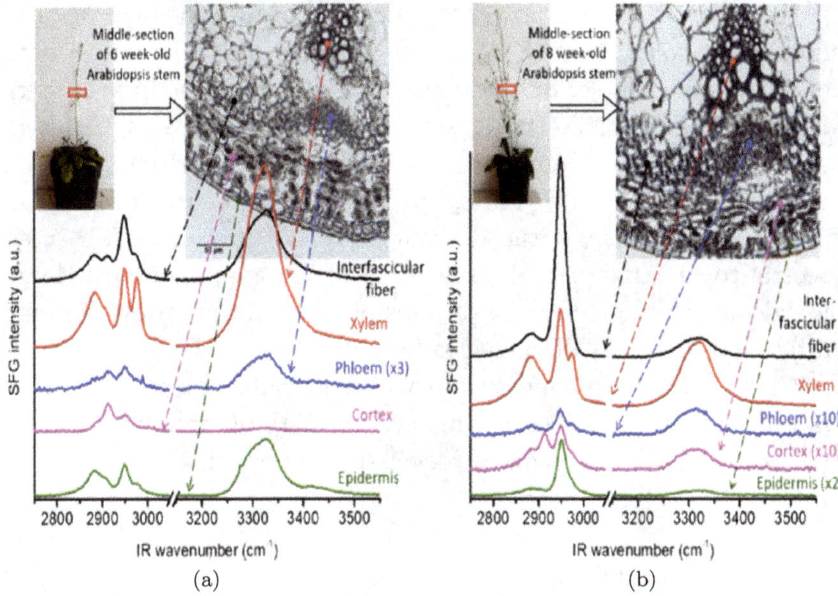

Fig. 11.23. SF vibrational spectra in the CH and OH stretch range taken by SFM from five different tissue regions of (a) 6-weeks old and (b) 8-weeks old influorescence stems: interfascicular fiber, xylem, phloem, cortex, and epidermis. The photos at the left top corner of (a) and (b) show where the stem samples came from, and the cross-sectional images were obtained from the samples by linear optical microscopy (after Ref. [61]).

compressive sensing scheme that scans only a fraction of the whole set of pixels randomly without losing much of the information.[62] It saves the collection time for an image by orders of magnitude. Moreover, they used machine learning/neural network algorithm to help identify molecules and microstructures and greatly improved the image quality.[63] Ultrafast SFM can be readily implemented in the experimental setup of Fig. 11.14(e) or Fig. 11.20(b) with fs broadband IR and narrowband visible input pulses by simply adding an fs excitation pulse on the sample ahead of the SF probe pulses, and was recently applied to study H-bonding dynamics at water interfaces with biomimetic self-assembled lattice materials.[64] Near-field SFM has been demonstrated with essentially the same experimental arrangement as near-field SHM, although it may not be suitable for biological imaging.[55]

There are still challenges for further development of SFM. The information content of SFM images is huge; accordingly, the data analysis is not simple. Spectral range for SFM is currently limited. Most studies, especially those using cw mode-locked lasers with high pulse repetition rate, are restricted to the 2700–4000 cm^{-1} region because of insufficient IR energy per pulse at lower frequencies. On applications to biology, the difficulty also lies in relating observed spectra to the microscopic structure of a sample because large biological molecules often have multiple subunits with overlapping vibrational modes as discussed in Sec. 10.2.

Chiral SF microscopy for surface and bulk imaging of chiral materials was described earlier in Sec. 5.8. With proper input/output polarizations, all the technical discussions above in this section on SFM apply to chiral SFM.

11.4. Device Probing

SH/SF microscopy can be used for *in situ* probing of device structure and functionality even during operation. It has been well demonstrated by Iwamoto's group that SHM can map out the electric field, and hence the charge, distribution in an organic electronic or optoelectronic device. The idea is based on the electric-field induced second harmonic (EFISH) process that can occur in a material with inversion symmetry (or induced change of SHG in materials without inversion symmetry); the SH output field (or change of the SH field) is directly proportional to the electric field in the material:

$$\vec{E}(2\omega) \propto \vec{P}^{(3)}(2\omega) = \overset{\leftrightarrow}{\chi}^{(3)} : \vec{E}_{dc}\vec{E}(\omega)\vec{E}(\omega)$$

where $\vec{E}_{dc}$ is the electric field of interest, and $\overset{\leftrightarrow}{\chi}^{(3)}$ is the third-order nonlinear susceptibility of the material. Assuming $\vec{E}_{dc}$ is constant (or its variation with position is known) in a focus spot in a material with known $\overset{\leftrightarrow}{\chi}^{(3)}$, measurement of SHG allows direct deduction of $\vec{E}_{dc}$ in the spot. SHM imaging can then provide a distribution map of $\vec{E}_{dc}$ in the material. With the field distribution $\vec{E}_{dc}(\vec{r})$ known, the charge distribution can be found from the Poisson equation. In the transient dynamic case with charges flowing in response to applied voltage, the charge and field distributions can be obtained by solving the coupled Poisson equation and the drift-diffusion equation of charged carriers with proper boundary conditions. Compared to inorganic metals or semiconductors, organic materials have higher resistivity, lower charge density, and accordingly larger $\vec{E}_{dc}(\overleftarrow{r})$ that can be easily measured. (We note that current-induced SHG discussed in Sec. 4.3 is supposedly negligible compared to EFISH in organic materials.) We should note that linear optical measurements can also measure electric field distribution through field-induced refractive index and absorption spectrum change, but unlike EFISH, they always have to fight against a strong background, and are particularly inconvenient to probe buried structure.

To illustrate device probing by EFISH, Fig. 11.24 describes an experiment by Iwamoto's group on SHM imaging of the time-varying field distribution in an organic field effect transistor (OFET) following the application of a voltage pulse across the source and drain electrodes.[65] The OFET was composed of a 100-nm centrosymmetric pentacene $(C_{22}H_{14})$ film, which is a hole transport material, deposited on a poly-methyl-methacrylate (PMMA)/SiO_2-coated

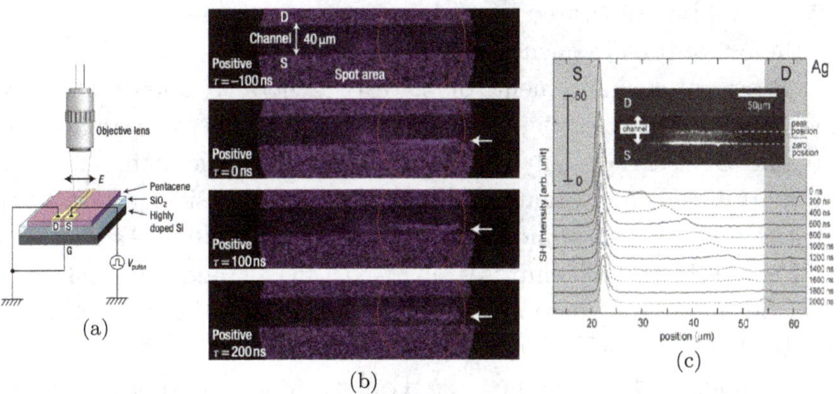

Fig. 11.24. (a) Second harmonic microscopy arrangement for EFISH to probe electric field distribution in an OFET device. (b) SH microscopic images in the OFET channel at a few pump–probe time delays (after Ref. [65]). (c) SH intensity variations across the OFET channel at various pump–probe time delays showing propagation of a field front from the source to the drain electrode after switch-on of a positive voltage at the source electrode (after Ref. [66]).

heavily n-doped Si wafer as the gate electrode; the source and drain electrodes on the pentacene layer were made of Au forming a long channel with a width of few tens of μm between them (Fig. 11.24(a)). When a voltage between the source and drain was switched on at t = 0, holes were injected from Au into pentacene, and the voltage front traveled across the channel from the source to the drain. Solution of the coupled Poisson and drift-diffusion equations yielded the associated electric field distribution that varied with time in the form of a propagating field pulse following the voltage front.[66] A short laser pulse impinging on the OFET channel could image the field distribution at selected time t. As displayed in Fig. 11.24(b), the SH image at a given t showed a narrowband parallel to the channel; as t increased, the band started from the source end and moved toward the drain. Matching of the observed field distribution with the calculation could yield the carrier mobility and carrier distribution in the channel. It was also possible for EFISH to measure the electric field distributions in both layers of a double-layer OFET, namely, a poly-3-hexylthiophene (P3HT)/pentacene film instead of a single pentacene film in the above example.[67] Because P3HT and pentacene have different visible absorption spectra, SH signals from the two layers could be separated by their respective resonant enhancements at the SH output frequency.

Resonant enhancement of EFISH generally allows selective measurement of electric fields in different layers of a multilayer organic device, not only OFET, but also organic light emission diode (OLED), and organic solar cell (OSC). Consider a four-layer OLED studied by Iwamoto's group as an example.[68] As sketched in Fig. 11.25(A), it comprises an α-NPD layer and an Alq3 layer that form a p-n junction sandwiched between the ITO and Al electrodes. Here, α-NPD is for N,N′-di-[(1-naphthyl)-N,N′-diphenyl]-(1,1′-biphenyl)-4,4′-diamine ($\underline{C}_{44}\underline{H}_{32}N_2$), Alq3 for tris(8-hydroxy-quinolinato) aluminum(III)($C_{27}H_{18}AlN_3O_3$), and ITO for indium tin oxide. The chemical structures of α-NPD and Alq3 and the equivalent circuit of the device are also described in Fig. 11.25(A). With the voltage V_{ex} switched on, holes and electrons were injected into α-NPD and Alq3 layers, respectively. They met at the interface

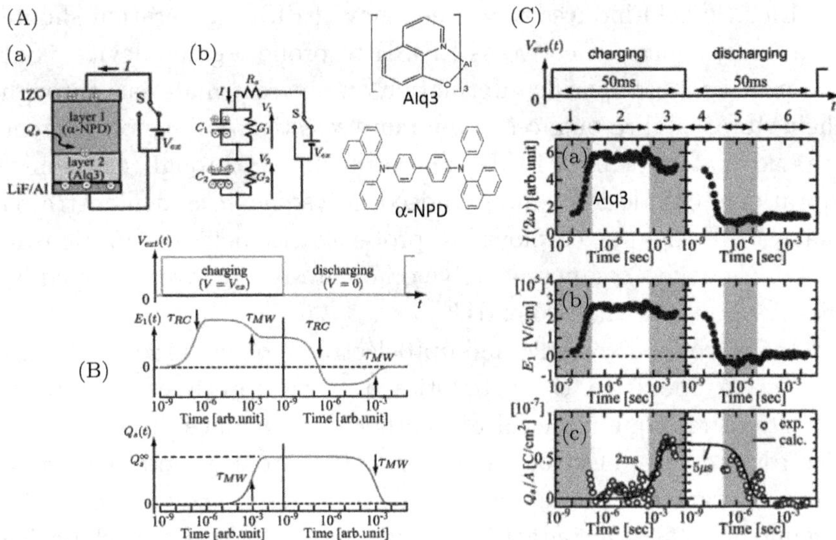

Fig. 11.25. (A) Structure of a double-layer OLED device composed of Alq3(layer 1) and α-NPD(layer 2) in (a) and its equivalent circuit in (b). The chemical formulae of Alq3 and α-NPD are given on the right. (B) Transient electric field in layer 1 and charges accumulated at the two electrode interfaces in response to switching on and off the square voltage obtained from a model calculation using the equivalent circuit in (a–b). (C) Experimental result on transient SHG induced by the dc electric field in the α-NPD layer (top frames, a), the deduced electric field in the α-NPD layer (middle frames, b), and the deduced charge density at the α-NPD/Alq3 interface (bottom frames, c) during charging and discharging of the device. Solid lines in the bottom frames are theoretical fits with $\tau_{\mathrm{MW}} = 2$ ms (after Ref. [68]).

between α-NPD and Alq3 and emitted light when recombined, and the interface became charged (the Maxwell–Wigner effect). From the equivalent circuit, the charge at the interface, $Q_s(t)$, and the fields in the two layers, $E_1(t)$ and $E_2(t)$, during charging and discharging, could be found, as described in Fig. 11.25(B) ($E_2(t)$ not shown), where $\tau_{\mathrm{RC}} \equiv RC_1C_2/(C_1 + C_2)$ and $\tau_{\mathrm{MW}} \equiv (C_1 + C_2)/(G_1 + G_2)$. Experimentally, Taguchi *et al.* used resonantly enhanced EFISH to probe the OLED. As seen in Fig. 11.25(C), $E_1(t)$ and $Q_s(t)$ deduced from the measured SH output intensity $I(2\omega,t)$ agreed more or less with the calculation in Fig. 11.25(B).

Electric-field-induced sum frequency (EFISF) generation should be at least equally effective as EFISH to probe organic devices, but can provide more specific information on different materials through their vibrational resonances. Even more selective probing of different organic materials by EFISF can be achieved through electronic-vibrational double resonances. Indeed, Miyamae *et al.* demonstrated that EFISF could be employed to probe electric field distributions in an OLED device composed of five different organic layers capped by two electrodes at the two ends.[69]

Interfaces of electronic and optoelectronic devices are important for device operation characteristics as they govern carrier transportation from one material to another. For instance, in an organic electronic device, electron or hole injection from a metal electrode into an organic material layer may depend strongly on how the organic molecules in contact with the metal are oriented. Such device interfaces are usually buried and difficult to access, especially during operation. Obviously, SF vibrational spectroscopy (SFVS) being surface-specific in media with inversion symmetry is uniquely suited for interrogation of device interfaces.

Perovskite solar cells are most promising next-generation solar cells because of their low cost, high efficiency, light weight, good flexibility, and other merits. They usually have a multilayer structure comprising perovskite as the active layer (absorbing light) sandwiched between electron and hole transport material layers and capped by suitable electrodes. The device performance, such as efficiency, stability, degradation, etc., depends critically on material structure, and interfacial structures between layers particularly may play a crucial role. For example, organic materials are often used as electron or hole transport layers; molecular orientation at an interface can strongly affect electron or hole transfer between perovskite and an adjacent organic layer. As we discussed in Sec. 6.3, SFVS provides an effective means to probe molecular structure of such buried interfaces. Polythiophene (PT) is one of the effective hole transport materials for organic devices. In a study of a halide perovskite photovoltaic cell composed of a three-layer structure of TiO_2/perovskite/PT (polythiophene) capped by ITO and Ag

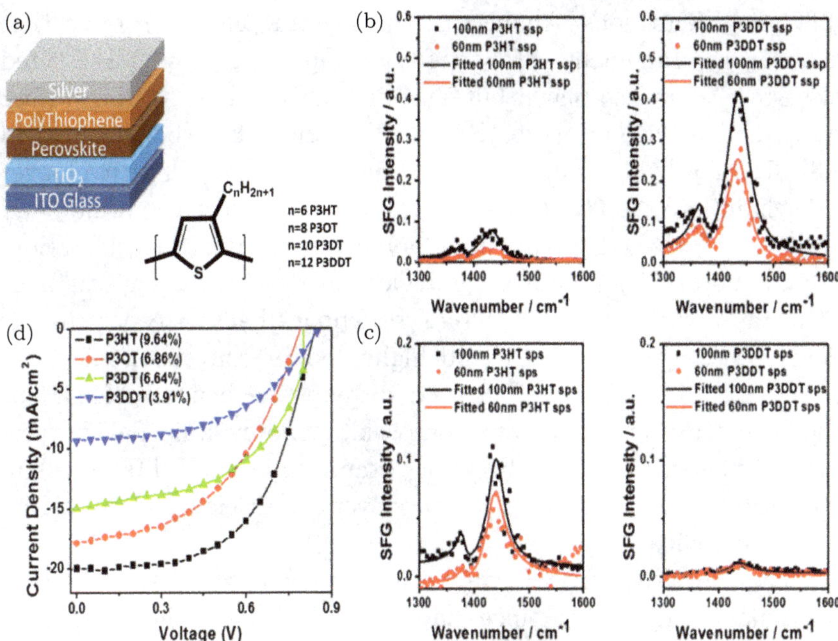

Fig. 11.26. (a) A perovskite solar cell comprising five layers with halide perovskite as the active layer, poly-thiophene (PT) and TiO$_2$ as the hole and electron transport layers, and silver and ITO as electrodes. The chemical formula of PT under study is also shown. (b) and (c) SF spectra of C=O stretches of P3HT and P3DDT obtained with SSP and SPS polarizations, respectively, from PT/perovskite interfaces with 60 and 100 nm layers of P3HT and P3DDT. Solid lines are theoretical fits. (d) Measured current density versus voltage generated upon constant illumination, in four perovskite solar cells with the same structural parameters but different PT materials. The measured power conversion efficiencies of the four cells are noted (after Ref. [70]).

electrodes with PT as the hole transport material, Z. Chen and coworkers showed from SFVS measurement that the orientation of PT molecules at the PT/perovskite interface could affect the photovoltaic efficiency example.[70] Figure 11.26(a) displays the C=C stretch spectra of P3HT (poly-3-hexylthiophene, PT with a 6-carbon side chain) and P3DDT (PT with a 12-carbon side chain) obtained from the PT/perovskite interfaces with SSP and SPS polarizations. They exhibit very different polarization dependence, which is an indication that P3HT and P3DDT interfacial molecules had very different orientations. Analysis of the spectra (Sec. 6.3) found that the transition moment of C=O of the thiophene ring of P3HT tilted close to the surface plane, but that of P3DDT closer to the surface normal. They also measured P3OT (PT with an 8-carbon chain) and P3DT (PT with a 10-carbon chain), and found that PT with a longer side chain had the C=O tilted more toward the surface normal, and accordingly, the PT backbone adopted a more upright orientation. For the perovskite photovoltaic devices made of these four different PTs, those with more upright PT backbones had a steeper current density rise against voltage and higher power conversion efficiency (Fig. 11.26(d)). Zhang *et al.* used dithiophene benzene (DTB) as the hole transport layer of a perovskite photovoltaic device, presumably having the DTB backbone stand up at the DTB/peroskite interface;[71] they obtained a power conversion efficiency close to 20% from the device.

Time-resolved EFISH can be used to monitor charge transfer dynamics at an interface since charge transfer leads to the appearance of an electric field across the interface. Zhu and coworkers first applied the technique to the study of hot electron transfer from colloidal PbSe nanocrystals to a TiO_2 substrate.[72] In their experiment, an fs pump pulse shining on a $PbSe/TiO_2$ assembly excited electon–hole pairs in PbSe. Subsequently, the e–h pairs partly relaxed back in PbSe and partly underwent electron–hole separation with electrons moving over to TiO_2 and creating a dc field across the interface. The field variation with time could be monitored by EFISH with a time-delayed probe pulse. Their result is presented in Fig. 11.27. It is seen that the signal is weak with the sample at room temperature because of the fast relaxation of excitons within PbSe, but at 80K, the signal shows clear oscillation of the field reflecting the dynamics of electron–hole separation under the influence of the time-varying field created at the interface. Zhu and coworkers also studied ultrafast dynamics of charge transfer as well as formation and dissociation of interlayer excitons across a model photovoltaic junction formed by copper phthalocyanine (CuPc, donor) and fullerene (C60, acceptor).[73] From their time-resolved pump-probe EFISH measurement with the help of two-photon photoemission, they could monitor how the pumped

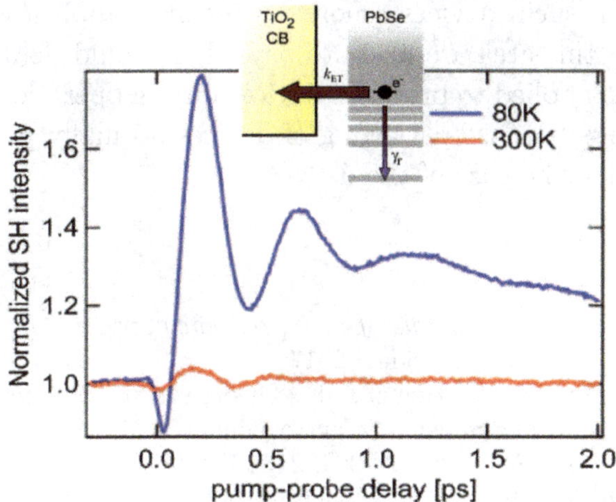

Fig. 11.27. Time-resolved SHG from a TiO_2 substrate coated with 1.5 monolayer of hydrazine-treated 3.3 nm PbSe nanocrystals excited by 50-fs pump pulses at 800 nm. Results at two different temperatures are presented. The initial rise of the SHG signal is due to electric field created at the interface by electron transfer from PbSe to TiO_2. The inset describes the electron relaxation and transfer processes. At 300K, the relaxation rate is significantly faster than the electron transfer rate, resulting in little electron transfer across the interface and hence a weak SHG signal. At 80K, the relaxation rate is slower than the transfer rate and the oscillation of the signal in time reflects the dynamics of electron–hole separation at the interface (after Ref. [72]).

excitons decayed and dissociated into electrons and holes that move oppositely away from the junction and create the E field at the junction. It was found that interlayer excitons formed at high energy levels would decay first into excitons at the lowest-energy level and then rapidly dissociated into electrons and holes moving away from the junction.

Time-resolved SF vibrational spectroscopy should be more informative on ultrafast interfacial dynamics of devices involving both electrons and phonons (or molecular vibrations), but it has not yet been well explored. As we discussed above, SHG/SFG can provide information on both bulk and interfaces of an organic/polymeric electronic or optoelectronic device that can be useful for improved design of the device. Since SHG/SFG is the only technique that can be employed to learn about microscopic structure about buried interfaces of such devices, more systematic applications of the technique in this area are likely to pay off. It would be more helpful if it could be applied to practical devices during operation. This may be a challenge because packaging of devices normally prevents light from accessing the core of the devices.

References

1. Shen, Y. R. *Fundamentals of Sum Frequency Spectroscopy*. Cambridge University Press, Cambridge, 2017.
2. Dadap, J. I.; Shan, J.; Heinz, T. F.: Theory of Optical Second Harmonic Generation from a Sphere of Centrosymmetric Material: Small-Particle Limit. *J. Opt. Soc. Am. B* 2004, 21, 1328–1347.
3. Pavlyukh, Y.; Hubner, W.: Nonlinear Mie Scattering from Spherical Particles. *Phys. Rev. B* 2004, 70, 245434.
4. de Beer, A. G. F.; Roke, S.; Dadap, J. I.: Theory of Optical Second Harmonic and Sum-Frequency Scattering from Arbitrarily Shaped Particles. *J. Opt. Soc. Am. B* 2011, 28, 1374–1384.
5. Gonella, G.; Dai, H. L.: Second Harmonic Light Scattering from the Surface of Colloidal Objects: Theory and Applications. *Langmuir* 2014, 30, 2588–2599.
6. Wang, H.; Yan, E. C. Y.; Borguet, E.; Eisenthal, K. B.: Second Harmonic Generation from the Surface of Centrosymmetric Particles in Bulk Solution. *Chem. Phys. Lett.* 1996, 259, 15–20.
7. Wang, H. F.; Yan, E. C. Y.; Liu, Y.; Eisenthal, K. B.: Energetics and Population of Molecules at Microscopic Liquid and Solid Surfaces. *J. Phys. Chem. B* 1998, 102, 4446–4450.
8. Yang, N.; Angerer, W. E.; Yodh, A. G.: Angle-Resolved Second Harmonic Light Scattering from Colloidal Particles. *Phys. Rev. Lett.* 2001, 87, 103902.
9. Gonella, G.; Dai, H. L.: Determination of Adsorption Geometry on Spherical Particles from Nonlinear Mie Theory Analysis of Surface Second Harmonic Generation. *Phys. Rev. B* 2011, 84, 121402(R).
10. Srivastava, A.; Eisenthal, K. B.: Kinetics of Molecular Transport across a Liposome Bilayer. *Chem. Phys. Lett.* 1998, 292, 345–351.
11. Yan, E. C. Y.; Eisenthal, K. B.: Effect of Cholesterol on Molecular Transport of Organic Cations across Liposome Bilayers Probed by Second Harmonic Generation. *Biophys. J.* 2000, 79, 898–903.
12. Liu, Y.; Yan, E. C. Y.; Eisenthal, K. B.: Effects of Bilayer Surface Charge Density on Molecular Adsorption and Transport across Liposome Bilayers. *Biophys. J.* 2001, 80, 1004–1012.

13. Pons, T.; Moreaux, L.; Mertz, J.: Photoinduced Flip-Flop of Amphiphilic Molecules in Lipid Bilayer Membranes. *Phys. Rev. Lett.* 2002, 89, 288104.

14. Zeng, J.; Eckenrode, H. M.; Dounce, S. M.; Dai, H. L.: Time-Resolved Molecular Transport across Living Cell Membranes. *Biophys. J.* 2013, 104, 139–145.

15. Gh, M. S.; Wilhelm, Dai, H. L.: Azithromycin-Induced Changes to Bacterial Membrane Properies Monitored in Vitro by Second Harmonic Light Scattering. *ACS Med. Chem. Lett.* 2018, 8, 569–574.

16. Wilhelm, M. J.; Sheffield, J. B.; Sharifian, G.; Wu, M.; Spahr, C.; Gonella, G.; Xu, B.; Dai, H. L.: Gram's Stain Does Not Cross the Bacterial Cytoplasmic Membrane. *ACS Chem. Bio.* 2015, 10, 1711–1717.

17. Roke, S.; Roeterdink, W. G.; Wijnhoven, J. E. G. J.; Petukhov, A. V.; Kleyn, A. W.; Bonn, M.: Vibrational Sum Frequency Scattering from a Submicron Suspension. *Phys. Rev. Lett.* 2003, 91, 258302.

18. Shank, C. V.; Yen, R.; Hirlimann, C.: Femtosecond-Time-Resolved Surface Structural Dynamics of Optically-Excited Silicon. *Phys. Rev. Lett.* 1983, 51, 900–902.

19. Sitzmann, E. V.; Eisenthal, K. B.: Picosecond Dynamics of a Chemical-Reaction at the Air-Water-Interface Studied by Surface Second Harmonic Generation. *J. Phys. Chem.* 1988, 92, 4579–4580.

20. Castro, A.; Sitzmann, E. V.; Zhang, D.; Eisenthal, K. B.: Rotational Relaxation at the Air-Water-Interface by Time-Resolved Second Harmonic-Generation. *J. Phys. Chem.* 1991, 95, 6752–6753.

21. Zimdars, D.; Dadap, J. I.; Eisenthal, K. B.; Heinz, T. F.: Anisotropic Orientational Motion of Molecular Adsorbates at the Air-Water Interface. *J. Phys. Chem. B* 1999, 103, 3425–3433.

22. Harris, A. L.; Levinos, N. J.: Vibrational-Energy Relaxation in a Molecular Monolayer at a Metal-Surface. *J. Chem. Phys.* 1989, 90, 3878–3879.

23. Harris, A. L.; Rothberg, L.; Dhar, L.; Levinos, N. J.; Dubois, L. H.: Vibrational-Energy Relaxation of a Polyatomic Adsorbate on a Metal-Surface — Methyl Thiolate (Ch_3S) on Ag(111). *J. Chem. Phys.* 1991, 94, 2438–2448.

24. Morin, M.; Levinos, N. J.; Harris, A. L.: Vibrational-Energy Transfer of Co/Cu(100) — Nonadiabatic Vibration Electron Coupling. *J. Chem. Phys.* 1992, 96, 3950–3956.

25. Schmidt, M. E.; Guyot-Sionnest, P.: Electrochemical Tuning of the Lifetime of the CO Stretching Vibration for CO/Pt(111). *J. Chem. Phys.* 1996, 104, 2438–2445.

26. Guyot-Sionnest, P.; Dumas, P.; Chabal, Y. J.: Lifetime of an Adsorbate-Substrate Vibration Measured by Sum Frequency Generation: H on Si(111). *J. Elec. Spec. Related Phenomena* 1990, 54–55, 27–38.

27. Morin, M.; Jacob, P.; Levinos, N. J.; Chabal, Y. J.; Harris, A. L.: Vibrational Energy Transfer on Hydrogen-Terminated Vicinal Si(111) Surfaces — Interadsorbate Energy Flow. *J. Chem. Phys.* 1992, 96, 6203–6212.

28. Wang, Z. H.; Carter, J. A.; Lagutchev, A.; Koh, Y. K.; Seong, N. H.; Cahill, D. G.; Dlott, D. D.: Ultrafast Flash Thermal Conductance of Molecular Chains. *Science* 2007, 317, 787–790.

29. Smits, M.; Ghosh, A.; Bredenbeck, J.; Yamamoto, S.; Muller, M.; Bonn, M.: Ultrafast Energy Flow in Model Biological Membranes. *New J. Phys.* 2007, 9, 390.

30. Bonn, M.; Hess, C.; Funk, S.; Miners, J. H.; Persson, B. N. J.; Wolf, M.; Ertl, G.: Femtosecond Surface Vibrational Spectroscopy of CO Adsorbed on Ru(001) During Desorption. *Phys. Rev. Lett.* 2000, 84, 4653–4656.

31. Hsieh, C.-S.; Campen, R. K.; Verde, A. C. V.; Bolhuis, P.; Nienhuys, H.-K.; Bonn, M.: Ultrafast Reorientation of Dangling OH Groups at the Air-Water Interface Using Femtosecond Vibrational Spectroscopy. *Phys. Rev. Lett.* 2011, 107, 116102.

32. Hsieh, C.-S.; Campen, R. K.; Okuno, M.; Backus, E. H. G.; Nagata, U.; Bonn, M.: Mechanism of Vibrational Energy Dissipation of Free OH Groups at the Air-Water Interface. *Proc. Nat. Acad. Sci.* 2013, 110, 18780–18785.

33. McGuire, J. A.; Shen, Y. R.: Ultrafast Vibrational Dynamics at Water Interfaces. *Science* 2006, 313, 1945–1948.

34. Lock, A. J.; Bakker, H. J.: Temperature Dependence of Vibrational Relaxation in Liquid H_2O. *J. Chem. Phys.* 2002, 117, 1708–1713.

35. Zhang, Z.; Piatkowski, L.; Bakker, H. J.; Bonn, M.: Ultrafast Vibrational Energy Transfer at the Water/Air Interface Revealed by Two-Dimensional Surface Vibrational Spectroscopy. *Nature Chem.* 2011, 3, 888–893.

36. Eftekhari-Bafrooei, A.; Borguet, E.: Effect of Electric Fields on the Ultrafast Vibrational Relaxation of Water at a Charged Solid-Liquid Interface as Probed by Vibrational Sum Frequency Generation. *J. Phys. Chem. Lett.* 2011, 2, 1353–1358.

37. Piontek, S. M.; Tuladhar, A.; Marshall, T.; Borguet, E.: Monovalent and Divalent Cations at the Alpha-Al_2O_3(0001)/Water Interface: How Cation Identity Affects Interfacial Ordering and Vibration Dynamics. *J. Phys. Chem. C* 2019, 123, 18315–18324.

38. Tuladhar, A.; Dewan, S.; Pezzotti, S.; Brigiano, F. S.; Creazzo, F.; Gaegeot, M. P.; Borguet, E.: Ions Tune Interfacial Water Stucture and Modulate Hydrophobic Interactions at Silica Surfaces. *J. Am. Chem. Soc.* 2020, 142, 6991–7000.

39. Singh, P. C.; Nihonyanagi, S.; Yamaguchi, S.; Tahara, T.: Ultrafast Vibrational Dynamics of Water at a Charged Interface Revealed by Two-Dimensional Heterodyne-Detected Vibrational Sum Frequency Generation. *J. Chem. Phys.* 2012, 137, 094706.

40. Guyot-Sionnest, P.: Coherent Processes at Surfaces — Free Induction Decay and Photon Echo of the Si-H Stretching Vibration for H/Si(111). *Phys. Rev. Lett.* 1991, 66, 1489–1492.

41. Hellwarth, R.; Christensen, P.: Nonlinear Optical Microscope Using Second Harmonic Generation. *Appl. Opt.* 1975, 14, 247–248.

42. Florsheimer, M.: Second Harmonic Microscopy — A New Tool for The Remote Sensing of Interfaces. *Phys. Status Solidi Appl. Res.* 1999, 173, 15–27.

43. Dombeck, D. A.; Kasischke, K. A.; Vishwasrao, H. D.; Ingelsson, M.; Hyman, B. T.; Webb, W.W.: Uniform Polarity Microtubule Assemblies Imaged in Native Brain Tissue by Second Harmonic Generation Microscopy. *Proc. Nat. Acad. Sci.* 2003, 100, 7081–7086.

44. Florsheimer, M.; Paschotta, R.; Kubitscheck, U.; Brillert, C.; Hofmann, D.; Heuer, L.; Schreiber, G.; Verbeek, C.; Sohler, W.; Fuchs, H.: Second Harmonic Imaging of Ferroelectric Domains in $LiNbO_3$ with Micron Resolution in Lateral and Axial Directions. *Appl. Phys. B* 1998, 67, 593–599.

45. Freund, I.; Deutsch, M.: Second Harmonic Microscopy of Biological Tissue. *Opt. Lett.* 1986, 11, 94–96.

46. Rivard, M.; Popov, K.; Couture, C. A.; Laliberte, M.; Bertrand-Grenier, A.; Martin, F.; Pepin, H.; Pfeffer, P.; Brown, C.; Ramunno, L.; Legare, F.: Imaging the Noncentrosymmetric Structural Organization of Tendon with Interfeometric Second Harmonic Generation Microscopy. *J. Biophotonics* 2014, 7, 638–646.

47. Smolyaninov, I. I.; Liang, H. Y.; Lee, C. H.; Davis, C. C.; Aggarwal, S.; Ramesh, R.: Near-Field Second Harmonic Microscopy of Thin Ferroelectric Films. *Opt. Lett.* 2000, 25, 835–837.

48. Zhuo, S.; Chen, J.; Xie, S.; Hong, Z.; Jiang, X.: Extracting Diagnostic Stromal Organization Features Based on Intrinsic Two-Photon Excited Fluorescence and Second Harmonic Generation Signals. *J. Biomed. Opt.* 2009, 14, 020503.

49. Nadiamykh, O.; LaComb, R. B.; Brewer, M. A.; Campagnola, P. J.: Alteration of the Extracellular Matrix in Ovarian Cancer Studied by Second Harmonic Generation Imaging Microscopy. *BMC Cancer* 2010, 10, 94.

50. Boyd, G. T.; Haensch, T. W.; Shen, Y. R.: Continuous-Wave Second Harmonic Generation As A Surface Microprobe. *Opt. Lett.* 1986, 11, 97–99.
51. Li, Y. L.; Rao, Y.; Mak, K. F.; You, Y. M.; Wang, S. Y.; Dean, C. R.; Heinz, T. F.: Probing Symmetry Properties of Few-Layer MoS_2 and h-BN by Optical Second Harmonic Generation. *Nano Lett.* 2013, 13, 3329–3333.
52. Yin, X. B.; Ye, Z. L.; Chenet, D. A.; Ye, Y.; O'Brien, K.; Hone, J. C.; Zhang, X.: Edge Nonlinear Optics on a MoS_2 Atomic Monolayer. *Science* 2014, 344, 488–490.
53. Zimmermann, J. E.; Li, B.; Hone, J. C.; Höfer, U.; Mette, G.: Second Harmonic Imaging Microscopy for Time-Resolved Investigations of Transition Metal Dichalcogenides. *J. Phys. Condens. Matter* 2020, 32, 485901.
54. Schaller, R. D.; Saykally, R. J.; Shen, Y. R.; Lagugne-Labarthet, F.: Poled Polymer Thin-Film Gratings Studied with Far-Field Optical and Second Harmonic Near-Field Microscopy. *Opt. Lett.* 2003, 28, 1296–1298.
55. Shen, Y. Z.; Swiatkiewicz, J.; Winiarz, J.; Markowicz, P.; Prasad, P. N.: Second Harmonic and Sum-Frequency Imaging of Organic Nanocrystals with Photon Scanning Tunneling Microscope. *Appl. Phys. Lett.* 2000, 77, 2946–2948.
56. Schaller, R. D.; Johnson, J. C.; Saykalay, R. J.: Time-Resolved Second Harmonic Generation Near-Field Scanning Optical Microscopy. *Chem. Phys. Chem.* 2003, 4, 1243–1247.
57. Hoffmann, D. M. P.; Kuhnke, K.; Kern, K.: Sum-Frequency Generation Microscope for Opaque and Reflecting Samples. *Rev. Sci. Instr.* 2002, 73, 3221–3226.
58. Han, Y.; Raghunathan, V.; Feng, R. R.; Maekawa, H.; Chung, C. Y.; Feng, Y.; Potma, E. O.; Ge, N. H.: Mapping Molecular Orientation with Phase Sensitive Vibrationally Resonant Sum-Frequency Generation Microscopy. *J. Phys. Chem. B* 2013, 117, 6149–6156.
59. Florsheimer, M.; Brillert, C.; Fuchs, H.: Chemical Imaging of Interfaces by Sum Frequency Microscopy. *Langmuir* 1999, 15, 5437–5439.
60. Cimatu, K.; Baldelli, S.: Chemical Imaging of Corrosion: Sum Frequency Generation Imaging microscopy of Cyanide on Gold at the Solid-Liquid Interface. *J. Am. Chem. Soc.* 2009, 130, 8030–8037.
61. Huang, S., Makarem, M.; Kiemle, S. N.; Hamedi, H.; Sau, M.; Cosgrove, D. J.; Kim, S. H.: Inhomogeneity of Cellulose Microfibril Assembly in Plant Cell Walls Revealed with Sum Frequency Generation Microscopy. *J. Phys. Chem. B* 2018, 122, 5006–5019.
62. Zheng, D.; Li, L.; Kelly, K. F.; Baldelli, S.: Chemical Imaging of Self-Assembled Monolayers on Copper Using Compressive Hyperspectral

Sum Frequency Generation Microscopy. *J. Phys. Chem. B* 2018, 122, 464–471.

63. Li, H.; Kelly, K. F.; Baldelli, S.: Spectroscopic Imaging of Surfaces — Sum Frequency Generation Microscopy Combined with Compressive Sensing Technique. *J. Chem. Phys.* 2020, 153, 190901.

64. Wang, H.; Wagner, J. C.; Chen, W.; Wang, C.; Xiong, W.: Spatially Dependent H-Bond Dynamics at Interfaces of Water/Biomimetic Self-Assembled lattice Materials. *Proc. Nat. Acad. Sci.* 2020, 117, 23385–23392.

65. Manaka, T.; Lim, E.; Tamura, R.; Iwamoto, N.: Direct Imaging of Carrier Motion in Organic Transistor by Optical Second Harmonic Generation. *Nat. Photonics* 2007, 1, 581–584.

66. Manaka, T.; Liu, F.; Weis, M.; Iwamoto, N.: Studying Transient Carrier Behavior in Pentacene Field Effect Transistor Using Visualized Electric Field Migration. *J. Phys. Chem. C* 2009, 113, 10279–10284.

67. Shibata, Y.; Nakao, M.; Manaka, T.; Iwamoto, M.: Probing Electric-Field Distribution in Undercover of an Organic Double-Layer System by Optical Second Harmonic Generation Measurement. *Jap J. Appl. Phys.* 2009, 48, 021504.

68. Taguchi, D.; Inoue, S.; Zhang, L.; Li, J.; Weis, M.; Manaka, T.; Iwamoto, M.: Analysis of Organic Light-Emitting Diode As a Maxwell-Wagner Effect Element by Time-Resolved Optical Second Harmonic Generation Measurement. *J. Phys. Chem. Lett.* 2010, 1, 803–807.

69. Miyamae, T.; Takada, N.; Tautsui, T.: Probing Buried Organic Layers in Organic Light-Emitting Diodes under Operation by Electric-Field-Induced Doubly Resonant Sum-Frequency Generation Spectroscopy. *Appl. Phys. Lett.* 2012, 101, 073304.

70. Xiao, M.; Joglekar, S.; Zhang, Z.; Jasensky, J.; Ma, J.; Cui, Q.; Guo, L. J.; Chen, Z.: Effect of Interfacial Molecular Orientation on Power Conversion Efficiency of Perovskite Solar Cells. *J. Am. Chem. Soc.* 2017, 139, 3378–3386.

71. Zhang, L.; Liu, C.; Zhang, J.; Li, X.; Cheng, C.; Tian, Y.; Jen, A. K. Y.; Su, J. B.: Intensive Exposure of Functional Rings of A Polymeric Hole-Transporting Material Enables Efficient Perovskite Solar Cells. *Adv. Mater.* 2018, 30, 1804028.

72. Tisdale, W. A.; Williams, K. J.; Timp, B. A.; Norris, D. J.; Aydil, E. S.; Zhu, X. Y.: Hot-Electron Transfer from Semiconductor Nanocrystals. *Science* 2010, 328, 1543–1547.

73. Jailaubekov, A. E.; Willard, A. P.; Tritsch, J. R.; Chan, W. L.; Sai, N.; Gearba, R.; Kaake, L. G.; Williams, K. J.; Leung, K.; Rossky, P. J.; Zhu, X. Y.: Hot charge-transfer excitons set the time limit for charge separation at donor/acceptor interfaces in organic photovoltaics. *Nat. Mater.* 2013, 12, 66–73.

Review Articles

Colloidal Particles.

- Roke, S.; Gonella, G.: Nonlinear Light Scattering and Spectroscopy of Particles and Droplets in Liquids. *Ann. Rev. Phys. Chem.* 2012, 63, 353–378.
- Gonella, G.; Dai, H. L.: Second Harmonic Light Scattering from the Surface of Colloidal Particles: Theory and Applications. *Langmuir* 2014, 30, 2588–2599.
- Wilhelm, M. J.; Dai, H. L.: Molecular Membrane Interactions in Biological Cells Studied with Second Harmonic Light Scattering. *Chem. Asian J.* 2020, 15, 200–213.

Ultrafast Surface Dynamics.

- Nihonyanagi, S.; Yamaguchi, S.; Tahara, T.: Ultrafast Dynamics at Water Interfaces Studied by Vibrational Sum-Frequency Generation Spectroscopy. *Chem. Rev.* 2017, 117, 10665–10693.
- Inoue, K.; Nihonyanagi, S.; Tahara, S.: Ultrafast Vibrational Dynamics at Aqueous Interfaces Srudied by 2D Heterodyne-Detected Vibrational Sum-Frequency Generation Spectroscopy. In Cho, M.(editor): *Coherent Multidimensional Spectroscopy.* Springer Series in Optical Sciences 2019, Vol. 226, pp. 216–236.

SH/SF Microscopy:

- Mazumder, N.; Deka, G.; Wu, W. W.; Gegoi, A.; Zhuo, G. Y.; Kao, F. J.: Polarization Resolved Second Harmonic Microscopy. *Methods* 2017, 128, 105–118.
- Pavone, F. S.; Campagnola, P. J. (eds): *Second Harmonic Generation Imaging.* CRC Press Taylor & Francis: Boca Raton, 2nd ed., 2019.
- Mizuguchi, T.; Nuriya. M.: Applications of Second Harmonic Generation (SHG)/Sum-Frequency Generation(SFG) Imaging for Biophysical Characterization of the Plasma Membrane. *Biophys. Rev.* 2020, 12, 1321–1329.

- James, D. S.; Campagnola, P. J.: Recent Advancements in Optical Harmonic Generation Microscopy: Applications and Perspectives. *BME Frontiers* 2021, Article ID 3973857.
- Cisek, R.; Joseph, A.; Harvey, M.; Tokarz, D.: Polarization-Sensitive Second Harmonic Generation Microscopy for Investigations of Diseased Collagenous Tissues. *Front. Phys.* 2021, 9, 726996.
- Sun, W.; Wang, S.; Han, X.: Biointerface Characterization by Nonlinear Optical Spectroscopy. In Zhao, X. and Lu, M. (eds.), *Nanophotonics in Bioengineering.* Springer Nature, Singapore, 2021, Chapter 5.
- Wang, H.; Xiong, W.: Vibrational Sum-Frequency Generation Hyperspectral Microscopy for Molecular Self-Assembled Systems. *Ann. Rev. Phys. Chem.* 2021, 72, 279–306.
- Parodi, V.; Jacchetti, E.; Osellame, R.; Cerullo, G.; Polli, D.; Raimondi, M.T.: Nonlinear Optical Microscopy: From Fundamentals to Applications in Live Bioimaging. *Front. Bioeng. Biotechnol.* 2020, 8, Article 585363.
- Maekaw, H.; Kumar, S. K.; Mukherjee, S.; Ge, N. H.: Phase-Sensitive Vibrationally Resonant Sum-Frequency Generation Microscopy in Multiplex Configuration at 80 MHz Repetition Rate. *J. Phys. Chem. B* 2021, 125, 9507–9516.
- Li, H.; Kelly, K. F.; Baldelli, S.: Spectroscopic Imaging of Surfaces — Sum Frequency Generation Microscopy Combined with Compressive Sensing Technique. *J. Chem. Phys.* 2020, 153, 190901.

Organic Devices.

- Manaka, T.; Iwamoto, M.: Second Harmonic Generation Spectroscopy in *Organic Semiconductors for Optoelectronics*, (ed. Naito, H.; J Wiley, 2021) Chapter 9.
- Xiao, M.; Lu, T.; Lin, T.; Andre, J. S.; Chen, Z.: Understanding Molecular Structures of Buried Interfaces in Halide Perovskite Photovoltaic Devices Nondestructively with SubMonolayer Sensitivity Using Sum Frequency Vibrational Spectroscopy. *Adv. Eng. Mater.* 2020, 10, 1903053.

Chapter 12

Prospects

Among nonlinear optical spectroscopic techniques for material characterization, SHG/SFG plays a central role. After nearly 40 years of development, SHG/SFG has become arguably the most versatile and powerful surface analytical tool. It has impacted many areas of different disciplines, and has begun to find more applications on problems of practical interest. One may wonder whether the technique can be further developed and where the field is going. The following describes the author's personal view on the subject.

Currently, SHG/SFG spectroscopy suffers technically on three fronts. First, resonances below $\sim$600 cm^{-1} ($\sim$80 meV) are not accessible by usual experimental setups for lack of coherent IR source with sufficient energy. Although free electron lasers can provide the desired IR pulses, they are not easily available. This problem is likely to be solved in the near future by the development of broadband THz generation from wave mixing.[1] Second, phase-sensitive SHG/SFG has not yet been developed for all spectroscopic measurements. Ordinary SHG/SFG measures only intensity spectra that allow deduction of $|\chi^{(2)}|$ spectra of a medium, but not those of the complex nonlinear susceptibility $\chi^{(2)}$. This means that half of the spectral information is lost. It is not inconceivable to set up phase-sensitive spectroscopy measurement in all cases, but for different interfaces, the obstacles to be overcome are different. Third, the current SF spectroscopy systems are too complex for novices. They are also not easily movable. In order to attract more practitioners, they have to be more user-friendly and compact. The solution is not easy as we have to wait for advances of laser and optics technology. Hopefully, in the future, AI-assisted computerized compact systems can be built with high-power, high rep-rate, femtosecond fiber lasers, and optical lenses and filters made of 2D metamaterials.

On characterization of bulk inorganic materials, nonlinear optical spectroscopy has not attracted a great deal of interest in the past. Researchers seem not to care much about structural properties of materials beyond what can already be obtained by conventional spectroscopic techniques. There are however specific areas where SHG/SFG spectroscopy is capable of providing unique valuable information. *In situ* probing of material symmetry and anti-ferromagnetism are examples. Recently, SHG has been employed to interrogate 2D and other novel materials, but SFG spectroscopy can be more informative. For instance, through doubly resonant excitations, electron-phonon and magnon-electron couplings can be studied. Being surface-specific, it can be used to probe interlayer excitons and phonons, interlayer coupling associated with surface conductivity, multiferroics, interfacial superconductivity, and inter-layer surface dynamics. SFG spectroscopy to probe topological insulators could also be interesting.

SF vibrational spectroscopy (SFVS) as a tool to monitor atomic and molecular adsorption at various interfaces has been well developed. In competition with infrared reflection absorption spectroscopy, it has some obvious advantages, especially with respect to *in situ* probing of adsorbates and their orientations at buried interfaces. The technique is already at the stage that it could be applied to more practical problems such as monitoring surface contamination, catalysis, and surface reactions in real environments. It will be exciting if this actually happens although there is the lingering issue that we may still not know the surface structures of substrates during reactions.

One of the unique features of SF spectroscopy is its unique capability to probe surface structures of bulk materials in real environments. Unfortunately, because of the limited IR range SFVS can address, studies have been limited to solids and liquids composed of light to elements.[1] So far, only a small number of different solid and liquid (fluid in general) interfaces have been investigated even though the technique has become more mature. Selective applications of the technique to practically relevant surfaces can be anticipated in the near future. Special interest will be on buried liquid/solid interfaces since SFVS is the only viable tool that can be used for *in situ* interrogation of surface structures of both liquids and solids at a buried interface.

A major contribution of SF spectroscopy to surface science is that it provides an effective means to probe liquid interfaces. It is probably fair to say that essentially all our current knowledge about microscopic structures of liquid interfaces has come from SFVS results. This is particularly true for water, the most important liquid on Earth. Gas/water interfaces have been most extensively studied. Recent research interest has shifted to solid/water interfaces, which are scientifically and technically important in many disciplines. At such interfaces, SFVS has shown that the interfacial water structure varies with composition of water solution, but the surface structure of the solid is generally not probed and is assumed known when interpreting the interfacial spectra of water. Recently, it was demonstrated that SFVS can detect surface vibrations/phonons of oxides in water, thus allowing *in situ* probing of both oxides and water at an oxide/water interface and providing a more complete picture of the interfacial structure. Emphasis of SFVS studies of solid/water interfaces in the coming years is likely to be on learning how both solid and water structures at the interfaces change when the water solution is adjusted.

A particularly important, technologically relevant area of water science deals with charged water interfaces. Yet, despite extensive studies, still not much is known about their microscopic structures. SFVS has been able to record the OH stretching spectrum of interfacial water of a charged water interface and its changes with pH and salt concentrations in water, but interpretation often disregards the existence of a water layer in the immediate vicinity of the opposing medium with a hydrogen bonding structure very different from that of the bulk. This layer is most important as it directly controls the functionality of the interface. Probing the layer is challenging because the SF signal can be overwhelmed by contribution from the electric double layer created by surface charges. Recently, the situation has changed. The newly developed SFVS scheme allows separate measurements of vibrational spectra for the two interfacial water layers. In addition, SFVS can probe the structure of the opposing material at the interface and its change in response to changes in the water solution. These advances are likely to create a new paradigm of research on charged water interfaces. More works along the line are expected that will forge a much better understanding of charged water interfaces.

Electrochemistry is another most important branch of surface science, but little is known about its microscopic interfacial structure. SFVS is uniquely poised to probe electrochemical interfaces but has suffered from relatively weak signals due to strong attenuation of IR input on accessing such interfaces. In certain IR spectral range, it is still possible to identify and track interfacial species during electrochemical cycling. However, to have a thorough microscopic understanding of an electrochemical process, we generally need to know the microscopic structure of the electrochemical interface including surface charge density and interfacial water structure in correlation with the measured cyclic voltammogram. This unfortunately has not yet been realized. Recently, it was demonstrated that a properly designed patterned metal electrode can greatly enhance the fields at an electrochemical interface and hence the SFVS signal from the interface, making measurement of vibrational spectra of water at the interface feasible. If the technique is further developed, SFVS studies of electrochemistry are likely to create an exciting new frontier.

Polymer science is one of the fields SFVS has impacted most. As the only viable and versatile technique for *in situ* studies of polymer interfaces, SFVS has provided a plethora of hitherto unattainable information about various polymer interfaces. While the technique is gaining popularity in the field, there is still further room for development. Interpretation of observed spectra is always challenging in cases where polymers are complex, especially if vibrational modes are entangled. It is imperative that phase-sensitive SFVS be implemented to facilitate spectral analysis; MD simulations of observed spectra would help greatly in interpreting observed spectra. Otherwise, SFVS studies of polymers have already reached the stage of being able to address technological questions. It is time that researchers can collaborate more with industry to find more impactful problems to solve.

Being capable of probing buried interfaces, SFVS has the potential to become a powerful analytical tool for bio and medical sciences, but works so far have been largely limited to model systems to show what can be learned with the technique. As in the case of polymers, it is time to apply SFVS to problems biologists would care about, e.g., identification of biomolecules and their interactions with membranes and other biological entities, penetration of molecules through real cell membranes and subsequent effects, and many others. Most importantly, SFVS should be able to find niche areas that other techniques cannot explore. Microscopic bio-imaging could be such an area. Similar to coherent antiStokes Raman scattering (CARS) and stimulated Raman scattering (SRS) imaging, SFVS imaging requires no labeling, and can be expected to be more widely adopted to study bio-systems. Its unique ability to image both chiral and achiral biological interfaces should be noted. Chiral SF spectroscopy allows *in situ* probing of molecular chirality and related phenomena that can be of interest in chemistry and biology. It offers new opportunities to study, for example, chiral conformational changes of molecules, induced chirality, and chiral dynamics. So far, research in these areas has not attracted much attention. Hopefully, the situation will change in the coming years.

In short, we have witnessed the growth of SHG/SFG spectroscopy into a most powerful tool for material characterization, in particular, surface characterization. Yet, there is always space for further development. To broaden the scope of applications, ability to cover a wider spectral range is needed. To allure more practitioners, simplification of the system is vital. To attract more recognition, demonstration of how the technique can solve impactful problems is important. Finally, success of the technique depends on how observed spectra can successfully characterize materials. In this respect, theoretical help in interpretation of spectra is crucial. Overall, we can look forward to more progress in both technical development and valuable applications of SHG/SFG spectroscopy in the future.

Reference

1. Le, J.; Su, Y.; Tian, C.S.; Kung, A.H.; Shen, Y.R.: Bridging the Terahertz Gap by Resonant Four-Wave Mixing in Diamond", (To be published).

Index

CPSIA information can be obtained
at www.ICGtesting.com
Printed in the USA
JSHW051955090323
38637JS00002B/4